I0061799

Charles Jean Marie Letourneau

Sociology

Charles Jean Marie Letourneau

Sociology

ISBN/EAN: 9783742820136

Manufactured in Europe, USA, Canada, Australia, Japa

Cover: Foto ©berggeist007 / pixelio.de

Manufactured and distributed by brebook publishing software
(www.brebook.com)

Charles Jean Marie Letourneau

Sociology

SOCIOLOGY

BASED UPON ETHNOGRAPHY

BY

DR. CHARLES LETOURNEAU.

TRANSLATED BY HENRY M. TROLLOPE.

NEW EDITION.

LONDON: CHAPMAN AND HALL, LD.

PREFACE.

Social Science has never been so much talked about as in our own times. We all know now that the life of human societies, like everything else, is governed by rules, by laws, and may therefore become a question of science. This idea is far from new, for the "Politics" of Aristotle is a treatise upon Sociology, doubtless very incomplete, but nevertheless scientifically conceived. And in their way the "Laws" and "The Republic" of Plato are also sociological works, though in them the scientific method is that which is most wanting. Aristotle and Plato have also had many imitators or followers. To the former we may trace much of what we read in Machiavelli and in Montesquieu; on the other hand, still confining ourselves to a few names, Campanella and Rousseau may be counted as descendants of Plato. By the side of these two schools, a third has made for itself a place;—this we may call the systematic school. It is doubtless from the results of observation that systematic socialists start their theories. But they confine their field of observation; and they distort facts by so selecting them that to their questions they may find an answer which will agree with their preconceived opinions. Vico, Condorcet, Saint-Simon, Auguste Comte, are the most illustrious representatives of this school, and all these men possessed minds of the highest order.

But have all these thinkers succeeded in founding Sociology? —for this hybrid word was first brought into fashion by Comte. We can hardly think so, unless we wish to close our eyes very determinedly. We have the word without the thing, and it cannot be otherwise. The commencement of any science, however simple, is always a collective work. It requires the constant labour of many patient workmen, succeeding each other, each benefiting from the toil of his predecessor, and through a long series of generations. Isolated minds, let them be ever so powerful, cannot do more than promulgate speculative questions that may be more or less ingenious.

Again, the rise of any science is all the more laborious in proportion as it is vast; and what can be more complex than Social Science? Modern investigations have taught us the fundamental truth that everything in the universe must be governed by laws. We have therefore our sociological laws. But it becomes more difficult to discover a law in proportion as the phenomena which it governs are multiple, are variable and intricate; and social facts are numberless, their intricacy and their variability are extreme. By dint of observations and experiments made and continued during historic periods we have succeeded in formulating a few astronomical laws; and yet we are told that the isolated meditations of some few systematic minds can give us a ready-made scientific Sociology! Anyone who likes to believe in this illusion is of course free to do so.

It is difficult for us to look around over the whole vast field of Sociology; for we must take into account all the infinitely various manifestations of human activity, and also all the exterior agents that may in any way influence that activity. Will the evolution of societies always continue to unfold itself confusedly and spontaneously? Must we despair of ever finding and possessing a scientific Sociology? Assuredly not; but it is all-important that we should not think complete a work as yet hardly begun.

We now know how sciences first rise, and how they grow. It is before all things necessary—and this is a very long task—to bring together rich materials of well authenticated and carefully noted facts. We must then select them, group them, class them, and arrange them in order. For until then we are not entitled to make inductions; we cannot see correctly the links between the phenomena in past times; we ought not to risk observations as to their future evolution. As a matter of course all this elaboration will be of greater value in proportion as it rests upon a larger basis. In many sciences experience may assist us in our observations, and especially in controlling our inductions. This precious means of verification has hitherto been wanting to Sociology; but human societies, as they felt the need, have made many attempts, which in a great measure may serve as premeditated analytical experiments, perhaps to be undertaken at some future date.

Before this vast preliminary labour that we have just indicated can be accomplished centuries must first elapse; all hope of a scientific Sociology would otherwise be vain. All that we can now do is to make some few attempts; and it behoves us to define our objects, directing our efforts to each one in turn of the many sides of social life. Sociology must necessarily rest upon the groundwork of many sciences: natural history, anthropology, ethnography, demography, pedagogy, the study of climates, political economy, history, etc. etc. etc. The enumeration would be infinite, for everything which can, directly or indirectly, have influence upon human life has also its sociological importance.

This is the scientific method. It is doubtless long and laborious; but it is the only one that can do the work, and more than one pioneer has already begun to clear the way. It will be enough to mention the large historic pictures of Buckle and Draper, the ethnographical works of Lubbock, Tylor, and others, and lastly Herbert Spencer's book on *Sociology*. This latter is also mainly ethnographical, but it has in some way deceived the public; for more was expected of its author, than whom few men in our own time possess a larger or a more acute intellect, or a mind more richly furnished. Mr. Spencer's work no doubt gives evidence of much sound thought and nice perception; but his exposition of facts is singularly unmethodical, and he is often led astray by *a priori* systematic conceptions. We may mention his very strong belief in the doctrine of the Greek philosopher Euhemeros, according to which the pagan gods were superior men who had become deified by the people; and also his unwarrantable comparison between social and biological organisms. We may add, that in many of his conclusions Mr. Spencer has run directly counter to noticeable facts, and to those which have been already established.

For ourselves, our own views or opinions here in this volume are very confined. Such has been our intention. Our purpose was to write a chapter on Sociology—the ethnographical chapter—and we have endeavoured not to heap up our facts confusedly and without order. We have undertaken to describe the principal manifestations of human activity successively in the principal human races, connecting them as nearly as possible with similar phenomena that have been observed in animals. In nearly every case we have closed our short inquiry with an attempt at generalization, and even of induction; but the reader will at once distinguish our own personal views from the facts which in our opinion will justify them, and may himself draw any other conclusion that appears to him to be more sound.

After what has been said, no one will expect to find in this volume an enumeration of "Sociological Laws," drawn up with all the strictness of the laws of true science. Social science is yet in its infancy; to formulate its laws is therefore beyond our power. But scientific laws do not spring suddenly as from spontaneous generation. The way is first prepared by extracting from the chaos of minute information some few general facts. This has been our endeavour: we hope we may have succeeded.

CH. LETOURNEAU.

CONTENTS.

ETHNOGRAPHICAL PROLEGOMENA.

BOOK I.

Nutritive Life.

BOOK II.

Sensitive Life.

BOOK III.

AFFECTIVE LIFE.

BOOK IV.

SOCIAL LIFE.

BOOK V.

INTELLECTUAL LIFE.

ETHNOGRAPHICAL PROLEGOMENA.

SOCIOLOGY.

CHAPTER I.

ENUMERATION OF THE HUMAN RACES.

When we attempt to enumerate and to classify the various races of man, the anthropologist and the ethnographer are at once beset with difficulties,—so changeable, so multiform, and so various are the human mammalia. Let it not be thought for a moment that we wish, as do some over-metaphysical anthropologists, to perch man up in the clouds, to pretend that our puny vertebrated body has upon this earth a separate existence of its own—a gulf dividing us from all other animals. Though man is incontestably a mammiferous animal of the highest order, he differs, nevertheless, very widely from his more humble congenerous creatures; for with him the higher nervous centres, and their use, which is shown in his intelligence, are susceptible of a relatively enormous development. Again, as far as scientific data will allow us to judge, the origin of man is multiple. It may very well be that the now existing human race descends, coming down through a long series of metamorphoses, from monkey-bearing breasts. But these early progenitors of man were very numerous, and even from the first of very various kinds. Starting from this low primitive state, the earliest types that were even nearly human must have been subject to changes from the ordinary habits of life; for during very long geological cycles man was necessarily obliged to live in various climates, to which he was constrained to adapt himself in order to maintain his own existence. And, in its turn, this labour of accommodating himself to places more or less inclement has become a cause of organic change. Everywhere on the face of the globe man has formed for himself a separate existence, and he has made for himself a civilisation that is more or less intelligent; both of which have served to protect him from rough contact with the surrounding elements, and have also either tended to stifle in him certain preponderating energies, or to foster some latent capabilities. Now there is no functional modification which is not both the sign and the effect of a corresponding modification in the organs. Owing to the combined influence of the diversity of his origin, and to the disparity of civilisation,—to both of which we must add the effect, ever various, of the innumerable ethnical unions which took place during the long night of the prehistoric ages, when his last thought was to write his own annals,—man modelled himself after many and different types. In one place these different types were clearly determined, in another they were so closely joined together by intermediary mixings, and were effected so gradually, that further gradation is no longer possible. In a word, that has happened to man, but upon an infinitely larger scale, which has happened to our domestic animals—for instance, to our canine creatures. The greyhound and the bulldog, the spaniel and the Newfoundland, are all canine mammalia, but how different are they one from the other!

But before we begin to speak of Sociology itself we must endeavour, as well as we can, to unravel this chaos of ideas; we must divide the human species into different kinds, and extricate some of the principal types, more or less homogeneous, from out of the confused mass of the human races. We shall now attempt to do this as briefly as possible.

If the anatomical characters, now so carefully studied by contemporary anthropologists, were well classified and well arranged in their proper order, our task would then be simple enough. As it is the special object of this book to speak of sociology, we might pass over small anatomical details, tracing out the principal groups and showing the peculiarities of each. But unfortunately the study of anatomical anthropology is as yet in its early phases. It enables us only to prove certain facts, not to classify them. One anthropologist will base his classification of races on the form of the cranium, on the quantity and the formation of the brain, another will satisfy himself by examining the hair. In this still confused state of human taxonomy we must proceed somewhat at random, taking groups of characters for our guide, so as to lessen as far as possible the chances of falling into error. No doubt that the actions of men and women form the principal study of the sociologist; but he is bound nevertheless to connect these actions as far as possible with certain anatomical facts, or at all events to show the connection existing between them,—for between the labourer and his work the tie is very close. An inferior anatomical race has never created a civilisation superior to itself. Over such a race hangs an organic malediction, the weight of which can only be reduced by millenary efforts, and by a struggle for improvement constantly going on during geological cycles. As to the nobility of organisation, we see the greatest variety in men, and these differences are so strongly marked, that any idea of close and gradual progression is at once excluded. Nevertheless, taking into consideration only the very large and important features, we may group, both anatomically and sociologically, the existing types of man under three main heads:—

I. *The black man*, whose brain is small, especially in the frontal region, which with him is narrow and receding. His cranium is elongated and oval-shaped. Correlatively, his jaws are prognathous, that is to say, the rudimentary organs are projecting. His nose is more or less flat. The skin is also more or less black, and the hair woolly, except with the Australian negro, who seems to be a man of mongrel breed.

II. *The yellow man*, the Mongolian, or Mongoloid, is still farther separated from the animal form. His brain, more developed in the Asiatic Mongolians, but very small with the American Mongoloid, is better shaped. The forehead—where the intelligence mainly lies—is with him less sacrificed, it is even relatively largely developed in the case of the Asiatic Mongolians. His cranium is large and short, brachycephalous; the prognathism is here much less strongly marked than in the preceding type. His eyes, or rather his palpebral openings, are very elongated, scarcely open; they are contracted, and often raised obliquely, both inside and outside.

His hair is always black and straight; his skin is yellow, or yellowish in colour.

III. *The white man* has ascended a few degrees higher in the organic hierarchy. His brain is developed; his forehead has expanded and become straight; his jaw-bones have become reduced, and in him we no longer find the prominence of a blobber-lipped mouth. His eyes are set straight, are open, sometimes dark and sometimes light in colour, whilst with the two preceding types they are nearly invariably black. And so with his hair, instead of being always black, we see it of very various colours, from quite fair to jet black. His skin is more or less white, and his hair is sometimes straight, sometimes curly, but never woolly.

From the sociological point of view these different types of human beings are far from possessing the same dignity; there are also many subdivisions in each class of very various kinds, and these differ very widely one from the other. The negro will generally be found to belong to the most inferior type. The black man, guided by no instincts but his own, in whom there is no admixture of blood of a superior race, has never been able to form for himself any sort of cultivated civilisation. In this way the yellow man—the Mongolian—is a much superior creature. From a very early date, the Asiatic Mongolians, who are the best specimens of this type, formed for themselves large and well-organised societies, which, as well as the Chinese, are actually maintaining a rivalry with the white races, and in some respects may serve for them as a model. Even the American Mongoloids, who of the Mongolian class are the most mongrel, the most inferior in race, and whose brain is the feeblest—of whom some of the most humble specimens are always to be found dwindling away in the lowest stage of intellectual and social existence—even they, by their better and superior types, have been able in days gone by to show in Mexico and Peru some remarkable examples of social progress.

In spite of its imperfections, its weaknesses, and its vices, the white men still maintain their foremost position in this steeple-chase competition among the human races. It is in the heart of the ethnical groups of the white races that intellectual energy has shown its finest and widest developments; it is there that art, moral nobility, science, and philosophy, have spread themselves the most widely. And as we proceed we shall see, with some degree of detail, that the white race in all its forms is now the least opposed to progress.

The following table will enable the reader to see at a glance the hierarchy of both the principal and the subordinate races of men. As some of the negro races differ in essential points from the most common of the negro types, we have given to them a separate place by the side of the principal black races with which they are in fact actually connected:

THE PRINCIPAL TYPES OF THE HUMAN SPECIES.

I. BLACK MEN	OCEANIC NEGROES (Melanesians)	Tasmanians.	
		Australians.	
		Papuans.	
	PURE NEGROES ...	Veddahs, and Indian black men.	
		Andamans.	
		Negroes of small stature (Negritos).	
		Hottentots.	
	AFRICAN NEGROES	Inferior Africans.	
		Superior Africans	Mandingos, etc.
			Nubians.
			Abyssinians.
II. YELLOW MEN	MONGOLIANS ...	Tartars, Chinese, Japanese.	
	MONGOLOID AMERICANS ...	Fuegians.	
		Southern Americans.	
		Central Americans.	
		Northern Americans (Red Skins).	
	MONGOLOIDS OF VARIOUS TYPES	Esquimaux, Kamtschadales, etc.	
		Lapps.	
		Carolines, etc.	
		Malays.	
		Polynesians.	
III. WHITE MEN	CAUCASIAN RACES	Indo-Europeans.	
		Semites.	
		Berbers.	

CHAPTER II.

DISTRIBUTION OF THE HUMAN RACES ON THE FACE OF THE GLOBE.

Statistics are even now very far from enabling us to enumerate exactly, or even with tolerable accuracy, the population of the world. Only some few civilised states endeavour more or less to keep an account of the number of their inhabitants, a registry of their births, deaths, and marriages. We are obliged, therefore, when we wish to make an enumeration of each race, to confine ourselves to the most general means of calculation. In round figures we may say that from twelve to thirteen hundred millions of human beings are struggling for their existence on the surface of our little planet. It may be taken as a matter of course that the many coloured and multifarious members of humanity maintain their existence with very unequal chances, according as the temperature in which they live is more or less clement; or more especially according as they expend more or less intelligence and energy for their use and for their protection.

If we were to judge only by the extent of area occupied, and by the number of the occupants, the Mongolians, or the yellow races, would hold the first rank in the world as it now stands. Their dominion in Asia is very large indeed. A frontier line drawn, starting from the south of the Caspian sea, just touching the south of Afghanistan, running along the southern side of the Himalayas, then turning southward towards the Indian ocean, and afterwards joining the Irrawady river—such a line would indicate roughly the Asiatic boundary between the vast Mongolian territory and the much smaller territory inhabited by the more or less Caucasian races. The people to the north and to the east of this line are Mongolian, with a few exceptions here and there—in Burmah, for instance; on the other hand, the people on the south and west, in the angle formed by this broken line, are Caucasian, and advancing westward, the Caucasian blood becomes purer and more free from admixture. Let us not forget, too, that the Mongolians have very widely spread themselves in European Russia, in Turkey, in Hungary, that they have swarmed in many directions, peopling the Malay archipelago, occupying the Philippine islands, the Caroline islands, and many other islands in the Pacific, mixing themselves with the inhabitants and gradually moulding the aborigines of these islands into the large Mongoloid race. Also that the lowest specimens of their type—the Kamtschadales, the Esquimaux, the Samoyedes, and the Lapps—wander all along the arctic shore of Asia, America, and even of Europe. And it is also very difficult to avoid recognising a large admixture of Mongolian blood in the indigenous American and in the Polynesian. When we recollect all this, we may then have some notion of the position, morally as well as materially, which the yellow race holds upon the earth.

The only serious rival to the Mongolian is the white man; as to whom contemporary anthropology contests the Caucasian origin. Yet, if we consider the place and climate occupied by the white man on the old continent, we shall see that the Caucasian chain shows very plainly the centre of the region which we may call white; that it is in the Caucasus and the neighbouring countries that the purest specimens of the type are to be found. Starting from the Caucasian Alps we see that the white races spread all round about on every side. Towards the north and the west they cover Europe; on the south they occupy Asia Minor and Arabia; and on the east, Persia and Afghanistan. Then, going down into India they mix with other races, and gradually lose their own distinctive features as one gets nearer towards the Deccan. There are many linguists who would have us believe that all the Indo-European races come from the Caucaso-Indian Hindoo. There are certainly some representatives of the white races left there even now; but they are, so to speak, the sentinels chosen out and put there to mark and define the limits of their empire. Let this be said, without wishing for a moment to insinuate anything against the admirable systematisation of Indo-European dialects, from which, however, we must be careful not to expect more than it can in reality afford to us.

Taking into account only the Caucasian population of Asia and of Europe, it will be found that in point of numbers the white man bears a very secondary part by the side of his Mongolian rival; and if, as it seems probable, the yellow races awake up in right earnest from their long sleep and wrest from us that which our superior strength gives us—that is, our industry and our homicidal instruments—the white races will not find that their powerful auxiliaries, the new people which are gradually growing in America, in Oceania, etc., are at all too powerful to resist the attack.

In point of numbers the black man holds the third place; and he, with a greater or less mixture of Aryan blood, specially towards the north and towards the east, occupies the whole of Africa, excepting the Mediterranean shore, the territory of Barbary, and the region of Sahara. And here we find mixed up together the Berbers, the Semites, and a few even of the early instances of European colonisation. What is the number of the black population in Central Africa? No one can tell, but we know that these numerous hordes, savages though they be, are nearly always stationary, and are agricultural in their pursuits. This has lately been shown to us by Speke, Baker, Schweinfurth, Stanley, and others. The African negro is the principal representative of his type, and he also is a serious rival to the superior races, for he can live and work where other races, especially the white men, languish and die away. By the side of the African negro we ought to mention, that they be not forgotten, his inferior brethren the Oceanian negroes, the Melanesians. Of these the Tasmanian has already disappeared in face of the stronger European force. The same fate is reserved for the Australian. The Papuans of New Guinea only, have some chance of lasting, though they have not yet advanced beyond the Polished Stone age.

If we take away the white men, and the black men who have emigrated, the existing American is a man of much less importance than the African negro. The vast and singular empires of Mexico and Peru, original civilisations, are now for ever wiped out; the Aztecs and the Quichuas who peopled them at one time are now destroyed, or have degenerated, or have become merged with their

Spanish conquerors. As regards the Red Skins of the north, the Guaranis, the Pampeans, the Patagonians, the Fuegians, etc., of South America, people rebellious for the most part against European civilisation, their number is very small, and it is destined to become even smaller still. A future seems to be denied to these people, who are both very far backward and very badly provisioned.

We may say as much, too, for the Polynesian races—probably the offspring, or at least a part of them, of the ancient Peruvians, —the people also inhabiting Tahiti, the Sandwich islands, the Tonga islands, the Marquesas islands, New Zealand, etc. These people are interesting from more than one point of view, and later on we shall have to speak of their indigenous, rudimentary, but original civilisation. It can only be from mixing with European colonists that this very curious specimen of the human type can leave behind any lasting impression of his stay upon the earth.

As we have said at the beginning of the chapter, it is impossible as yet to draw up a table showing the number of inhabitants belonging to each of the different large human races. Nevertheless, we may venture to make a few approximate calculations. In putting together the Mongolians, and Mongoloid people, the Chinese, who are pure Mongolians, the Thibetans, the Japanese, the Malays, and the Indo-Chinese, we can hardly estimate their number at less than six hundred millions of inhabitants. As regards the white population spread over the world, it is perhaps going too far to reckon it at four hundred millions. But for the negroes in Africa, in Oceania, the indigenous Americans, and other small races scattered about everywhere, no computation with any sort of accuracy would be possible. Not taking into account the African negro, a race which certainly will number many millions of inhabitants, the other human types only form some poor ethnical groups destined to disappear, or be absorbed by the large battalions of the three great races—the yellow, the white, and the black.

Our preliminary sketch is now completed. We have traced

very broadly, as was suitable to our object, which is mainly sociological, the enumeration of the most important of the human species. The task now before us is more minute in detail, and is more difficult. We must now take each of these types singly, and examine it psychologically and ethnographically. We must bring them all before us, from the lowest to the highest, try them, measure their faculties by the work they have done, and at the same time say something of the various forms of civilisation which they have attempted or which they have realised. The most important considerations are not the numbers of individuals belonging to each race, though this element will very often be found in proportion to the actual value of the race. We shall have to consider principally the originality and the dignity of the industrial, moral, social, and intellectual labour which has come from human effort.

BOOK I.

NUTRITIVE LIFE.

CHAPTER I.

FOOD.

I.

Let us imagine an observer placed somewhere high up in air above our terrestrial equator, far enough from the globe on which we live to take in a whole hemisphere at one glance, and yet close enough to distinguish, with the aid, if need be, of a magnifying-glass, the continents and the seas, the great ranges of mountains, the white frozen tops of the polar regions, etc. etc. To anyone so watching us, the earth would appear to be a system made up of two hemispherical mountains joined together at their base, and at the top covered over with ice. At the foot of the mountains, in the middle region, which we will call *the equatorial zone*, between the tropics, our aerial observer will perceive on the continents or on the islands the existence of a rich country, where both the vegetable and the animal kingdoms are represented by organic forms of the largest and most various kinds. He will see there a sort of outburst of life which gradually becomes weaker and poorer as he draws his eye nearer to the desolate countries, more or less approximatively bounded by the polar regions. Beyond this the organic poverty becomes absolute barrenness; the vegetable kingdom is only seen by a small number of objects of small stature; the boned animals of the earth become rare; here and there a few deep caverns that are not frozen over may serve as a sort of protection to fish and other amphibious creatures, to whom the vegetable world is quite useless, and who live only by devouring each other. There is no doubt that the whole organic system of the world has not always been as we now see it. The temperature was formerly higher than it is at present, and this is proved by the vegetables in the antarctic regions reduced to a fossil state, but which to live require a warm, or at least a temperate climate; on the other hand, the length of the nights in the polar regions has always been very unfavourable both to animal and to vegetable life.

It is therefore in the middle terrestrial zone, in the clement portion of the globe, that the first dwelling-place of the great mammalia—and not excepting man, the foremost of the kind—must, with the greatest sense of probability, be placed. But man has not always been what he is now; his most distant ancestors, his precursors, if we wish, at a time when they were very little different from large monkeys, did not escape from the general law. The large monkeys now alive, our more or less distant cousins, whom we rightly term anthropomorphites, are frugivorous animals; and man also, judging by his teeth, even though we look at the formidable row of teeth of the Australian, is a frugivorous animal. We may, therefore, rationally suppose that the dwelling-place of the earliest specimens of the human type was placed, as is nowadays the case with the large monkeys, in the humid and warm regions, where fruit-producing plants, containing feculant and albuminous matter, will grow easily and without culture. There is every probability that it was in the torrid region of Asia and of Africa, perhaps on the large islands or continents now submerged, that man was first born and had his being. It is in this warm zone that we now find the big monkeys; and again, the fossil fauna in Europe is poor in the remains of ape-like form, and those of America and Australia appear to be totally destitute of them. It is in this zone that man, still savage, whose industry is yet so entirely rudimentary, now supports most easily a tolerable existence; it is here too that the precursor of man can only have lived before he knew the properties of fire.

But from this cradle, from this Eden, man gradually spread out towards the north and towards the south. At first he went into the more temperate regions, where here and there he was able to live and even to develop; but only under the condition that by his skill he could correct the parsimony of nature. All the great civilisations of historical antiquity, even now but ill provided, were first produced in warm or in temperate climates. Long afterwards, races who came later, the inheritors of previous societies, having acquired a more scientific education—among whom were many animal kinds and domestic creatures—succeeded in forming, less far away from the polar circles, large human agglomerations. And even now in the arctic and antarctic regions man can with difficulty find only a precarious existence. In a word, nature's banquet, as the poets say, seems to be very unequally spread over the earth, when the productive power of the globe, altogether careless of human interests, is left to take care of itself. We shall soon become assured of this if we look at the different countries and see what are the different alimentary resources which each offers to the many types of the human race.

II.

Food in Melanesia.

Excepting the desolate regions of the arctic wastes, and the island known as Terra del Fuego, no country has been so ungracious to the human biped as that which is now occupied by the Oceanian negro, including Tasmania, Australia, New Guinea, New Caledonia, and a few of the mixed archipelagoes—though they are rather better provided—such as the Solomon islands and the Fiji islands. In these last-mentioned islands, and also occasionally in New Guinea, the introduction of certain Asiatic vegetables has given some assistance to the black inhabitants; but in Tasmania and in Australia man's natural stock of provisions was, and still is, exceedingly small. There the precious family of palm trees, and other gramineous fruit, were not represented by any feculiferous plant. The poor Australian woman is obliged to dig up the earth with a pointed stick, as the Tasmanian woman had been obliged to do before her race had died out, to unearth the roots, the tubercles, and specially a sort of big truffle (*mylitta Australis*) growing at the feet of the dead trees, in places where the soil is rich with organic decay. And she cannot afford to disdain the juice of the eucalyptus, the sap of certain ferns, the gum of the acacia, certain berries, and even the algæ, which, when slightly cooked, help in some small way to augment the daily fare. These are the accessories; but it is from the animal kingdom that the customary provender is mainly supplied.

But in Australia and in Tasmania the animal kingdom is hardly more generous than the vegetable; these are the countries in which the marsupial fauna still hold their sway. Man must hunt there, and capture the different kinds of kangaroo, the wombat, the opossum—whose size is hardly double that of a rat—the emu, or the Australian cassowary, and various kinds of birds. By the sea-shore the molluscs form the principal article of food, and at the different resting-places the people manufacture for themselves a rough sort of cookery, as prehistoric man used to do in Europe in ages gone by. A few fish, and occasionally (a rare and happy godsend!) the dead body of a whale cast on shore, will furnish them with a more substantial nourishment; and as a last resource they have the dingo, the Australian dog, and even man himself.

New Guinea affords to the Papuan, its inhabitant, a somewhat less scanty fare. Not that the flora there will be found to satisfy all that man desires. According to O. Beccari, there is not an indigenous fruit to be found there at all tolerable; but the Malays have introduced a few cereals, especially rice, and also a few articles of the vegetable kind, notably sago, belonging to the beneficial tribe of palm trees, have been spontaneously brought on to this land, which still shows itself to be Australian from the majority of its organic productions. On the other hand the fauna are not there wholly marsupial. The wild boar roams in the forests, and has been domesticated by the Papuan, though indifferently enough, because the people are obliged to blind the young to prevent them escaping from the human yoke. But the dog, the habitual companion of man on the greater part of this world's surface, lives there a completely domesticated animal, and poultry also is common. The forests of New Guinea are full of birds, and on the sea-shore the fishing assists the Oceanian negro very materially with articles of food. The sea is also one of the principal resources of the Papuan of the New Hebrides, and of his analogous companion, the New Caledonian. These countries are altogether destitute of domestic animals, even of the dog. People there keep themselves alive with what fish they can catch, and with their agricultural industry, which is in a fairly advanced state, especially in New Caledonia; and they show some care and intelligence in cultivating the ignams and the taro. And these people too, before the arrival of the Europeans, were the first instances of mammalia who have more or less often furnished themselves with a dish of meat. Man himself would furnish the first dish;—afterwards would come large bats, vampires, of which we find different kinds in nearly all the Polynesian islands, and also in India.

To sum up: of all the Oceanian negroes, the Tasmanians, and afterwards the Australians, are the most destitute and the least liberally supplied by Providence; the Papuans of New Guinea and the adjacent islands have larger resources given to them; so have the people of New Caledonia, and the negro of the New Hebrides. As regards the other tribes of the Papuan race, they inhabit more fortunate islands: the Solomon islands, New Ireland, the Fiji islands, where, strictly speaking, the Polynesian flora abound, poor in kind, but rich in alimentary vegetables. Even in New Caledonia we find the cocoa-tree, the banana, the papyrus, the sugar-cane, etc. Let us now see how man is provided in the blessed region of Polynesia.

III.

Food in Polynesia.

We must, in the first place, observe that, with the exception of New Zealand, all the large Polynesian archipelagoes are situated in the tropics, that they all have a marine temperature, which is both very mild and very little changeable. In November and December, the hottest months, the thermometer at Noukahiva marks from 29 to 33 degrees centigrade in the shade, and from 20 to 27 degrees in the night. The minimum of temperature in February is 24 degrees. As we may perceive, in this happy country clothes are almost useless, and if it never rained so would be houses.

All these islands are madreporical, that is, calcareous. They have been slowly raised up above the level of the sea, sometimes by volcanic eruptions. The mineral kingdom here is extremely poor. From the nature of the soil there are no beds of metal, few hard stones, except a sort of serpentine and jasper, out of which the people have learnt to make themselves arms and other utensils. The sea shells, the teeth of sharks, etc., have also served them for the same purposes. Soft water was often scarce with them, and, in order to get a regular supply, they have been obliged to dig for themselves deep pits.

The organic kingdoms are hardly more varied than the mineral kingdoms. As these islands have assuredly never been joined to any continent, at least, not since the tertiary period, their plants and their animals must have been imported there by chance or by accident. As a matter of fact the flora in Polynesia came originally from India and from America, and the fauna there is remarkably poor. Except in the few instances when they have been imported by man, animals have not come there except by swimming or by flying. The animal kind are, therefore, very few: a few lizards, some birds, specially parrots, which become more common as one gets less farther away from Australia and New Guinea, but not going beyond the island Rouroutou (153° 6' East longitude). There are no savage mammalia except the rat and a huge bat vampire, which is found also in New Zealand and in New Caledonia. We may add a few domesticated animals, evidently brought there by man, poultry, which we find even as far distant as Easter island, and specially the dog and the hog; this latter animal is generally of very small stature. The dogs and hogs, used generally as articles of food, are found in all the large archipelagoes; but many of the smaller islands were deprived of them, and the pig, known only by tradition in the western island of New Zealand, was altogether unknown in the southern island. The dog, a dog-fox, was the only kind of domestic mammalia of the New Zealanders; he used to live in a wild state when left in the forests, but with the islanders he became domesticated.

In all these archipelagoes the flora is rich in aspect but poor in kind. From twenty-eight to thirty only were counted in the two groups the Society islands and the Marquesas, and they were the most rich in vegetables; but of these kinds there were some infinitely precious. First must be mentioned the cocoa-tree, which constitutes whole forests, grows even on the small islands, and furnishes of itself to the islanders their drink, their food, their wood for building purposes, and their rope. In the larger islands, we see growing by the side of the cocoa-tree the bread-fruit tree (*artocarpus incisa*), the providence of these countries, and of which tree may be reckoned no less than thirty-three kinds. This vegetable is so completely domesticated in Tahiti, that it no longer produces grains, and it is only reproduced by grafting. Its presence, therefore, indicates an agricultural people. The bread-fruit tree was flourishing in Tahiti, and was wanting in the other islands. It was wanting for instance in Easter island. After the bread-fruit tree comes the ignams (*dioscorea*), of which there are four domestic kinds (*alata, bulbifera, pentaphylla, aculeata*), and a wild plant. In the Fiji islands the natives count fifty different kinds of this plant. The roots, which may be kept for six months, sometimes attain to the weight of fifty pounds, and in some few localities the fruit has been picked twice in the year. Let us also mention the rhizome of the *colocasia esculenta*, a hectare of which might provide food for fifty-eight persons, and would demand only the labour of three men. And last, the papyrus (*carica papaya*), the bananas, the sweet potato, the sugar-cane, etc. The greater part of these vegetables were not to be found in New Zealand, where agriculture was unknown. But this archipelago, as large in size as Italy, possesses hardly more than a thousand vascular plants, among which there are one hundred and fifteen ferns; the roots of some of these were eatable, and formed one of the great alimentary resources of the country. These magnificent countries were both too richly and too poorly endowed to allow man to make much progress there. They were too richly endowed, because with the assistance of the sea, alimentation was abundant and was obtained with little labour; for want, the great spur towards progress, did not press man to work for himself, and to use his intellect. They were too poorly endowed, because the total absence of metal set a bar against the development of any perfected industry. And also the uniformity of benign nature, the contiguity of the greater part of the islands, offered but a very narrow field for any sort of investigation. It was literally an Eden in which man was living, as did Adam and Eve, stupidly, in their earthly Paradise.

IV.

Food in America.

In America human habitation is very much more varied in its kind than in the lukewarm, fertile, and uniform Polynesia. We have here a vast continent stretching nearly from one pole to the other, in which a great and lofty chain of mountains, with immense plateaus, run down all the length of it, from north to south, and whose height is often nearly equal to their breadth. Thanks to these horographical conditions, the Mexicans and the Peruvians were able, under the tropics, to found large, complex, and curiously-organized societies.

As in Polynesia, the food of the Mexicans and of the Peruvians before the conquest, was mostly vegetable. None of our European cereals were known to them, not even the banana, which, according to Humboldt, now affords a too facile nourishment to the inhabitants of the *terra caliente*. According to this author the produce of the banana is to corn as 133 is to 1, and to potatoes as 44 is to 1. Among the cereals, maize formed the principal article of food in Mexico, and it was cultivated also in Peru. To this were added the cassava (*jatropha manihot*) and cocoa. For other food they had the game that they killed and fish that they caught, and also a sort of dog, the only domestic animal of the country. According to Hernandez, there were three kinds of canine animals in Mexico, which have now disappeared, having been driven out by their stronger European rivals. The eatable kind, small, stubby, and dumb, was not unlike the dog which the Polynesians used to eat. B. Dias saw this dog, and even ate him.

The Peruvian civilisation, which seems to have remained quite different to that of Mexico, was based upon the cultivation of the potato and the *quinoa*. Large flocks of lamas and of alpacas were at once beasts of burden and beasts of slaughter. Plants and animals were looked after with great care, as we shall have occasion to point out when we describe the interesting social organisation of ancient Peru.

Outside the favoured zone where the great Peruvian and Mexican empires flourished, there were not in America, before the coming of the Europeans, other inhabitants than the tribes of more or less wild savages. Here and there, in the basin of the Mississippi and also in that of the Orinoco, etc., a few attempts at agriculture were made; but the vegetable kingdom, which the people did not sufficiently understand, and which indeed was less bountiful in alimentary plants, bore only a secondary part in the furnishing of food. Hunting, and fishing in the large rivers, were the principal occupations of the Indians. Everywhere, in fact, these employments have sustained man in his primitive state. They have offered him an occupation according to his tastes, and they have maintained him in a state of savage wildness. In uncivilised countries the alimentation always becomes less and less vegetable in its kind according as one gets farther distant from the equator. In the vast prairies of Western America man lived almost altogether upon one animal, the bison, which he used, and still continues to slaughter in very large quantities. The deer and the roe, no doubt, furnished their quota, but it was to the bison that man was mainly indebted for food, for clothing, and for string for his bow, etc. According to the Abbé Domenech, the town of St. Louis alone, in the year 1849, consumed 110,000 skins of buffalo, of deer, of roe; and the different American companies used then to buy annually about 90,000 skins of buffalo. In Southern America the Araucanians and the Patagonians, etc., used before the conquest, to lead the same sort of life as did their brothers the Red Skins, at the expense of the vicunia swandown, which they could hunt as far as the coast of the Magellan straits. At the present time the European horse, the bovine races, so very plentiful in the pampas, and also the *guanaco*, a known animal domesticated long before the discovery of America, and the American ostrich, or nandou (*rhea Americana*), are the animals which furnish food to the nomad inhabitants of these extensive plains.

In the extreme regions of the American continent, where the vegetable kingdom descends from poverty to indigence, and where no large terrestrial mammalia live in large numbers, the nourishment is nearly exclusively animal, and it is looked for specially in the ocean. In the north, the seal and the whale are the large game of the Esquimaux. In default, people have to content themselves with fish and with shell-fish when they cannot get the domestic reindeer.

The Peacherais of Terra del Fuego, or the Fuegians, much inferior in intelligence to the Esquimaux of North America, and having no other domestic animal but the dog, feeds upon animal food almost entirely. He piles up the remains of the shell-fish in kjøkkenmøddings, as prehistoric man used to do in Denmark; these will form the basis of his food, which he may improve or vary whenever he can capture a seal or a penguin.

V.

Food in Asia.

Leaving Terra del Fuego for Asia, is like leaving the desert to go into the promised land. In the regions that we have hitherto considered, man's table has always been more or less insufficiently found. Here animals were wanting, there vegetables. Taken as a whole, the vast Asiatic continent may be considered a blessed region, offering to man the produce of every climate in abundance. It constitutes a great reservoir of animal and vegetable kinds, all useful to the human race. But the distribution of these kinds will vary naturally with the degree of latitude and with the climate. In India, and in the Indian half of the Malay archipelago (Java, Sumatra, Borneo, etc.) we find the cocoa-tree, the bread-fruit tree, the banana, the sweet potato (*convolvulus batatas*), all the kinds of palm trees, and notably the sago palm tree, whose edible trunk is grown with so little labour in Malay and in New Guinea. To convert into food the trunk of this tree, which is often twenty feet high and four or five feet in circumference, the work of two men and two women for five days will be found sufficient; and the result will give food enough to maintain a man for a whole year. But the Malay peninsula often possesses also the cereal, which we find still more frequent in Southern Asia; rice forms the greatest part of the nourishment of the Hindoo, and it is an important addition to many of the Chinese. The peaceful and laborious Chinaman does not content himself with vegetable food alone, however good an agriculturist he may be. In reality he eats of everything; and though he makes use of pork, he does not disdain rats, monkeys, nor alligators, nor cuttle-fish, nor holothuriæ, nor even swallows' nests. The pastoral life of the ancestors of the ancient Mongolians is so thoroughly forgotten by the present existing Chinaman, that he looks upon milk with horror—excepting however woman's milk, by the aid of which old men endeavour to remedy their decrepitude.

A near relation of the Chinaman, the nomad Mongolian, wandering over the vast plains and plateaux of the north of Asia, over the *country of herbs*, has a much less varied diet. As agriculture is nearly unknown to him, he lives mainly upon the flesh and the milk of his flocks, specially mares' milk, prepared in different ways, and upon butter, and like substances.

As regards the northern Mongolians, the Kamtschadales and the Samoyede races, they live like their congenerous brethren, the American Esquimaux, on the flesh of seals and of whales, on fish, and sometimes on the milk and flesh of the reindeer, as do the European Lapps. The vegetable kingdom furnishes them only with roots, some berries, sometimes the young bark of the willow and of the birch tree, and a poisonous mushroom from which they make intoxicating liquor. They are essentially a fish-eating people, and the food of which they are the most fond is caviare.

In Asia the white race, Aryan or Semitic, is far from having the preponderance. As we have already said, it occupies only the north of India, Afghanistan, Persia, Asia Minor, and Arabia.

Over the other three quarters of the vast Asiatic continent the yellow man is predominant. Now, in that portion of Asia which we may call white, man lives upon our European cereals, and upon our domestic animals. Nothing is more natural than this, since the division between Europe and Asia is purely geographical. To say the truth, our little Europe is only a sort of Asiatic peninsula, in a great measure peopled by emigrants from the East, who brought with them their plants and their domestic animals, to which we may add the kinds indigenous to Europe; the *bos urus*, the small ox of the peatmoss, the reindeer, and a sort of wattle-fenced barley. Our domestic animals come, as we do ourselves, from many crossings; but there are still a certain number of horses, oxen, goats, etc., still in a wild state on the high plateaus of Thibet, where M. Préjévolsky has quite lately come across the wild savage. Starting from Afghanistan and coming westward one sees everywhere, not only our fruit trees, our cereals, our vegetables, our domestic animals, but even the greater number of articles common in our forests. Everywhere in these regions man is stationary and agricultural, living both upon the crops that he has sown and upon the cattle that he has reared. In Kashmir we find all the plants, the trees, and the flowers that we find in our own climates.

In Asia the white man is generally a husbandman, and is fixed to the soil.

We must, however, make an exception of Arabia, where the nomad life has not yet been completely abandoned, and where the people cultivate little but the date tree, though they do not disdain barley and rice when they can procure them.

This short enumeration will be enough to show that Asia, where the principal human types still reside, is certainly one of the great matrices of the *genus homo*. The greater portion of humanity has always lived there, and in that country have been founded some of the greatest civilisations that are known to have existed: for instance, the civilisations that have been created by the Indians, the Chinese, the Assyrians, and the Persians. These large social expansions have certainly been confined in their extent, though we cannot quite tell the reason why. They have conquered many people who have always remained conquered, but who have not lost their vigour. They have diffused themselves farther and farther west, and have allowed the more innovating European races to attain to their present state of civilisation.

VI.

Food in Africa.

Up to the present we have seen the number of animals and of plants that are useful to man grow or die away in proportion as we advance towards or go farther away from Asia, the real storehouse of plenty for the human race. And this general fact is confirmed if we take a bird's-eye view over Africa in an alimentary point of view. It is in the north-east of the great African continent, on the side which joins it to Asia, that animals and plants will be found to flourish in the greatest quantity.

It is evident that ancient Egypt had received much assistance of this kind from Asia; and these precious gifts have by degrees become more or less spread over the interior of the continent. There are certain kinds whose introduction is of recent date: for instance, it was only in the third century of our era that the camel became naturalised in Egypt. Rice, the great cereal of Asia, and specially of India, which is now grown in lower Egypt, was not known there before the time of the Caliphs. Corn, which was cultivated by the ancient Egyptians, is still little known in Central Africa. But the African *gramina par excellence* is the *sorgho* (*holcus spicatus*), a sort of millet. This plant would seem to be indigenous to the country, for it forms the principal nourishment in three-quarters of the African continent. The Asiatic-European grasses, and barley especially, become more and more known as one gets farther away from the northern and the north-eastern countries. In Barbary, on the shores of the Mediterranean, we find nearly the same fauna and flora that we see in Southern Europe. In the oasis of Sahara, and more generally in tropical Africa, the date tree is one of the great alimentary resources; but towards the equator the vegetable food consists mainly of bananas and of ignama.

In all the portions of Africa that are well watered, the vegetation is luxuriant; on the banks of the great watercourses, and of the lakes, the country is extremely fertile, and will generally recompense bountifully the slightest labour. Therefore the population of these favoured regions is often very close. According to Schweinfurth, in the district of the Shilluks, on the White Nile, there are as many as six hundred, or six hundred and twenty-five inhabitants to a square mile.

Everywhere among the native-born inhabitants the domestic animals are the same in Asia and in Africa; but, like the domestic plants, they are distributed very unequally. The horse, originally an Asiatic animal, is rarely found but in the northern half of the Asiatic continent—in the countries which have been more or less inhabited, or perhaps subdued, by the Egyptians in the first place, then afterwards by the Romans, the Arabs, and the Moors, or which at any rate have indirectly felt their influence. On the banks of the Niger Mungo Park occasionally saw the horse in his wild state hunted as an animal of food by the aborigines; but in Southern or in Western Africa the horse was, and is still, little or even totally unknown. In the north the cattle, specially the goat tribe, are very common, and sometimes, as in Sennaar, oxen and camels both serve as an article of food. But in the centre, and in the south of Africa, meat and milk are mainly provided by a race of black cattle. And nearly everywhere the people breed poultry of various kinds. But all over Africa one of the most important elements to the chase is still wanting; for the fauna of Africa, especially in Central and Southern Africa, is often exuberant. We find there in large quantities, the gnu, the antelope, the zebra, the giraffe, the hippopotamus, the elephant, and the ostrich—which latter animal has become domesticated by the colonists at the Cape.

And occasionally, for instance on the banks of the higher Nile and its tributaries, fish is an important article of food.

But it is to be remarked, that nowhere in Africa is dog eaten; and the animal is fairly common all over the country.

In short, life is not too hard for the African. But we must

except, first the Hottentots, who do not practise agriculture, and who live very poorly upon their cattle, and also the Bushmen. They both ignore the art of breeding cattle and also of agriculture. Their food consists principally of roots, locusts, and the larva of ants, which the Boers comically call "Bosjesman rice." These hungry people nevertheless sometimes manage to catch a small head of game, some bird hit by one of their poisoned arrows, or an animal which has fallen into a ditch dug specially for this purpose. Such are the hours of happiness to these poor creatures, which we must class by the side of the Tasmanians and the Fuegians, of whom we shall again say a word when we speak of the nutritive inebriation in the human races.

VII.

From what has been said we may gather some general views.

There is no doubt that man is everywhere, more or less an omnivorous animal; but, speaking roughly, we may say that the vegetable kingdom provides him with the greatest portion of his food. This is especially the case in the tropical regions, as regards the savage as well as regards the civilised man. All the great civilisations have had for their principal support one or several alimentary plants, domesticated or multiplied by culture. Mexico had maize; Peru, the potato and quinoa; Africa still grows the sorgho; India, China, and Malay have rice; the rest of Asia and of Europe live principally upon corn, barley, and rye. To the domestic plants we may add the domestic animals. It is even probable that, by domesticating certain kinds of animals, man was able to begin to form important ethnical groups; but no great society has been able to form itself by remaining merely in a pastoral state. Not that the pastoral state is one of necessity, but that agriculture is always and everywhere the sign and the cause of a superior civilisation. The primitive Mexican, the Polynesian, the Papuan, have never been shepherds; the New Caledonian is an agriculturist, and yet he had no domestic animal. As a matter of fact man lives as he can, turning to his use, with more or less good effect, everything which comes to his hand, and in the succession of his different modes of existence he has not been guided by rule, nor by any absolute laws.

It is likely that the most primitive man began by being frugivorous; then he became omnivorous, specially after he learnt the properties of fire. Now that the human race is spread everywhere, all over the earth, the savage in the tropical zone is from choice frugivorous, but we find that he desires animal food the more as he lives in countries farther north. Yet the taste for vegetable substances rarely leaves him, and even in the arctic regions the Esquimau will gladly regale himself upon the residuum of vegetable matter found in the stomach of the reindeer that he has killed.

It has been affirmed that the first civilisations worthy of being so called, developed themselves only in those regions where the vegetable kingdom afforded an easy means of subsistence, and without labour. There is truth in this statement; but even though such conditions be granted, they will not be found all sufficient. What could be more fortunate from this point of view than the happy islands of intertropical Oceania? But yet we see that man there lives still in his rudimentary state. Can we say that in Polynesia man has not developed because of his isolation, because his field for experiment or for emigration was too narrow? In Asia and in Europe the movement of civilisation would seem to agree with this interpretation of facts, but in Africa it is very different. The Kaffir is hardly superior to the Shilluk of the White Nile, and the Hottentot is very much inferior to him. In America, the influence of migration and of a temperate climate is still more doubtful. The only attempts at civilisation, however little advanced, are in the tropical regions. The people have remained inactive; and in the vast temperate regions of Northern and of Southern America they have grown stagnant in their savage state to such a point that the Red Skin has even no idea of domesticating the bison, which he spends his whole life in hunting.

The ways and means, the conditions under which man lives, will do much, but they cannot do all. The question of race is more important. There has never been any great civilisation among black men. Ancient Egypt was only negroid and mongrel; the Asiatic and Berber races have certainly brought their contingent towards furnishing the country.

There is a hierarchy in human races. Humanity owes to the pure Mongolian race the large and interesting society of Chinese. Among the Mongolian branches we may mention the Malays, and the people of Central America, who have endeavoured to raise themselves above the savage state. But still, not considering the Arctic races, the Mongolian, or a race sprung from the Mongolian, was in a quite savage state in the Mariana islands at the time of Magellan; in the Caroline islands, at the time of the travels of Duperrey; and in the steppes of Asia, large populations of pure Mongolian race have not yet passed the pastoral and nomadic life.

The white races only, whatever their origin may have been, have altogether abandoned their primitive savagery, at least as far as concerns their societies.

Race, therefore, has a larger influence than the ways and means upon sociological development. Whatever his dwelling-place may be, man is ill provided for progressive movement as long as he does not possess, and has not had transmitted to him, certain faculties, knitted and bound together, slowly and laboriously acquired in the struggle for existence. These are sociability, which unites and strengthens individual action; intelligence, which directs his efforts towards an object useful to the community at large; and a patient disposition, which makes him resolute and capable of endurance.

CHAPTER II.

COOKERY.

THE use of fire is of all things the most necessary to the civilised man. It is not only indispensable for all our industries, but without it we should not be able to bear even a single winter;

without it we should not be able to make use of the greater number of our articles of food. The savage thinks less of fire than we do. The Australian uses fire to warm himself, to drive away unclean or dangerous animals during the night; but the Fuegian goes nearly naked, and, like an animal, is able to brave the attacks of a rigorous climate. He hardly ever warms himself. Between the tropics man is far from wanting fire. There man does not need to warm himself; and if people are still in the Stone age, as they are in Polynesia, fire is only useful to them for cooking purposes, and even for this they can do without it without much effort. But we must not despise cookery. The art of preparing food by means of fire is undoubtedly one of the greatest primitive inventions. Thanks to it, man has been able to increase in strength, and in numbers; for, on the one hand, his table has become enriched with many dishes before unknown to him, and he has also been able to use the others more fully and to better advantage. There are nevertheless many people who still do without cookery. The Fuegians, at least in the time of Wallis, used to crunch their fish raw, just as they came out of the water. They would first kill them by biting them close to the gills, and then devour them whole from head to tail. Or, when they got the chance, one of them would tear with his teeth pieces from the body of a putrefied whale, and pass them on to his companions. Even in these coarse ways of life the man shows himself, for wolves and crows do not hand to each other their bits of food. The Fuegian women and children emulate with each other in devouring raw birds. They all of them share with their dogs the raw flesh of seals. At the time of Magellan's passage through the straits which now bear his name, the Patagonians were not more delicate than their neighbours, the Fuegians. The Portuguese saw them devour an ostrich (nandou) without taking the trouble to cook it. At the other extremity of the continent, with the Esquimaux, we find the same disregard for cookery. Ross has seen the Esquimaux enjoy themselves by dancing round long slices cut out of the flesh of one of their musk-smelling oxen. He has seen them devour raw salmon, etc., sprinkling it over with the oil from the seal. Between the Behring and the Kotzebue straits the Esquimaux begin their meal by slitting open the stomach of a seal; they then, one after the other, force their heads into the opening to suck the animal's blood; then they cut off a piece of flesh and swallow it greedily.

The Kamtschadale, without being much more delicate, does make a sort of apology for cookery. He can make a sort of paste with smoked and dried fish; he is very fond of caviare, and sometimes he mixes with it the bark of the willow or of the birch tree. According to Father Huc, the Thibetans, who are people much more civilised, eat their meat either raw or cooked.

We all know of the hideous repasts at which Bruce was present when he saw the Abyssinians take delight in cutting thongs out of the flesh of an ox, who was alive and bellowing with pain. No sooner is a hippopotamus killed than the Bushmen rush at him, open his belly, and fight for his entrails, as though they were dogs. Thompson tells us of similar facts observed amongst the Griquas Hottentots. Here we find man, *flagrante delicto*, turning himself into an animal. It is not a poetical sight, but it is nevertheless instructive. How little are these voracious human bipeds like the man of fantasy described to us by our psychologists now in fashion!

Culinary art, like everything else, has grown by degrees. We find it in an embryo state with the Tasmanian, who because he knows no other kind of vessel than such as may be made of large leaves fastened at the edges with pins, and can therefore have no idea of boiling water, usually contents himself with broiling his meat or his fish. He cooks his eggs or his shell-fish on the hot cinders. Sometimes, however, he prepares his fish by laying it down on hot stones. We have now an idea of the commencement of the oven of the Polynesians, and of their most complicated form of cookery. The Australian, a little more advanced, would sometimes make use of an underground stove—a custom no doubt imported from Polynesia, where it seems to have been general, starting from the Sandwich islands, in the north, down as far as New Zealand. But the Papuans of New Ireland do not trouble themselves with taking the skin off the animal—dog, hog, bird, or lizard—whatever they are going to eat. They cast the animal upon the burning coals, and when he is nearly roasted, they gluttonously devour him. With the Polynesians, who are relatively civilised enough, pottery was as much unknown as it was with the Australians; and their most delicate way of cooking their meat or their fish—which, indeed, they would as often as not eat quite raw, dipping it into the sea as though into their sauce-dish—was by digging a hole and putting into the hole stones that had been heated by fire. On these stones they would place the dog, or the hog, etc., enveloped in aromatic herbs, then they would put more hot stones over him, and again earth on the top as a general covering. In three or four hours' time the meat was well baked, and if we may believe Cook and other navigators, it was extremely good. The New Zealanders used to do the same with the roots of their feculent and edible ferns. But, ignorant as he may have been of the potter's art, the Polynesian knew how to boil water. His plan—a somewhat complicated one, which has been used in different parts of the world—was to throw hot stones into the water. For this vessels are not absolutely necessary. Very often he would simply pour some water into the hollow place in a rock. Some of the Polynesian, and certain American tribes, who were more advanced, made use of wooden vessels for this purpose. Herodotus tells us that the Scythians had but leathern bottles made of skins which they used for boiling; and this was also the custom with the people in the Hebrides as late as the sixteenth century. At Amboyna and Ternate, in New Guinea, people used formerly to cook rice and sago by putting them into the fire in a cocoa-nut on a fragment of bamboo, so destroying a vessel at each meal; and this is also now the practice in the Papuan islands. But cooking by means of water was not introduced until the invention of pottery, which, according to Goguet, people in some countries began to make by covering a wooden vessel with earth or with clay.

When man has learnt how to boil and to roast his food, he has learnt the two most important features in the art of cookery. The rest is only a question of becoming more and more perfect, and every stage of progress in cooking, in making it more various, makes it also more complicated. For instance, with the pastoral people, melted butter was a most useful adjunct. The Aryans of the Rig Veda esteemed it so much that they burnt it in their sacrifices; and even now the Arabs hold melted butter to be a great delicacy. From milk we get articles of food as well as many kinds of drink—cheese, and the koumiss (mare's milk) which the Tartars drink. The coarse Lapp can ice the milk of his reindeers, and so preserve it for his nourishment during his severe winter.

The greatest of culinary inventions, that of making bread, has been extended to many countries. The Egyptians were known to make bread of the lotus grain (*nymphæa lotus*). In certain countries even, people make bread before they have learnt agriculture, by extracting the lees from wild fruits. The Tahitians, or rather the Polynesians, had already got as far as this, for the fermented paste (mahei-popoi) which they prepared from the fruit of the *artocarpus incisa* was a sort of bread.

The ethnography of cookery supplies us, therefore, with another instance that no step in the general march of the progress of humanity is isolated. If by nothing else than by their food and their manner of preparing it, we might draw out a table of the human races, showing their advancement from savagery to civilisation. At first, man, hardly out of his animal state, devours without preparing it almost everything of an eatable nature that comes to his hand; then he broils, first the flesh of animals, and afterwards of fish, which is more tender. By degrees he learns to prepare and to preserve, by means of fire, certain fruits and certain roots. Then the rudimentary intelligence of the savage rises to the idea of cooking in water, and that induces him to invent the precious art of pottery. At this standpoint he is in a state of relative civilisation. Culinary progress gets no further check, and it is the means of bringing about important social events.

Henceforward there is a hearth around which the family unites and governs itself, and around which the ties which bind us together are formed and become strengthened. Man no longer feeds as a beast of prey; he eats humanly, at first only in company with his parents and his male friends. Woman, inferior creatures, have to wait, or else to eat apart. This is a barbarous usage, dating as far back as the primitive times in which man used to kill and seek his food in the forest, and would swallow it without preparing it, as do the wild animals. As kindly feelings developed themselves, women and children became the guests of man, or of men; the family was then constituted.

CHAPTER III.

PSYCHOLOGY OF THE NUTRITIVE WANTS.

To complete the picture of nutritive life, it still remains to us to explain its psychology. It is not sufficient to enumerate the principal aliments of man, or to indicate roughly how he prepares them. We have also to see what are his nutritive needs—such wants as he actually requires—to appreciate the energy peculiar to each race; to take note of, by some characteristic signs, the manner in which he shows his wants, and the degree of pleasure he feels in satisfying them.

We have abundant evidence showing that men differ very widely in their ability and in their energy for consuming nutritive matter. Speaking generally, the more coarse civilisation is, digestive life will be found to hold a larger place; for then not only is the animal need stronger, but it is also less abundantly satisfied.

In the life of every living creature there is ever going on an incessant work of oxydation, which wears away the anatomical elements molecule by molecule. Inside the organised tissues an exchange of matter, which is itself the essence of life, is ever taking place. The injured molecules are perpetually being driven out, and new molecules replace them. In plants and inferior animals this everlasting change of demolition and re-edification works unconsciously; but it is not so with superior animals and with man. With them the molecular mechanism of nutrition awakens in their nervous centres a conscious echo—the feeling either of hunger or of satiety. The hunger of the wild animal—a roaring, bellowing hunger—is almost or quite unknown to most civilised animals. These latter feel little more than appetite, its agreeable forerunner. But it is very different with primitive man, whose larder is always more or less badly furnished. The life of the savage, and especially of the savage who is neither pastoral nor agricultural, is very different indeed from that of certain well-fed townsfolk, whose tissues are overcharged with adipose tissues, with alimentary reserves, and who often, vainly endeavouring to awaken in them a simple appetite, sit down several times in the day to a too plentiful table with the most mechanical regularity. The meal of the savage will depend upon a thousand chances. Nature, as one used to say formerly, serves him very unpunctually. In this kind of life, so much akin to that of the animal, man must eat when he can and how he can, compensating as far as possible his hours of famine by hours of gluttony. To know what he will eat is then the great affair of life; it is the all-absorbing care. All the faculties of his intelligence are absorbed, and often to no purpose, in looking after his daily food. For nearly every other object man's thought is dormant; the cry of the empty stomach makes itself heard before every other. Man needs to eat nearly always, to eat enormously; and he feels extreme pleasure in satisfying this starving want.

We will quote a few instances among a thousand of typical examples that have been related by different travellers.

What a holiday it must be to them in Australia when a dead whale is cast upon the sea-shore! That is their beau-ideal; that is their most perfect bliss! "Fires, lighted instantly, tell afar the news of the happy event. The Australians rub their bodies all over with grease, and make their favourite wives go through the same toilet. They cut through the fat into the lean meat, which they eat, sometimes raw, or sometimes grilled upon pointed sticks. By degrees, as the natives arrive on the spot their jaws are soon well at work upon the whale. You see them, some here, some there, climbing over the stinking carcase, looking with anxiety after all the tid-bits During whole days they will stay by the body. They are rubbed all over from head to foot with fetid grease. They are stuffed to repletion with fetid meat. Their excess drives them to anger and to quarrelling amongst themselves; and they become affected with a cutaneous disease, brought on by their highly-flavoured food. Altogether they present a most disgusting aspect. There is nothing in the world more revolting than to see one of these young native women, a creature whose form is graceful, coming away from the carcase of a whale in a state of putrefaction."*

Analogous cases have been observed constantly among the Fuegians, the Esquimaux, the Lapps, the Bushmen, and others. We have already related, quoting Wallis, how the Fuegian devours live fish from head to tail, exactly as though the man were a seal. He and his companions swallowed down every kind of food that was offered to them, caring nothing whether it was cooked or raw, fresh or salt. Others would greedily eat of the flesh of putrefied whale, which one of them would tear in pieces with his teeth. A Fuegian woman and her children—her little ones rather—picked to pieces raw kids, while the blood was trickling down over their bodies.

The voracity of the Esquimaux is comparable only to that of famished wolves. Ross has seen them cutting long slices out of one of their perfumed oxen, just stunned. They would cut the pieces from about the nose, and stuff them into their mouths, greedily sucking in enormous mouthfuls; and at intervals, when they could eat no more, stop to take breath, complaining that they could not go on. They never let go their knife or the unfinished slice, which they would again begin to eat as soon as they had somewhat regained their strength.

In his diary kept upon his journey, Lyon has described to us one of these stomachic orgies: † "Koulittuak showed me another kind of orgy practised among the Esquimaux. He had eaten until he *was drunk*; and at every moment he went to sleep, his face red and burning, and his mouth open. Arnaloua (his wife) was sitting by his side to take care of him, and to thrust into his mouth as far

* "Explorations in the west and north-west of Australia," by Captain Grey.

† "Journal de Lyon," p. 161.

as she could with her forefinger, a large piece of half-boiled meat. When his mouth was full she bit off the meat that remained outside his lips. During this operation the happy man remained immovable, moving only his jaws, and not even opening his eyes; but every now and then, when the food allowed a free passage for sound, he would express his satisfaction by an expressive grunt. The grease from this savoury repast trickled down over his face and over his neck in such abundance that I became convinced that man became most like a brute when he ate or drank to excess."

Ross has seen other Esquimaux devour, each of them, fourteen pounds of raw salmon; and this was only by way of tasting it! Parry, too, tells with disgust how they would gluttonously swallow raw fat, and suck the rancid oil which remained on the skins of the seals. According to Egede, a young Esquimau girl can eat every day, and for several consecutive months, from ten to twelve pounds of meat, and also a great quantity of biscuits. Eyre has seen an Australian eat in one night six pounds and a half of boiled meat. In spite of the very great similarity of organization among all the human types, it is difficult for us not to admit that these races of enormous eaters have not a more energetic nutrition than other people, and also a greater rapidity of molecular change.

The Bushmen, and even their brethren the Hottentots, go through the same kind of performances. Burchell has seen some of the former make a rush at the bowels of a hippopotamus, "and every now and then wipe off the grease upon their fingers on to their arms, their legs, and their thighs. They would rejoice and make merry, each of them over the share that had fallen to his lot, while they were all bespleshed with blood, and were most disgusting objects, covered as they were with the filth." * Thompson says almost as much of the Hottentot Griquas; but he adds that the power of abstinence of the Bushmen is equal to their voracity. He says that "one of them has lived for fifteen days only upon water and salt." †

* Burchell, "Hist. Univ. des Voy." vol. xxvi. 349.
† Thompson, "Travels in South Africa," p. 88.

There are some races organically superior and much more civilised who are hardly less brute-like in the way they take their food. Cook has seen New Zealanders drink oil with a greediness that would be worthy of the Esquimaux, empty the lamps, swallow the wicks, and press round the boiler in which the fat of seals was being melted, with the enraptured face of children expecting bonbons. The same traveller has told us of the repast of a Tahitian chief: "A woman, seated beside him, was filling the mouth of this glutton by handfuls with the remains of a large boiled fish and with several pieces of bread-fruit, which he would swallow down with a voracious appetite. A perfect insensibility was marked upon his face, and I imagined that his whole thoughts were concentrated upon his stomach. He hardly deigned to notice us. If he did pronounce a few monosyllables when we looked at him, it was only to excite his nurse and his valets (who were preparing for him the paste of the bread-fruit) to do their duty more energetically." *

In Polynesia we are still in a savage country; but we find manners very analogous to those in Abyssinia, where the people are barbarous. At the time of Bruce's travels,† the Abyssinians, after they had cut their beefsteak from the ox, who was standing straight upright, would sit down to table, each man between two women, and they would eat as follows: "Their hands placed upon the knees of each of their female neighbours, their bodies bent forward, their heads projecting and their mouths open, they looking like idiots all the while, and turning constantly to the hands that gave them their food and stuffed them. This went on with such rapidity that they ran near risk of being choked. That was a sign of their greatness. He who swallowed the largest slices and who made the most noise in eating was regarded as the best brought up, and who was most practised in the usages of good society." We see, therefore, that in communities still coarse in their manners, there is no disgust attaching to the satisfaction of nutritive wants. Many a petty king in Central Africa makes his wives

* Cook, Deuxième Voyage ("Hist. Univ. des Voy." vol. vii. 388).

† Bruce, "Hist. Univ. des Voy." vol. xxiii. 362.

chew his food for him; the happy man has then only the trouble of swallowing it. In the Marquesas islands, to take a piece of chewed food out of your mouth and offer it to a friend is considered an act of civility.

The foregoing instances, to which it would be very easy to add others, lead us to contemplate with some degree of attention the life of man who is as yet hardly removed from the animal state. In him the noble faculties of the brain are wanting, or are barely in existence. He has too little intelligence to force his nature for his own benefit; and, above all, he has no foresight. The privations he has undergone leave but a vague recollection in his disfigured memory, for his brain is hardly yet able to retain any durable impression. On the other hand, the pleasure of the moment, and the pleasure of eating above all others, is supreme. Primitive man does not know what subsistence means; he will immoderately consume at one time the provisions that ought to last him for several days. His nutritive faculties must therefore acquire a great elasticity. Sometimes he will adopt this habit from choice. The New Caledonian, when he eats, eats enormously, but he does not pretend to eat every day. The Kafirs, when threatened by famine, impose voluntary fasts upon themselves; they do not eat every day. Even the frugal Bedouin, who can at a push satisfy himself with a few dates and a little milk, devours great quantities of food when by chance he finds his table well served. In some countries famine is so common that the people have accustomed themselves to fill their stomachs with non-edible substances: for instance, the New Caledonians of the Loyalty islands eat the aluminous earth loaded with organic detritus, which they pick up in the excavations of rocks filled with humus. Humboldt saw the Otomacs on the banks of the Orinoco eat after this fashion.

We can, therefore, now see clearly to the bottom of the mind of the primitive man. The noble intellectual faculties, described in our treatises upon psychology as being essential to man, shine by their absence in the brain of the savage. In this conscient life, as yet hardly more than an outline sketch, the care of digestive needs dominates over every other. To get food to eat, to enjoy the happy pleasure of digestion, are the ends and aims of psychical life. We should strangely abuse ourselves if we supposed it was otherwise with men so-called *civilised*. In the societies of modern Europe, how many savages are there not still existing? With the greater number of us, to find our daily provender is ever the all-important matter; and even with many who imagine that they belong to the governing classes, the cream of their pleasure is still that of eating, when it is increased by a few pleasant savours. To tell the truth, when we look under the brilliant surface of our so-called civilised societies, we find that the brute has the upper hand over the angel; and in taking humanity in general as it now stands, we may say that its affective and intellectual needs of a superior kind are only an epiphenomenon.

CHAPTER IV.

INTOXICATING SUBSTANCES.

In his book entitled "The Last of the Tasmanians," the Rev. Mr. Bonwick somewhere ridicules those who have imagined it to be a sign of inferiority in the Tasmanians that they were ignorant of any intoxicating substance or liquor. His satire does not really touch the matter; for in this case, as in many others, there is an apparent analogy between over civilisation and extreme savagery, but for reasons diametrically opposite. The truly civilised man, who is both soft-minded and intelligent, has a horror of drunkenness. He asks for pleasures of a higher kind. On the other hand, the most ignorant savages, the Tasmanian and the Australian, who in spite of their communication with Europeans, did not intoxicate themselves, still showed no aversion to drunkenness. They simply remained in ignorance of it; only to become passionately fond of it as time went on. And so it has happened nearly everywhere among the rare instances of people who have not themselves invented the art of learning how to get drunk. The

New Caledonian has until now remained the only instance of a man who has not succumbed to intoxication; and this exception is very singular.

In the bestial life of the quite inferior races the taste for drunkenness supposes a certain state of progress. In order to be able to make oneself drunk, one must first have been ingenious enough to manufacture, or to acquire, a substance having the necessary properties. And also we must begin by disentangling ourselves from the animal life, we must accustom ourselves to the nutritive pleasure; we must ask from aliments more than the mere satisfying of a want, something more than the happy stupor which accompanies the long process of digestion. Drunkenness, in a sort of way, may be said to be the poetry of digestive life. It at once excites the cerebral life, and lifts man up out of the commonplace routine of his daily existence. It is therefore a pleasure all the more precious as man's life grows rougher, more dangerous, or more weighted with cares. For a poor creature ever fighting against the anguish of hunger, leading an existence often like that of game constantly being pursued, it must be a great happiness to feel, if even only for a moment, a joy in life, and to have the idea of worldly welfare without alloy, not to know the torments of his physical and moral condition, but to imagine himself a god to rule over the beasts and the wild men among whom he is surrounded. The short hours of drunkenness are taken for what they are worth. The drunken man cares little for what may follow; the savage, whether he be Australian or Parisian, never gives a thought for the morrow.

The human race has nearly everywhere sought after and found the coarse pleasures of drunkenness. We can hardly count the few countries in which man has not invented some means of voluntarily losing the small share of reasoning power which he possesses. The means may be different, but the end is always the same. Alcohol is the substance most frequently used, or to speak more exactly, alcoholic drinks. The drunkenness which they produce is generally agreeable, and the means of manufacturing them are abundant, for it is sufficient to have at one's command sweet or feculent substances.

A brief enumeration of the means employed in different parts of the earth to deaden or to excite our conscient life may not perhaps be without some interest.

In the first class of substances that are conducive either to drunkenness, or towards causing an agreeable emotion of the nervous centres, we must place the multiple group of various drinks. Every human race, so to say, has its alcoholic liquids; but the typical drink is the wine from the grape. We are not quite justified in agreeing with Pictet, on the strength of some etymologies of which even we are not altogether certain, that wine was not ignored by the mythical Aryans; and though the ancient Egyptians knew what wine was, though probably at second hand, it is most likely that in the Greco-Roman civilisation wine first became the favourite drink of a notable portion of the white race of man. The Greeks were so delighted with the properties of this gladsome liquor, that they commissioned the god Bacchus to represent it on Mount Olympus. Putting aside alcohol properly so called, and the innumerable intoxicating liquors of which it forms the essence, the alcoholic drinks succedaneous with wine are to be found in every country. At the time of Marco Polo the Chinese knew how to manufacture a wine from spiced rice, which the old chronicler tells us was both generous and excellent.

The Tartars, a wandering and pastoral people, and kindred with these Chinese, conceived the idea of fermenting their mares' milk, and so making an intoxicating liquor, which they called koumiss; they learnt also how to distil this koumiss, and extract from it a sort of brandy, arrack, a drink of which they were very fond. In Arabia, and wherever the Arabian civilisation has taken root, specially in the Fezzan, and with the Moors of Central Africa, the wine of the palm tree is held in high esteem. For the Mahomedan casuistry, supple and nimble like that of her Christian sister, has learnt how to distinguish between the heterodox drunkenness stigmatised by the Prophet, and the orthodox drunkenness produced by drinks of which the cursed wine forms no part. In Persia, for instance, the people manufacture from aromatic substances, from different fruits, specially oranges, a liquor extremely alcoholic (mâ-el hiâs), and this the religious-minded people will drink without any scruple. "It was very funny," Fraser says, "to see Mirza-Reza take the flask in his hands, and as he turned round towards me, explain in a most puritanical tone of voice, the very wide difference between this precious liquor of life and those prohibited and abominable things called *wine* or *brandy*, of which, he assured us, he was never allowed to taste." *

In certain districts of Arabia, and in the northern regions of Central Africa, in the Fezzan for instance, and even with the Timmanis, and also with the Ibo, on the banks of the Niger, where the Moors have more or less introduced the manners and the religion of the Arabs, the people intoxicate themselves as often as they can with wine from the palm tree, which has probably hitherto escaped the sacred ban.

But the African drink *par excellence*, that of the African negro, is a sort of beer made of sorgho; this may be found from the region of the higher Nile as far down as the Kafirs. But the Hottentots, who are not an agricultural people, have replaced this sorgho beer with a sort of hydromel made with fermented honey and various kinds of roots.

The American aborigines, at least those who had any notions of agriculture, have not failed to fabricate for themselves their alcoholic drinks. The Indians of Guiana have learnt how to make an intoxicating liquor from their cassava. But the fermented drinks most common in America seem to have been invented by the ancient Mexicans and Peruvians. In Mexico it is the pulque, or the fermented juice of the maguey—the American aloe—which is held in highest honour. The use of this was so general that, in the time of Humboldt's travels, the duty paid upon the pulque, brought annually into the public treasury, and only from the two Mexican towns Puebla and Tolluca, a sum of 800,000 dollars. The ancient Mexicans doubtless also knew of a beer made of maize, or chicha; but this is more widely spread in Peru and Bolivia, etc. In certain districts this blessed drink, the joy of the Indians of Central America, is prepared by means of mastication,

* Fraser, "Hist. Univ. des Voy." vol. xxxv. 295.

like the kava of the Polynesians, of which we shall have again to speak later on. This practice, disgusting as it appears to us, is only another proof of how far the nutritive pleasures are held in honour with the primitive man, and also how far removed he is from understanding the delicacies and the polished manners of him who is civilised.

In our brief enumeration we naturally confine ourselves to the typical drinks, those which in a certain measure characterise the people and the race. In addition to these there are many others, less general and less important, which we have not now time to mention. The reader, if he wishes to know, may find a list of them in an interesting work by Professor Mantegazza.*

Generally, though not always, as we have seen by the instance of the Tartar koumiss, the use of fermented drinks supposes a certain degree of agriculture. But as the drunkenness produced by alcohol is the pleasantest that man can enjoy, nearly all men of every race eagerly addict themselves to it, as soon as they have learnt its value. The inferior races, too, rush into it generally with an animal eagerness that knows no intellectual or moral bounds. This is generally the first thing they borrow from European civilisation, and they quickly pay for it at the price of their degradation and their rapid extinction.

But touching this matter, we must mention the New Caledonian Kanaks as a singular exception; for they, according to M. de Rochas, are not to be tempted by fermented drinks.

Unfortunately we cannot say as much of the Polynesians. The Tahitians, who at the time of Cook's travels refused all alcoholic drinks, have since become too well reconciled to them. Their speedy and bloody conversion to Christianity was accompanied with a more sincere and more thorough conversion to drunkenness. Distilleries were set up in every portion of the christianised island. During a certain time the islanders, who were nearly always drunk, debased themselves to a lower condition than that of the brute. The women rushed into prostitution, calling out at every moment for more rum.

* "Quadri della natura umana," etc. vol. ii. Milano. 1871.

There is hardly an inferior race with whom the passion for drink is so strongly developed, or with whom it produces analogous effects. For rum, the Australian would willingly prostitute his wife or his wives. The Malays do their best to get drunk with the arrack or with Java rum, which they procure by means of exchange when they cannot manufacture it for themselves. The people of the Arou islands only, consume annually three thousand casks, each cask holding fifteen bottles, and each bottle containing half a gallon. We all know the important part that *fire water* has upon the life, or rather upon the death, of the Red Skin of North America. And the delight coming from drunkenness is equally sought after over the whole of South America. The tribes in the equatorial or the sub-equatorial regions fabricate for themselves their chicha, or some similar drink. And the cavaliers of the Pampas will procure it for themselves as often as they can, giving in exchange leather, ostrich feathers, etc. With the Puelches, the Araucanians, the Patagonians, drunkenness is supreme happiness, they will sacrifice everything for it; D'Orbigny has seen an Indian woman sell her son to procure for herself and her family an orgy lasting for three days.

As regards drink, and the effects it produces, there are many Europeans who are still savages; we know the fact only too well. In our societies, so brilliant on the surface, a true civilisation, which enlightens and ennobles, is yet far from having gone very deep. There is also much suffering, and food is not always assured to everyone. It is known that alcohol in every form will deaden pain and will appease hunger. Still we must not necessarily conclude that because the consumption of alcohol in Europe is always on the increase, drunkenness necessarily follows in the same proportion. The use of alcohol increases certainly much more than the abuse of it. The daily fare of the labourer is gradually ameliorating, and the luxury of alcoholic drinks therefore becomes introduced with it. Drunkenness is a legacy of past times, which ought to disappear with the moral and intellectual progress of the human kind.

As regards the desire for drunkenness among Europeans, we may make an interesting remark. It will be found that this brutal passion is most frequent in the Germanic and in the Slav countries. The people in the south of France, the Italians, and the Spaniards are usually sober. The reason commonly alleged, that of climate, is not sufficient to explain this great difference in the habits of the people. For we may recollect that in the tropics the Tahitian drinks outrageously, and that the Esquimaux in the arctic regions do not possess fermented liquor. If the Slav and Germanic races, otherwise so richly endowed, are so strongly addicted to drunkenness, it would seem that as they entered relatively later into European civilisation, they are, therefore, less far removed from a barbarous life than the Latin races, and that they struggle more unequally against old hereditary instincts, which are destined one day for ever to disappear. The Chinese, who have been civilised for very many ages past, use their alcoholic drinks in moderation; and we find this to be the case all over the empire, though the winters in the northern part of China are extremely cold. On the other hand, in tropical Africa, on the banks of the Niger, there are towns in which all the inhabitants, the governor, the priests, the laity, and even the women, drink to excess.

CHAPTER V.

STUPEFYING OR EXCITING SUBSTANCES.

So addicted as nearly every human race appears to be to drunkenness produced by alcohol, yet that has not been sufficient for man, for he has invented drunkenness of other kinds. To a certain extent, in all quarters of the globe, man appears to have believed what many rose-water pessimists are now showing to us, that life is an evil, and that we ought to abstract ourselves as much as possible, to prevent ourselves from seeing it in its true colours.

Some day, when psychological physiology is further advanced, we shall doubtless know more or less exactly what are the effects of tobacco, of opium, of coffee, etc. upon the conscient cells of the human brain. Upon this interesting point we possess at present hardly more than the incomplete or the fantastical results of labour, and we are obliged to divide these substances, these nervous *aliments*, as they have been very wrongly called, into narcotic agents and exciting agents. This is a commonplace and false division; for the most narcotic, such as tobacco and opium, first begin by exciting more or less the cerebral activity; on the other hand, the most exciting, such as coffee, tea, and cocoa, end by producing mental depression, as they exhaust the cerebral reserves.

The principal narcotic stimulants, as we may call them, are tobacco, opium, hashish, Java betel, and Polynesian kava.

We need not say much upon each of these agents, but we must characterise them shortly. Tobacco, betel (*piper betle*), kava (*piper methysticum*), are all stupefying narcotics. Cook, Forster, and others agree in saying that kava, even if taken in a small quantity, throws one into a state of torpor, during which every noise has a painful effect. Our knowledge of the action produced by betel is as yet far from perfect. It is a plant belonging to the same family as the kava, but the effect it produces seems to be much less powerful. As regards tobacco, it would seem that its ultimate effect produces a sleepy state, more or less lightly marked, accompanied with a feeling of a nutritive advantage—a precious condition, which, since the discovery of America, has been propagated all over the world. People smoke on the banks of Behring's straits, and Schweinfurth has seen various kinds of *nicotiana tabacum* cultivated in central Africa, near to Bahr-el-Ghasel.

With tobacco we may connect the poisonous mushroom. The stupid Kamtschadales use this for making an intoxicating liquor, which at first produces a little gaiety of heart, and afterwards delirium and convulsions.

The narcotic of opium and of hashish (*cannabis indica*) is of a more elevated kind. This is the narcotic of a man of refined tastes, who tries to invest his imagination with fairy dreams, and to produce at the same time a delicious calm state of some inexpressible happiness. The brain only is active; the rest of the nervous system is so deadened that he is hardly conscious that he has a body, and the feeling of heaviness is, so to say, altogether extinguished.

We find the same debauch of imagination in the drunkenness of the American who has accustomed himself to chew the leaves of the coca-tree. He has present to his mind a display of imaginary fireworks; and this is very strong with the Peruvian or Bolivian aboriginal, when he has made himself drunk with the leaves of the *erythroxylon coca*. All this has been well described by Mantegazza.

The effects produced by these three well-known substances are so powerful and so agreeable that the desire to taste them easily degenerates, as does that of alcohol, into a passion and into an irresistible want. In the eastern towns one finds sometimes opium-eaters wandering through the streets crying: "Opium, opium! give me some opium, or I shall die!" We know the desire that many of the Chinese have for this precious drug; and their nomadic neighbours, the Mongolians, imitate them in this habit, and so, through pleasure, lose the qualities of their race. In the same way the American coca-eater, without any hesitation, will abandon his duty, will sacrifice his family and everything, for the sake of his darling leaves.

Tea and coffee, by the side of which, according to Mantegazza, we must put maté (*ilex Paraguayensis*) and Brazilian guarana (*paullinia sorbilis*), have quite another character. Man has here found substances of which the utility is of the highest order, and as to which it would be difficult to deceive himself. They are real exciting stimulants, which give to the mind a greater clearness and more vivacity, a prolonged use of them, generation after generation, must surely refine the brain of any race of people. But nevertheless, we have seen in Paris a fit of madness, which degenerated into suicidal monomania, brought on from swallowing a litre of coffee; whereas in Hedjaz, a man may drink with impunity as many as twenty or thirty cups of coffee every day.

If, as we have already said, alcoholic drunkenness is the poetry of digestion, the delirium produced by coca, and, more strongly still, the cerebral excitement brought on by too much coffee, must assuredly be held to be the poetry of drunkenness. For, far from drowning the conscient personality, such stimulants will bring it out into an exaggerated and deceitful relief.

And even from this point of view, most of the substances of which we have just spoken have their beneficial side, because for a certain time, more or less long, they create a cerebral stimulation. This passing over activity is often intemperate; the use of alcohol, opium, hashish, etc. is generally paid for very dearly. But still humanity need not altogether condemn these kinds of drunkenness; for human life is essentially rough, and is so all the more strongly in primitive communities where the ways and means of life are poor and scanty. It is something, therefore, for man, in the midst of the deluge of evils which assails him on all sides, to be able at will to find a moment of forgetfulness and some appearance of a shelter. We must also admit that from this factitious excitement man has often become imbued with ideas which would not have come to him in his normal state, and that he has also been capable of efforts which he could not have accomplished in his natural weakness.

From the specially sociological point of view, man has also reason to be grateful to the nervous aliments. The use of these stimulants is both the sign and the cause of an ever-advancing sociability. Ordinarily, and among most people, we meet together without any hostile second thought, to enjoy these precious aliments—often indeed to prepare them. In many Indian hamlets in Guiana, piwari, a liquor made from cassava, is made in a trough scooped out of the trunk of a tree, and all the inhabitants help themselves from the same bowl. In Polynesia the roots of the *piper methysticum* are chewed in common, when people wish to make kava. The Indians of Balsopuerto, on the banks of the Maranon, prepare in the same way an intoxicating liquor from the yucas that they have chewed. In Polynesia, a branch of *piper methysticum* is used often to serve as a symbol of peace. To drink kava together was a strong mark of confidence, for it was giving oneself up, more or less drunk, to the mercy of others. In South America, wherever rudimentary agriculture exists, fermented drinks form the basis of all the feasts; with the people it is the joy and the recompense of labour. Coffee plays a similar part in Arabia, and tea in Central Asia.

To sum up: many of these substances, specially the alcoholic substances, put all the sensitive part of man's body into emotion. They excite us to joy, to dancing, to song, to music, and to poetry. Wherever fermented drinks are consumed, more or less of bacchic literature will be found to exist. The stolid Chinese are no exception to this rule; they have sung of their melancholy sorgho beer,* of their brandy made from grain, and of their wine made from rice.

There is surely a wide difference between the heavy digestive satiety of the Australian or the Esquimaux and the gladsome drunkenness of wine, or the coloured visions produced by coca. On the side of drunkenness the nutritive life in man borders upon the sensitive; and upon this we now purpose to speak.

* "Poésies de l'époque de Thang," p. 106, D'Hervey Saint-Denys, 1862.

BOOK II.

SENSITIVE LIFE

CHAPTER I.

OF SENSITIVE LIFE IN GENERAL.

We have just seen the nutritive life expand, by means of a sort of coarse idealisation, into an exaltation of the senses and of the imagination. In the first place to find food, then to feel, and afterwards to think, is the law of organic development in the animal kingdom as well as in the life of an individual man, and of all the human race. We need not disdain or despise the nutritive life, since it forms the large basis upon which rest all the habits of our conscient life, but the phenomena of sensitive life show a higher degree of complexity in our organisation. They result, as does everything else, from our nutritive acts; they are more noble, and we are fully warranted in assigning to them a higher place. In short, with man progress consists in making his conscient life richer and richer, in extending the limits, in freeing it as much as possible from the yoke of the nutritive life.

And from this point of view, the various fractions of humanity will offer many divergences, many singularities, many degrees of development. But that we may better appreciate all these peculiarities, we must divide our conscient life into sections. Each one of these necessary subdivisions will form the subject of a separate chapter.

CHAPTER II.

OF GENESIC WANT, AND OF SHAME.

As Schiller says somewhere: "While the philosophers are learning how to govern the world, hunger and love are performing the task." The motive power *par excellence* of progress assuredly is hunger—hunger, which will never for a long time remain asleep. We have seen how, all over the world, man has exercised his ingenuity to appease his hunger, and to it he is indebted for many conquests. Hunting, fishing, agriculture, and all the industries, and even the social institutions connected with them, have no other reason for their existence than the sharp and ever-piercing needle of hunger.

The energy of man will place the genesic want immediately after the nutritive wants, properly so called. Like them, it is one of the principal factors of the animal and human societies; since in all animals, even in those of a slightly superior kind, its satisfaction necessitates an association of a longer or shorter time. However short our life may be, it obliges every organised and conscient being, whoever he may be, to take account of his companion or companions, to take care of them, and often to obtain their consent. From this necessary commonalty, specially when young people of the two sexes are together, affective sentiments are born, likewise moral ties and social habits.

In order to see in man anything higher than his being the chief of earthly animals, we must be drunk with metaphysical wine; and nothing will cure us of all pride on this head more surely than comparative psychology. All the human sentiments are found in a more or less rudimentary state in the animal kingdom. An idealised love, such as some poets have conceived it, and such as is felt by a few of the choice specimens of the human race, is certainly quite unknown to the brute. But coarse love, the genesic need, as far removed from an ideal state as it is possible to imagine, such as is felt by many people who call themselves civilised, and by the greater number of the savages—this altogether degrades man to the level of or below the other inhabitants of the animal kingdom; for there are some animal kinds who know how to clothe their sexual intercourse with a poetical covering.

There are many birds who try to captivate their female by the sweetness of their song, as for instance our nightingale. Others appeal to her sight; they strut about before her, and show off their bright colours. The albatross of the southern hemisphere (*diomedea exulans*) with his beak touches the beak of his female, they both swing their heads to and fro, keeping time together, and thus remain for a long while looking at each other.* Our turtle doves and our pigeons actually kiss each other. But in this kind of way the palm belongs to certain birds of paradise, and specially to the *amblyornis ornata*, who is certainly more delicate in his æsthetic love-making than his neighbours the Papuans. This curious bird, in order to carry on his love-making, builds a little conical shaped hut, and before the door of it contrives to make a lawn carpeted over with moss; he further shows off his green sward by planting different objects variously adorned with bright colours—bog-berries, grain, flowers, stones, and shells. And he takes care when his flowers are faded to replace them with fresh ones. These singularly constructed nests are strongly made; they last for several years, and probably they serve for several birds. Now these are marks of refinement of which the inferior human races are quite incapable: for instance, the Fuegians, the Tasmanians, the Australians, etc.

With these races, and even with others a little more advanced, shame is a feeling altogether unknown. The Fuegians and the Australians go about naked, and clothe themselves only against the cold, without any thought of decency. It would seem that the feeling of shame arises more frequently in woman. Sometimes the Tasmanian woman, who is ordinarily quite naked, takes care when she sits down to use one of her feet as a sort of covering. It is generally the woman who first thinks of clothing herself from

* For all these matters on animal psychology we have made much use of the conscientious work of M. J. C. Houzeau: "Étude sur les facultés mentales des animaux," etc. Mons. 1872.

feelings of decency. But often the means which she employs will only very indifferently effect her object. We may mention the little apron worn by the Hottentot women; the belt with a fringe to it worn by the women of New Caledonia, and even this is forbidden to young girls, married women only having the right to wear it. In many savage tribes the young girls are obliged to go naked, even among tribes in which some clothing has become common. Amongst many instances where this custom prevails, we may quote the Ashiras of Equatorial Africa, and the Chaymas of Central America. Columbus observed the same custom on the coast of Paria, etc. But sometimes we find the opposite to be the case. In 1498, when Christopher Columbus landed in Trinity Island, the women were completely naked, whilst the men wore their guayuco, a sort of narrow bandelet. In the town of Lari, in Central Africa, all the women go about quite naked, though this country is barbarous rather than savage.

The notion of shame amongst savages is quite relative. In Polynesia, where the women were habitually clothed with two bits of stuff, the upper one with holes in it and worn in poncho, the other rolled over the loins, they would undress themselves at a moment's notice; one might see them swimming round about the vessels, climbing up on to the deck, and even about the masts, in a state of absolute nudity. The women in the Sandwich islands, already half civilised according to European notions, would swim quite naked towards the European vessels, and as they swam carry on their heads their silk dress, their shoes, and their parasol, so that they might dress themselves decently when they got on board. At Tahiti, the women would uncover themselves from their belt downwards out of pure politeness. A young princess, making a short journey in Cook's long-boat, wished to assure herself *de visu* that Europeans were made like the men of her own country. And later, when the missionaries had christianised New Cytherea, the women used to perform the most delicate part of their toilet by the banks of the sea, in places where the water was not more than a foot deep; they would also take care to choose the places where the greatest number of strangers would pass by.

Facts of this kind are numberless. In Africa, the young queen of the Apingi tribe, to whom Du Chaillu gave a bright-coloured piece of Indian stuff, immediately undressed herself in presence of the donor to try on the gift. In Kamtschatka, where on account of the climate the women must be warmly dressed, they will, without the least shame, allow their child to be born while on their knees before any ostrogoth, without distinction of age or sex. This disregard for clothing is not peculiar to savages only. In Mendoza, a Spanish town on the confines of the Pampas, at the foot of the Andes, the ladies every evening and every morning will bathe together with the men, quite naked, in a brook running along by the side of the Alameda of the town.

Sometimes the feeling of shame will show itself in most singular ways and in the most fantastic forms. In China a woman ought not to show to a man her small deformed foot; painters will avoid showing it in their pictures, and it is not considered proper to speak of it in conversation. At Basra, on the Euphrates, if a woman was surprised when taking her bath, it would be her duty to turn her face, without caring about anything else. And this too was the habit of the fellah women in Egypt.

All the foregoing facts, and one might continue to enumerate others almost indefinitely, prove superabundantly that the feeling of shame is altogether artificial. Like all other delicate sentiments, it is a moral ornament which man has slowly and tardily acquired. And even nowadays it soon disappears when danger or malady is close at hand. It is peculiarly a feminine sentiment, and no doubt arose in the mind of woman from the idea of menstruation and of childbearing. Men, even civilised men, feel it little; it is unknown to the majority of those who lead a savage life. There are people, as the Dinkas, who glory in their absolute nakedness. To them clothing conveys a feeling of dishonour; it is the exclusive appanage of the woman. They ironically spoke of Schweinfurth as the Turkish lady.

We need not therefore wonder that the fragile screen of shame becomes easily broken when genesic want makes itself felt.

CHAPTER III.

INTERCOURSE BETWEEN THE SEXES.

Like animals, man at first attached no idea of shame to sexual intercourse, and later on we shall have to establish that in primitive societies promiscuity generally preceded marriage. In the most savage hordes and tribes, even where conjugal union exists in some gross form, the chastity of a married woman is quite a relative matter, and the husband, who owns a property in his wife, has the right to lend her to whomsoever he pleases. But in every savage country, in the Andaman islands, in Australia, in New Caledonia, in Polynesia, in black Africa, etc. young girls were, or still are, perfectly free to dispose of their person at their own will. The same liberty of manners exists even in much more highly-governed countries, for instance in Cochin-China, in Japan, where parents, when they are poor, let out their daughters to hire to houses for the purpose of prostitution, and this noviciate will not at all bear prejudice against the girl in her future marriage. Our genesic instinct always has its share in our nutritive wants, it has almost equal energy, and it chafes at restraint or moral teaching. In insects it will survive the most fearful mutilation. Decapitation even will not prevent the male of the *mantis religiosa* from making his female become pregnant. We find nothing of this kind in superior animals who have a centralised nervous system; but nevertheless with them, and with man, who is their chief, the need of generation is very masterful, and in primitive societies he thinks little of controlling it.

As a matter of course the more inferior the race is, and the nearer it approaches to animal existence, the more the question of generation becomes unlimited. The life of the Australian woman is but a long state of prostitution.

From the age of ten she cohabits with boys of fourteen and fifteen. As she grows older it is her duty at night to offer herself to each guest made welcome by the tribe. The Australian woman who is married, or rather owned by some man, may be lent by her husband. If he is absent another man takes his place. If several tribes camp beside each other, the men at night go and debauch themselves in the camp next to them; for among the Australians, prostitution, like their marriage, is exogamic.

In New Caledonia chastity is not more esteemed or more respected than in Australia. M. de Rochas says that "we should want the burin of Juvenal, instead of the smooth brush of Forster, to paint these masculine savages carrying on their unholy practices upon young lads; to show how old matrons can become intent upon pointing out the road to vice to young virgins, while they themselves are directing the sacrifice," etc.*

In Polynesia we may say that the satisfaction felt in the desire for generation was one of the principal motive-powers of life. In all the archipelagoes the liberty of sexual union was absolute outside the marriage state; and even in marriage it was only restricted—the husband having an undoubted right to lend his wife, his property, to whomsoever he pleased; and infidelity on the part of the woman was punished with only a light correction. On this point the accounts of travellers are unanimous. Everywhere fathers, brothers, and sometimes husbands, would offer their wives to European sailors—that is, of course, if a price was paid. The price asked would vary according to the fashion prevailing at the time. Now it would be a red feather, again some bauble; in the early days nails were chiefly in request. Naked women would climb about the ship and climb up on to the deck. Canoes laden with female passengers would float about near to the vessels, and as they started to go, their fathers, their brothers, etc. would tell them the price they ought to put upon their favours.

These easy manners were the same all over Polynesia, including Easter island, where the women, who were not numerous at the time of Cook's travels, supplied their want in numbers by their activity, and far surpassed the exploits of the historic Messalina. In New Zealand the lubricity was less great; the women there did not prostitute themselves until they received the express permission

* "Nouvelle Calédonie," p. 235.

of the men, their masters; but this permission was easily given—for a nail, a shirt, etc.

In all these islands the feeling of shame did not exist, even in the most rudimentary state. The habitations were hardly more than a roof supported by posts, without other walls than a few rush mats attached from one side to the other, according to the chances of the inclemency of the weather. There was no seperate and individual life; everything passed *coram populo*. All the members of the family slept beside each other under the rush mats; the master of the house, with his wife or his wives by his side; the young boys and the young girls passed the night in the same manner, completely naked. The young girls were brought up to dance the timorodea, a most lascivious dance, and to accompany it with obscene songs. Sexual union was performed publicly, and without any sort of restraint. A princess named Oberea did not hesitate to instruct a young girl of eleven or twelve years to cohabit publicly with a young man.

To offer a girl or a woman to a visitor was, with the Tahitians, an act of ordinary politeness. Bougainville has described this kind of reception, so strange to our European customs: "Every day our people used to walk about the country, unarmed, alone, or perhaps a few together. They were invited to go into the houses, and food was set before them; but the civility of the master of the house does not end with a light meal. They offered to them young girls; the house was immediately thronged with a crowd of curious men and women, who made a circle around the visitor and the young victim, who was singled out for this act of hospitality. The ground was strewn with leaves and with flowers, and musicians were singing to the strains which were being played through the nose upon a flute, as a hymn of rejoicing. . . . . Every glad event is a festival for the natives. They were surprised at the embarrassment felt by our people."*

The conversion to christianity in Tahiti, effected at the cost of fearfully bloody civil wars, was quite apparent. Debauch became more general, more revolting, and it was covered over

* Bougainville, "Hist. Univ. des Voy." vol. iv. 230.

with the veil of hypocrisy. The women did not go on board the ships in the daytime—they went, or they were taken there at night. Amorous desires were so strong in these people that, in spite of the Anglican despotism, the conversation of the women and children was constantly upon the most obscene subjects, and they would talk in the most barefaced manner. Women were fattened with a paste made from the bread-fruit (*popoi*), from bananas, etc. During the time of the genesic impulse they could not walk except to go and bathe themselves, and before reappearing in public they were inspected by men and in a state of absolute nakedness.

To indulge in these amorous pleasures was the chief occupation and the great enjoyment of the Tahitians and the people of the Society islands. To vary their delights they would often travel from one island to another; and they invented the famous society of the Areois, as to which some few words must be said. Among the early Mexicans—those of the eleventh century—there existed a sect called the Ixcuinames, the members of which—though they were in a country where the women were obliged to feed apart from the men—used to hold their festivities and get drunk all together without any distinction of sex, and live in a state of absolute promiscuousness. The members of this society used repeatedly to give themselves up to orgies and to obscene practices, mingling with it all religious ceremonies and sacrificing human victims. This was exactly what the Areois people did; and the analogy is another argument in favour of the Polynesians being people of American origin.

In Tahiti, in the Marquesas islands, etc. the association of the Areois had a religious side to it. In many countries, and with many races of people, man has placed his pleasures and his passions into the safe keeping of heaven. The society of the Areois was a freemasonry, at once mystical and lewd, under the patronage of the god Oro, the son of Taaroa, the Polynesian Jehovah. Everyone was not at once, and without difficulty, admitted into the brotherhood. After a long noviciate the new member, painted red and yellow, was obliged to have a paroxysm of religious fervour. On his second trial, many months or even years after the first, he solemnly swore to put to death every child that might be born to him. He then belonged to the seventh and lowest class of the society. He learnt their songs, their dances, their sacred mimetic practices, which formed the ritual of the Areois. He could not go up the various classes in the brotherhood until at each step he had undergone new trials and new ceremonies; and a special tattoo marking distinguished the members in each category.

Now the object of this religious association was the satisfying without any restraint of their amorous passions; and for all their members infanticide was a duty. In this community all the women were common to all the men; the cohabitation of each couple rarely lasted longer than two or three days. They passed their lives in perpetual holiday-making. They feasted, they wrestled, they sang, and the women danced their lascivious timorodee. The first duty of every female member of the society was to strangle her children immediately after they were born; if the child lived but half-an-hour only he was saved. To entitle her to keep her child, a woman was obliged to find among the members an adopted father, but they were both expelled from the association, and the woman was branded with the name of "child-bearer."

It was considered to be a great honour to belong to the brotherhood of the Areois. A Tahitian, whom Cook brought over to England, declared that he considered himself equal to the king of Great Britain because he was an Areoi.

Facts of this nature though strange are still true, and render very difficult the position of certain belated metaphysicians commissioned to teach in our schools the innateness of the idea of moral worth; for it is very clear that the institution of the Areois had no other object than the fiery lust of the Polynesians, sanctified by religion, and strengthened perhaps by the wish to lessen the inconvenient increase of the population.

We have hitherto spoken mainly of Tahiti, which may be considered as the Polynesian metropolis, but the same amorous licence was common in all the Polynesian islands. There was everywhere absolute liberty in general intercourse; except perhaps a sort of restriction was placed upon married women, who legally required the consent of their husbands—whose property they were, and who had over them the power of life and death. At Nouka-Hiva, Porter says, many parents felt themselves honoured in the preference that was shown to their daughters, and showed their satisfaction by presents of hogs and fruits. The continence of the first English missionaries astonished the Noukahivans beyond measure, and one of their missionaries was obliged to escape to prevent an examination of his person; the natives were constrained to believe that he was made in some peculiar way, and they were resolved to satisfy their curiosity. In the Sandwich islands the great difficulty was in teaching chastity to the women; they were ignorant both of the word and the thing itself.

In Polynesia generally the people have no laws or no shame in anything that touches their amorous passion. But according to W. T. Pritchard, an exception must be made in favour of the daughters of the chiefs of Samoa, whose chastity, guarded by two duennas, is the glory of the tribe. Nevertheless, before marriage, the woman's innocence is verified by a most indecent examination held in the presence of the whole tribe.

This predominance of amorous desire, and the total absence of any scruple of shame with the Polynesians, coincides with an infantine nature in more than one way.

We shall have to speak of their passion for the colour red; and this passion showed itself in every brilliant object. Futtafaiha, prince of Tongatabou, was very much charmed by a tin plate which Cook gave him; he said that he intended that the plate should represent the traveller in his absence. At Tahiti, a princess named Oberea, a woman of about forty years of age, received with much gratitude a doll given to her as a present. Men and women showed an excessive mobility of disposition; for from laughing they instantly burst into tears.

Beside their amorous licence, perfectly innocent as it appears to be to them, there is scarcely place for the affectionate side of love. All travellers are agreed in saying that the feeling of jealousy is absolutely unknown among the men, who regard the women simply as property belonging to them, or as instruments for their own pleasure. This is not quite the case with the women. They do not pride themselves upon fidelity, but are not therefore less susceptible of jealousy. Volatile as they are, they often make some pretensions to the exclusive possession of their lover, whose instability sometimes drives them to despair. All over the world love with woman is a more important matter than with man. And it would seem, too, that woman has more than man contributed to ennoble it. Shame is more essentially a feminine than a masculine sentiment, and in the condition of the Polynesian promiscuous life it is in woman only that this feeling would have any small place in the domain of sensual love.

The same liberty of manners, more or less close to promiscuity, is seen all over the world wherever civilisation is still but slightly advanced. When any reserve, more or less, does exist, it is nearly always imposed by man upon married women; it is the right of the proprietor over an object possessed. But any show of moral continence, resulting from delicate sentiments, slowly acquired here and there from the mental evolution in the race, is a rare occurrence. We shall perceive this more fully as we continue to observe the peculiarities in the different races of the human kind.

Lichtenstein says that with the Kafir Kousas no feeling of love has anything to do with their marriage. "The idea of love, as we understand it," says Du Chaillu, in speaking of the tribe of the Gaboons, "appears to be unknown to this people." According to Mungo Park the Moors of Senegambia consider woman as a sort of inferior animal. They bring her up simply that they may apply themselves to sensual pleasures, without even asking from her any intellectual qualities, and they value her as with us a man values animals in the slaughter-house—by weight.

In Darfur every man is held bound to espouse the quarrels of the lover of his daughter or of his sister. In Abyssinia the courtesans often occupy a high rank in the court of the prince, and frequently they have given to them the government of a town or of a province. According to Bruce the Abyssinian women lived as though they belonged to everybody. The Abyssinian marriages were made and unmade without the slightest ceremony; at their banquets the most distinguished women would publicly give themselves over to their paramours without any idea of shame. "A couple of lovers come down from their bench to place themselves more conveniently; then the two men who were nearest to them raise their cloaks and hide them from the other guests." The ancient neighbours of the Abyssinians, the early Egyptians, were also very dissolute. At the time of Herodotus the lasciviousness of the Egyptian women was well known, and in ancient Egypt it was the rule not to give the dead bodies of the women to be embalmed until three days after their death. We may easily guess why this precaution was taken.

In America the passions of the people are quite as bad as in Africa. The Esquimaux, contrary to the modulating effect which it is said low temperatures have upon the genesic needs, are of all Americans the most shameless. The men and women sleep quite naked, all huddled up together under a deer-skin. They lie a little closer to make way for any guests; and even Captain Parry himself was once so made welcome. It is the duty of the master Esquimaux to offer his wife or his wives to his guest, and he also makes her over, lends her, hires her, or sells her without scruple. The women will prostitute themselves as often as they can when their husbands are absent. "My nation," an inhabitant of the Aleutian islands one day said to a Russian missionary, "in its pairings follows the example of the sea otters."

The Red Skins are less lewd, but not more scrupulous. With the Nadowessioux, a tribe of North America, a woman was held in great esteem because she had lodged, and treated as her husband, the forty principal warriors of her tribe. The men of this race seem to be but moderately inclined to these passions. But it used to be, and is now, very different in Central America. As we shall see later on, the ancient Mexicans were very much given to unnatural desires; and all over South America the Indians of Bolivia are celebrated for the energy and longevity of their manhood. Either from the force of example, or from the effect of mixed race, the Spanish Americans seem to have adopted these lax customs to a very large extent. It was usual for the creoles in the Antilles to offer a girl to their host at night when they gave him his candle. Not long ago at Santiago, in Chili, public girls might be seen in very large numbers. They occupied the ground floor in most of the houses, and would call to the passers-by from the doorsteps; and at the same time in their rooms a taper was burning before the holy images. La Pérouse here also remarked the extreme shamelessness of the people in Chili at the time of the Feast of the Conception of the Virgin.

In Asia, in Mongolian Asia, as well as in the country of the white man, there is great licence of manners, joined at the same time to a ferocious jealousy on the part of the proprietary husbands. In Malay the manners of the unmarried women are very free; but in certain localities, notably at Lombok island, adulterers are punished by being tied back to back and thrown to the crocodiles. In Cochin-China, in Japan, where the woman is considered bound to respect her conjugal ties, parents will readily hire out their daughters either to individuals or to houses for the purposes of prostitution. In China, rich men will, for their own use, buy pretty girls as soon as they have reached the age of fourteen. We cannot here describe the singular artifices practised in China for the purpose of increasing the pleasure of sexual intercourse; they show a degree of voluptuous refinement which has nowhere else been imagined.

It has been thought that lewdness in a race of people has a close connection with the latitude in which they live, and that their amorous desire grows less strong as they dwell farther away from the equator and come nearer to the polar regions. That is one of the superficial views, of which in ethnology we already see too many. The anatomical and physiological characters of races result from infinitely complex causes, which have been gradually forming themselves during the long course of innumerable centuries now past and gone, and which are now nearly impossible for us to unravel. Where have the existing races formed themselves? under what influences? how have they become crossed? what have been their surroundings? These are questions which we cannot answer; and yet before answering them we cannot determine accurately the influence and the conditions under which man lives. For instance, from the point of view of genesic energy the most marked differences are shown among different races and quite independently of climate. It is said that the Bolivian is very salacious; the Red Skin is much less so, and with him the development of the genital organs is very small. On the other hand the Esquimau is very erotic, in spite of his arctic temperature. And again, with other Mongolians of the North—the Lapps of Finmark—the amorous desires are very languid. A minister of a Lapland parish in Finmark, at the end of twenty years, met with only one instance of illegitimate birth among his flock; and in the families of the people there were rarely more than three or four children.

Although incontestably more advanced than other races, both in intelligence and in morality, the white races have shown and still continue to show a thousand instances of great freedom in their amorous passions. The chroniclers tell us that when Çakya-Mouni, the founder of Buddhism, first arrived in the Indian town of Vesali, he was received by the grand-mistress of the courtesans. And now the Brahmins bring up bayadeers in their pagodas, they teach them singing and music, and then the girls can advantageously let themselves out to hire. The holy town of Mecca is filled with public women, composed partly of Abyssinian slaves; they pay a tax to the sherif, and they exercise their profession for the benefit of their masters.

The white race, like all other races, most probably began by living promiscuously, of which practice they have slowly corrected themselves. For a long time the exclusive possession of a woman by a single man was considered as a sort of theft against the community. Hence there were laws which obliged women in different countries to prostitute themselves religiously at least once during their lifetime. On this subject the ancient writers give us abundant evidence. The Lydian women were held bound to acquit themselves of this obligation before their marriage in Hagnœu, the place where their chastity was sacrificed. The same custom was observed in the country of Akisilena, between the Euphrates and the Taurus. Once in their lives the Babylonian women were bound to prostitute themselves for money, in the temple of Mylitta or Venus. The Cypriot mothers sent their daughters to prostitute themselves on the seashore, offering up their virginity to Venus. Later, in Greece, the accomplishment of this amorous task on behalf of the community was confided to the priestesses, or hetæræ, who by their works exonerated the other women.

The Latin writers, historians, and poets have given us coloured pictures of the debauchery practised in Rome. Their descriptions are so well known that we need not now further dwell upon them. We will merely refer to the writings of Juvenal, Tibullus, Ovid, Petronius, etc., to Tacitus and to Suetonius. The Romans had a vocabulary of voluptuous words, for which the equivalents in our own language are wanting. On the theatre they would perform lascivious dances and pantomimic gestures that were very loose indeed. Juvenal tells us of the effect which these spectacles produced upon their contemporaries:

> Cheironomon Ledam molli saltante Bathyllo
> Tuccia vesicæ non imperat, Appula gannit,
> Sicut in amplexu, etc.
>
> Juvenal, Satyra vi.

And Boileau says:

> Le latin dans les mots brave l'honnêteté.
>
> Art Poétique, ii, 175.

The Roman people would cause decency of manners to blush also.

In default of so many facts the erotic, the phallic, and the vulvary religions are sufficient to show us how far the genesic need has beset and tormented humanity, and how also man has with difficulty freed himself from bestiality. A few words will make this last point clear to us.

CHAPTER IV.

GENESIC ABERRATIONS.

THE genesic aberrations of which we are going to speak very shortly are abnormal, but they are not unnatural, because we observe them among a certain number of animals. It is in the females when on heat that deviations from the amorous instincts have mostly been observed. Naturalists and rearers of cattle know that certain cows feel the desires common to the male beast, and that they endeavour to satisfy themselves in the same way. These cows, the *taurelières*, are often sterile, and so also are the hens, called *coquières*, who imitate the genesic mimicry of the cock. In truth, in the animal kingdom, when the powerful genesic instincts are thwarted, they will satisfy themselves as best they can.

And with man it is not otherwise. But man, being a more intelligent creature than the animals, thinks more of refinement, his aberrations tend rather to augment or only to vary his genesic desires. An *a priori* reasoning might lead us to suppose that man would be less scrupulous in this respect in proportion as his morality and intelligence are less developed. We find this to be the case. The voluptuous acts which are stigmatised as unclean among societies and individuals who are really civilised, cause no repugnance to the bestiality of savage races, or to individuals who are but little advanced.

That which we call an unnatural passion is not at all the result of an over-refined civilisation. On the contrary, it is one of the many signs of primitive savagery. It is practised in most savage societies, and provokes no sign of reprobation. The Kanak of New Caledonia is much less lascivious than the Polynesian, and he is satisfied with sexual intercourse once a month, after the very short passion of early youth is passed. But sodomy is his practice; at night the New Caledonians meet together in a hut, in more or less large numbers, purposely to give themselves up to the most foul debauchery.

In Polynesia people are not more scrupulous, and one of the Tahitian gods used to preside at meetings held for unnatural purposes. The New Zealanders abandoned themselves to this vice, even with women. Over all America, from the north to the south, we find similar habits, from the region of the Esquimaux down as far as the Rio de la Plata. According to Bernal Diaz, sodomy was a profession publicly practised among the ancient Mexicans: "Erant quasi omnes sodomiâ commaculati, et adolescentes multi, muliebriter vestiti, ibant publice, cibum quærentes ab isto diabolico et abominabili labore."* This practice of men disguising themselves as women for the purpose of sodomy was also a custom with the people of Illinois.

The Asiatic nations, so much more civilised, are not more moral in this respect. In China there used to be, and no doubt still are, houses of prostitution where sodomy is practised. In the town of Hebheb, in Mesopotamia, Buckingham has seen children walking about in public enticing the passers-by to this same sort of debauchery. In Mecca, the sanctuary of Mahommedanism, the same practices were committed, even in the mosque itself; and Palgrave has observed similar facts among the pious Wahabites of Central Arabia.

Everyone knows the licence practised in classical antiquity; it would be useless to quote instances. We may however call to mind that in his ethereal dialogues on love Plato was not speaking of love for woman.

At Rome the same vice was very common, as is attested by many writers from Virgil down to Petronius. With them it was not considered to be a vice, for not until very late was it condemned by public morality. It was scarcely until the early centuries of the Christian era, under Judaic influence, that this point of ethics became finally determined in Europe, and people truly began to feel the horror and the disgust at this unnatural excess which is now felt by nearly every individual in civilised countries.

This evolution in our moral senses is a most interesting study.

* B. Diaz, "Histoire véridique de la conquête de la Nouvelle Espagne." D. Jourdanet, 1st ed. ii, 594.

We have seen that from the age of primitive barbarism down to our own times, the delicacy of our moral senses has gradually become more and more refined. Our habits have gradually grown more disguised. Distastes have sown themselves in our imagination, they are transmitted to our descendants, in whom they grow stronger than with us; and now we cannot think, except with disgust, of the habits of our ancestors, who, more coarse than we, did not imagine that their daily habits were other than simple and commonplace. It is needless to say that if this progress is general in European civilisations, we do not sometimes find contrary instances. Often enough in our law-courts we are reminded that under the polish of our modern society—proud as it is of its advance of every kind—there is yet always remaining an old remnant of savagery which it is still our duty to abolish. *Memento quia animal es.*

CHAPTER V.

THE DELICACY OF THE SENSES.

WHAT must we understand by the delicacy of the senses? The expression is vague, for the senses are impressionable in many ways. The Père R. Salvado affirms that an Australian can follow the track of a carriage in a forest at night merely by feeling with his feet the impression made upon the ground; also that he can hear a horse's footstep a mile distant; and he tells us of a hundred other analogous facts, observed amongst savages of all races. Does it follow, therefore, that the sense of touch in the Australian or in the Red Skin is more developed than in a blind European well brought up—in a skilful pianist, or in an expert typographical workman? Evidently not. The senses of the savage are often delicate, though their field for active employment is small. For instance, the Australian, whose sight is ordinarily very piercing, is sometimes incapable of understanding the most simple drawing, he cannot always recognise his own portrait.

An old Japanese encyclopedia mentions the satellites of Jupiter; on an old Chinese map belonging to the Père Kegler were marked some stars of the seventh degree in size: and yet one is shocked by the imperfection of the Chinese and Japanese artists in their tints and in their notions of perspective.

We believe the reason to be that, in the perception of our sensations our intelligence plays more or less a large part. The sensation doubtless gives rise to the idea, but in its turn the idea makes the sensation more clearly felt. So that with a little practice we soon arrive at lessening the deviation at first necessary to allow the contact of the points in Weber's compass to give a double sensation.* The senses become more powerful when they are aided by a sustained attention and a well-developed intelligence. It is for this reason that the European when he adopts a savage life often ends by mastering it, even over the savages themselves, from the point of view of the delicacy of the senses, for with him the registry of conscience is larger and better kept. So long as man is reduced to depend uniquely, or nearly so, upon the chase as a means of getting his food, he accustoms himself from infancy to concentrate all his attention to the knack of following the track of man or of animals. It becomes with him a question of life or death; and on this point his mind becomes well stored with subtle observations. The Indian Red Skin will recognise the traces of footsteps; he may even be able to count the number of people that have passed, and to distinguish from which tribes rival to his own they belonged. In this way the European, absolutely cut off from the kind of life led by the savage, will necessarily be inferior to him, though he surpasses him in a hundred other things in which the delicacy of the senses are concerned.

Adventure and the mode of life will sharpen or will deaden this or that kind of sensibility after a whole series of generations. The savage is ordinarily less sensitive than the civilised man to the inclemency of his climate, to physical pain, etc. In the Magellan straits, while Darwin was shivering with cold over the fire, the

* Weber "De subtilitate tactus."

Fuegians, and even the Fuegian women with a child at their breasts, remained in a state of almost complete nudity, exposed to the wind, to the rain, and to the snow.

Taste and smell, modes of touch, as yet imperfect, appear to be not strongly marked with most of the inferior races. According to Humboldt the Peruvian Indians can distinguish at night the different races of human beings merely by the smell; they have three different words to distinguish the smell of the European, the American aboriginal, and the negro. But ordinarily the savage does not recognise odours except from a practical point of view; as regards perfumes and fœtid smells he is often indifferent. He has no æsthetic notions about smells; the olfactory delicacy hardly exists except in the white man. It goes to an extreme degree with the Arabs of Hedjaz, who, according to Burckhardt, cannot endure the least bad smell, and for this reason they dislike entering the towns.

The rudimentary state of cookery with the savage will alone prove how slightly the sense of taste is sharpened in a state of nature. The senses of taste and of smell are connected; they both become refined at the same time, or they are simultaneously obtuse. The Bongos, whom Schweinfurth saw near the tributaries of the Higher Nile, will make an excellent meal of the half putrefied remains of the body of a lion; and for this they have to compete with the flies and the vultures as to who will get the larger share. The Fuegian and the Australian also regale themselves on rotten flesh. And the Kafirs find this food to be excellent. The Kafirs of Natal call it *oubomi*; they use the word sometimes in a figurative sense to express some great pleasure. They explained this to the early missionaries who translated the Scriptures into their language, as being the equivalent for the expression "eternal happiness." This is very different from the sense of taste and smell in a European connoisseur, or in an expert taster, who can distinguish choice wines merely by their bouquet.

On the subject of hearing we may make similar remarks. The ear of the savage will often take notice of very slight noises, but it is always more or less unable to recognise and appreciate musical sounds. That is an interesting point to which we shall again have occasion to refer. Let us remark, however, that the Australian, who can hear the sound of a horse's hoof a mile off, has no instrument of music, that nearly all over the world the only musical instrument owned by the savage is the tom-tom, a sort of drum, more or less coarsely made. When Christopher Columbus first landed in Cuba the aborigines were so much charmed with the sound of his bells that they gave large pieces of gold in exchange for the precious instruments.

We find the same incapacity, we may say the same stupidity, in the sense of sight in the savage. The Red Skin is rarely near-sighted; his look is strong and piercing, but he is often incapable of distinguishing gray from blue. And even in many dialects in Central America there is only one expression used to mark the two colours gray and blue.

It may seem at first that the instances just mentioned would rest upon the too-celebrated theory of H. Magnus, on the evolution of the colour-sense. Must we agree with this author in thinking that at first man was able only to notice some luminous intensity, that he afterwards perceived the chromatic sensation of startling colours—red and yellow—and that later still he learnt to distinguish between colours whose luminous intensity was less clear? We should certainly fall into error if we were to interpret too literally this ingenious theory, based uniquely upon linguistic researches. It may be that the Rig-Veda, the Zend-Avesta, the poems of Homer, are wanting in expressions to designate certain shades of colouring; but direct experiments actually made upon the sense of sight among the different races of people can alone decide the question. Language may instruct us very well as to the condition of intelligence, or the measure of ideas in man's conscient life, but we need not apply to it to learn the degree of sharpness or the delicacy of our senses. Senses that are very subtle may co-exist with an understanding that is very obtuse. Then, though man may perceive sensations very plainly marked he will be unable to test them, or to decide as to what they are. If we were to push the arguments of M. H. Magnus to the extreme limit we should be obliged to conclude that animals, because they cannot talk, have therefore no sense of feeling. And yet the little crustaceous animals (*daphnia pulex*), noticed by M. P. Bert, could certainly distinguish between shades of colour, since they grouped themselves according to the different colours of the solar spectre, showing a marked predilection for yellow.

What is more certain is that bright colours, red especially, are much sought after by many of the inferior human races. The New Caledonian has a passion for red; he likes everything that is of this colour; he paints the posts of his huts, his ornamental objects, his statuettes, etc., all red. We find the same love for red in Polynesia. In New Zealand if any object was of a red colour, with the natives it was called "tabu." We learn from Marchand, from Porter, and from Cook, that red feathers were very highly esteemed in the Marquesas islands, in Tahiti, and in Tongataboo. In Tahiti the chiefs offered even their wives to Captain Cook in exchange for red feathers. We might easily multiply facts of this kind. We will only observe that with the Greeks purple was the royal colour, that in all catholic countries cardinals are clothed in purple, and that in different countries in Europe red is the dominating colour in military uniforms, and also in many popular costumes.

The conclusion of what has been said may be readily drawn. The man in whom intelligence is as yet but slightly developed has delicate senses as regards the exigences of savage life; but the mental side of his sensibility is still only rudimentary. He feels warmly and likes strong impressions, but he is not quick at observing, comparing, and classifying sensations and perceiving nice distinctions. In this way, as in many others, his conscious life reminds us of that of our own children. In studying the rudimentary phase of the æsthetic sense we shall see many other analogies of the same kind.

CHAPTER VI.

CLOTHING.

As we have noticed elsewhere,* two only of the human senses are artistic: the sense of hearing and of sight, and these we have called the intellectual senses. Only the impressions and sensations produced by hearing and seeing are revivifying; they only can be evoked by the imagination, spontaneously or voluntarily. Man may try to reproduce them objectively by external and artificial representations, varying them and combining them together in a thousand ways. Hence we get the beaux-arts. But there is another manifestation of the æsthetic sense more simple and more primordial—the taste for dress—which exists all over the world, even where we perceive no trace of the graphic or of the plastic arts, as for instance with the Fuegians. Before painting or adorning exterior objects, man begins by painting and adorning himself.

The desire to look well, that is, to produce upon oneself and upon others a sensitive and agreeable impression, both by the colouring and the formation of one's body, is not peculiar only to man. Many animals feel and show the same desires, specially during the period of generation. This fact is the more remarkable in many birds, who know how to make their feathers look glossy, to show off with grace, and make the most of their bright colours. In this way, the artifices of certain pigeons, of the turkey, of the peacock, etc., are typical. It does not appear, however, that any animal has yet had recourse to adorning himself with exterior ornament. But there is a bird in New Guinea (*amblyornis inornata*), described to us by O. Beccari, who is evidently on the way towards doing so. He does not exactly decorate his person, but during the period of generation he knows how to lay out a sort of garden ornamented with bright flowers and coloured stones; and in this Eden he builds for himself his little nest.

* "Physiologie des passions," 2nd ed. p. 101; "Biologia," 2nd ed. pp. 438-444.

Then when his earthly paradise is completed he brings his female there, so as to fascinate her by the pleasures of sight.

As man is the most intelligent of animals, in order to look well he uses artifices of more various kinds, and it is very curious to study this tendency in the different portions of the human race. The means used are manifold, and they may be classed into different categories.

When the state of nudity was habitual to primitive man, he first thought to paint or to tattoo himself, to embellish those parts of his body which would best lend themselves to ornamentation. This was ornamentation in its most primitive phase; and in the genesis of art it will correspond to drawing and to painting.

Man went much farther when he began to alter his own shape, when he cut and altered the shape of his human body by means of mutilations and deformations, some of which we shall have to enumerate.

With the progress of civilisation man has clothed himself more and more, and thus were lessened very considerably the outward surfaces on which painting or tattooing would appear visible. And also these deformations and mutilations grew less and less in favour, and by degrees were no longer practised. During this period the taste for dress manifested itself principally in temporary and portative ornaments: such as jewellery, which will still sometimes cause some slight mutilation; such as head-dress, more or less artistically arranged; such as clothes, on which man has ever exercised his ingenuity in varying the shape and in combining different colours. We shall notice these three phases of human ornamentation, which in actual practice will generally be found to be co-existing.

I.

Painting and Tattooing.

In the way of ornament we find that there is a very great similarity in all the Melanesian islands, in Tasmania, and in the islands of New Guinea; and we should be inclined to look for the cause in a community of origin, if many facts of every kind did not prove that the human mind, especially the human mind in its primitive state, often works in the same way in all countries and among all races.

As over all Melanesia the colour red is the one held in most honour, that is the colour usually chosen by the people to paint or to decorate themselves. Even now the poor Tasmanian will cover his body with grease from the wombat, from the seal, from the kangaroo, etc., with all of which he will mix red ochre. Before they go to one of their dances, or to pay their visits, the Australian dandies will trace or have traced on their chests and on their legs white and red lines crossing each other. So decked out they admire themselves, and strut about with a vanity that they cannot conceal. As it often happens with savages, the Australian women paint themselves less than the men. Before fighting, the latter will go through their toilet; they cover themselves with coatings of paint, in which yellow is often used concurrently with red. In the New Guinea islands and in New Caledonia red is less largely used by way of decoration. But yet the New Caledonians, who are passionately fond of it, will often try, by means of lime, to give a permanent red tint to their crispy hair.

To make himself appear beautiful, the Melanesian also tattooes himself, but he does so after the most sober and primitive fashion. In Tasmania and in Australia the men and the women used to, and perhaps do so still, cut long parallel lines upon their chests, upon their arms, their shoulders, and their legs, with a sharp stone, which would immediately produce a cicatrising swelling of a light clear colour. That is a kind of ornamentation of which the Australians are very fond. The Papuans in New Guinea tattoo themselves in the same way, except that with them the scarifications will often be made to cross each other; and in certain tribes tattooing is allowed to be practised only by the men.

But it is specially in Polynesia that the art of tattooing is practised upon a large scale, becoming also at the same time more artistic. And the process employed in Polynesia is very different. In Melanesia it is performed by cutting, and the people do not attempt to colour the cicatrices they have made upon themselves; but in Polynesia tattooing is effected by puncturing holes, into which the people afterwards introduce coloured matter. The first mode is done with a sharp stone, or a shark's tooth, etc.; the second with a small instrument with sharp teeth, something in the form of a comb. In New Zealand both systems are employed—that of the Papuans seems to have been with them practised first. Tattooing by means of cutting is still the method most employed by the New Zealanders; but the system of pricking enables them to adorn and enlarge upon the primitive custom. Much importance is attached to this form of ornamentation; it is a mark of distinction, shown chiefly upon the face. It is made by winding arabesques, showing off the different features of the face, and is often done with considerable skill. The practice is nearly altogether forbidden to women. The lines are coloured, and the tattooer is careful to draw them first only upon the skin, as is commonly done in Polynesia. In their dealings with the Europeans the facial tattooing of the New Zealanders has sometimes been exercised in a way that we should little expect. Some missionaries once bought from a chief a certain piece of land, and the tattoo marks upon the face of the seller of the land were copied on to the deed of sale, serving thereby as his signature to the contract.

But this ingenious form of tattooing does not prevent the New Zealander—after the fashion of the Melanesian—from anointing his face and his body with a pomade of red ochre; and, as this practice is not forbidden to women, they follow it more largely than the men, who often abstain from putting it upon their face for fear of disfiguring the tattoo marks.

In all the other Polynesian archipelagoes, tattooing by means of holes pricked was, or is still, the only one held in honour. The lowest classes of people and the children were alone exempted. Men and women practised it, especially the men. Women did not often tattoo their faces; but even amongst these creatures coquetry had its sway, for they would cover their thighs and their haunches with most wonderful drawings, which they would gladly show, and would do so with a certain amount of ostentation. In all these islands the tattooing was coloured, generally black, sometimes in blue. But the habit, which was common even as far as Easter island, began to disappear at the time of Kotzebue's travels. At this time the people of the Navigator islands confined themselves to painting themselves in blue, from their hips down to their knees, thus making for themselves a sort of trousers. But still we may observe that in Polynesia the practice of tattooing was not so universal as to have been very general. It was unknown, for instance, in the island of Rapa, and in Laivavai, an island belonging to the Pomotou archipelago, where the particular customs, the peculiar monuments (similar to the statues seen in Easter island) would seem to argue in favour of the American origin of the Polynesians, or at least of a portion of them.

Tattooing is not much practised in America; but we find it in South America, among the Charruas, among several tribes of the Grand Chaco, among some small tribes of the Guaranis, and among the most northern Pampas. Contrary to the habits of the Polynesian islanders, tattooing amongst these last-mentioned people is the appanage of the women; but it shows itself only in a few lines drawn upon the face, and it is used ordinarily as the mark of nubility.

The Red Skins of North America, so clever in drawing totems and hieroglyphics, do not tattoo themselves, or do so but rarely; and going farther still northward we do not find the practice to become general until we come to the Esquimaux. And even with them, as we learn from Ross, from Parry, and Beechey, the habit is left to the women, and with them it is seen only in a few lines drawn over the face; this is natural enough, for the Esquimaux are people who are well clothed. But everywhere Parry has seen women tattooed upon the face, upon the arms, the hands, the thighs, and sometimes upon the breasts. For tattooing, the Esquimaux have a special process of their own. They first draw upon the skin the figure they wish to imprint, then following these lines they pass through and under the skin a needle with a thread blackened with smoke and with oil.

In South America the custom of painting the body is more common than that of tattooing. The Fuegians, who are still so little removed from the animal life, paint white, black, and red images over their bodies and over their faces. The Patagonians have the same custom. At Balsapuerto, in the Peruvian Andes, the men and women stain their faces and different parts of their body with a purple rouge. On the banks of the Orinoco, when it is wished to give an idea of a man's extreme poverty, it is said of him "that he has not the means to paint the half of his body." Both sexes have a feeling of shame if they are seen unpainted. Painting with them serves as a sort of clothing; they have therefore some sort of modesty.

In Africa as in America tattooing gives way to painting, which sometimes is a sign of extreme savagery, and sometimes shows a step made towards civilisation. It is only amongst the Niam-Niams of Central Africa that tattooing is habitual. In this country it consists of four little squares, filled with punctured holes, imprinted on the forehead, on the temples, and on the cheeks; they add a sort of X drawn upon the chest, and sometimes various linear drawings are marked upon the arms. This tattooing is not merely ornamental; it serves a social purpose, indicating the tribe of which it acts as the coat-of-arms, and the totem. Remains of æsthetic tattooing still exist in certain districts of Senegambia, where the women endeavour to make their lips and their gums appear blue by pricking them with thorns, or with sharp iron points, touched with indigo; but upon the whole, as an article of clothing, in Africa tattooing plays only a secondary part.

If the people in Africa do not tattoo themselves much, they are very fond on the other hand of painting themselves. The favourite toilet of the Hottentot beauty is to anoint her body with grease and then to sprinkle over herself the dust of red ochre. Sometimes she will vary the colour, using a green powder upon her head and upon her neck. Among the Kafirs, who are neighbours to the Hottentots, the men also cover their body with grease, and then sprinkle over themselves a mineral powder. Farther northward, in Central Africa, on the banks of the Niger, near lake Tchad, in Soudan, we find a change, perhaps of Arabian origin, in the æsthetic ideas; the taste for red, so widely spread everywhere all over the world, gives way to a love for blue, or sometimes will be joined with it. Near to lake Tchad Denham and Clapperton saw a sultan whose beard was stained with a most magnificent azure blue. The women in Saccatoo stain their plaited hair with a blue colour, at the same time they paint their teeth, their hands, their feet, and their nails red. The Nyffe women look like an artist's palette, covered with colour of different sorts. They stain their hair and their eyebrows with indigo, their eyelashes are blackened with khol, their lips are coloured yellow, henna reddens their hands and their feet. This medley of colours is considered as the *ne plus ultra* of smartness. The people are passionately fond of bright colours; the whiteness of their skin, of which Europeans are so proud, with them only excites pity, or sometimes fear. This is one of the thousand facts which show us that the idea of that which is in itself beautiful exists only in the minds of a few metaphysicians in Europe.

The ethnical groups in Asia, all more or less civilised, and belonging for the most part to superior races, have in general long since outgrown the phase of tattooing; but, on the other hand, there are many who have not yet given up painting or staining themselves in a greater or less degree. We know that in Malay, in Indo-China, where the Malay-Mongoloids are predominant, the people consider it a point of honour to have their teeth black. A servant of the king of Cochin-China spoke contemptuously of the wife of the English ambassador (in 1821) because "her teeth were as white as those of a dog, and her skin was as rose-coloured as the flower of the patata." In Burmah the idea of what is beautiful is altogether different; there the women try to make themselves attractive by sprinkling over their face a fine powder from the sandal wood, and in staining red the nails both on their hands and on their feet.

In all these fashions that we have just mentioned we see always the same desire—to make oneself look well. But we must go into the clerical city of Lassa, the Thibetan Rome, if we wish to see disguise transformed into a means of mortification, into a moral practice. In that holy city, every woman who would be held in good repute ought, before she goes out, to smear her face over with a black glutinous varnish; this is not a fashion lately sprung up, for Rubruquis found it in vogue in the year 1352.

> Vous êtes donc bien tendre à la tentation.
>
> Molière, "Le Tartuffe," iii. 2.

If we make this exception, which is altogether of a sacerdotal character, we shall see that everywhere, all over the world, it is with bright colours that man tries to embellish himself; and the colour most preferred, that upon which humankind has generally placed its choice, is red. But there are, of course, exceptions—in Africa, as we have already seen, and also in Persia, and amongst the Arabs in Asia. Perhaps even in these countries the colour red is not altogether displaced, for the old men of Sari, in Persia, at the time of Fraser's travels, used to stain their beards bright red; but in other countries indigo was mostly used for this purpose; and in Bagdad the fashionable ladies seemed to be passionately fond of blue. They were not content with painting their lips azure blue, but upon their legs they drew circles and lines of the same colour, round their waist they painted a blue girdle, and round each of their breasts there would be a wreath of flowers painted blue. The same taste is prevalent, or used to be not long since, in Mongolia, where the women used to paint blue and black lines upon their faces. This desire for painting and tattooing was certainly everywhere very general, existing more or less among primitive individuals and primitive people of every race. Our European ancestors, even our historical ancestors, were not free from it. In Thracia the nobles used to distinguish themselves from the vulgar by painting their bodies. If we may believe Claudian, the Gelons on the banks of the Dnieper used to tattoo themselves. The Celts and the Illyrians used to tattoo themselves in black and blue. The Picts and the Britons used to paint their bodies blue, and so did also the Germans. Vegetius tells us that the Roman soldiers used to tattoo their skin. Pliny tells us that in the early days of Rome the conquerors used to paint their bodies red the day they had gained a victory. Even in our own times tattooing still exists to a small extent; we see a few drawings or some initial letters imprinted upon the arms; it is still used in our prisons and in our armies. And the use of paint of different colours is still far from being forgotten; it plays an important part in the toilet of many women. But the facts which we have just given—and we might give others almost indefinitely—show that the taste for tattooing and for painting exists in a large proportion amongst savage people, and that it is gradually disappearing as civilisation advances. And our next study, that of deformities and mutilations, will also bring us to the same conclusion.

II.

Deformations and Mutilations.

In quite the lowest stage of human development, when man is hardly superior to the brute beast, he, like the animal, does not think of changing the shape of his body for the sake of decorating himself, or for any other reason. The Tasmanians were satisfied with painting themselves, and the Fuegians continue to do so still. But the Australians pull out one and sometimes two of their incisor teeth from the upper jaw. Many of them pierce the *septum nasal* for the purpose of introducing a bony stick. This custom is common enough throughout all Melanesia; and specially with the Papuans, where this nasal ornament, called the *stigma*, is generally cut out of a shell, in the shape of a cylinder about six inches long, and it is often ornamented with red circular lines. The Papuans of both sexes make use of this, and they generally pierce holes in their ears large enough to admit of similar kinds of instruments. These auricular perforations are used for various purposes; it is not uncommon to see the Papuan put his cigar into the hole he has thus bored in his ear.

In ethnography we are often surprised to find the same usages among people who are very dissimilar in race and in language, inhabiting different countries, and between whom we cannot even suppose any sort of relationship. And this fact is doubly strong in the way that different people have mutilated and deformed themselves for what they have imagined to be aesthetic reasons. Ear ornaments, as we know, are used nearly all over the world, not excepting Europe of the nineteenth century; the few exceptions to this general custom are quickly counted. Putting aside the Tasmanians and the Fuegians, for the reason above mentioned, the only people who have not adopted it are the inhabitants of some thinly-peopled islands, amongst which, according to Cook, we may mention the Sandwich islands. Though the custom of piercing that portion of the nose which divides the two nostrils for the purpose of attaching some ornament is less common, it is still very frequently practised. We see it in the Melanesian islands, from Australia to New Caledonia. We find it in Nepaul in Hindostan, where large rings are substituted for the Papuan stigma. On the banks of the Niger the Cambri pass through the nasal septum a long piece of blue glass. The aborigines of Chili have the same custom. In North America the Natchez wear rings of bone in the nose. And the Americans on the banks of Behring's straits put into their nose pieces of string, of iron, of copper, and of amber.

The love for aesthetic beauty has also prompted many people to disfigure their lips. We know that the Botocudos of Brazil have made themselves famous because of this peculiar custom. According to Thevet they slit the lower lip, parallel with the mouth, and into the hole they put a disk of stone or of wood as thick as a man's finger, and as wide as a double ducat. The same observer tells us that they will sometimes take away this "botoque," and amuse themselves by putting their tongue in and out of their second mouth. This habit, inconvenient and ugly as it appears to our notions, was not peculiar to the Botocudos. We find it amongst many tribes in South America, notably amongst the large race of the Guaranis. It seems to us Europeans that the idea of so horribly deforming oneself could only arise in the brain of a madman, but the practice of the botoque is found in races very widely different from each other. It is common all along the north-western shore of North America, from Behring's straits to the banks of the Mackenzie river. Over the whole of this vast region parents are careful to perforate the lower lip of their children, then to put into the hole thus made, which at first is quite small, a piece of iron or of copper wire, which is soon replaced by a bit of wood or of bone, and this too is constantly exchanged for a piece of larger size. The labial orifice therefore grows larger and larger until it sometimes makes almost a second mouth, into which they introduce a disk as large as the hole will conveniently allow.

We may be quite sure that the aborigines of Sitka island, or of the neighbouring regions, have never held intercourse with the negroes of Central Africa; but, nevertheless, these latter people have adopted a custom similar to that which we have just described. We are told by many travellers that the labial ornament is worn, especially by women, from the banks of the Niger as far as the basin of the Upper Nile. In Kouka the women put into the lower lip a large silver nail, so long that to make way for it they are obliged to extract the two eye-teeth in their lower jaw. And the inhabitants of the town of Follindochie insert into each lip a piece of thin glass in the shape of a crescent. W. Baker tells us that on the Upper Nile also, the women of the tribe of Latoukas would dispute amongst themselves for pieces of a broken thermometer with which to adorn their lower lip. According to Schweinfurth, the same custom exists near to Bahr-el-Ghazal. There the women generally, but also sometimes the men, will insert into their upper or their lower lip either a nail or a disk of copper. The women of the Mittou tribe seem to try to give themselves a muzzle: they stretch their lower lip by means of a botoque, and then endeavour to prolong their upper lip symmetrically with it. We can only make guesses at the origin of this singular custom. It may be that the people wished to resemble certain animals; for the primitive man has never professed for animals, especially for those endowed with strength and ferocity, the same disdain which we see everywhere in Europe amongst the partisans of the human kingdom. Livingstone tells us that the negroes on the banks of the Zambesi kill an elephant with marks of respect, calling him "grand chief" as they do so. Many of the tribes in Southern Africa pretend to have as their ancestor either a crocodile or a lion. According to the Kirghiz, their race—that is the Mongolian race—offers the most finished type of human beauty, that which is absolutely beautiful, because the bony structure of their face resembles that of a horse—the greatest masterpiece in all creation.

We are also at a loss for any reason to ascribe to the very common custom of disfiguring the cranium. This habit, more or less practised even still in different localities in Europe, notably in the neighbourhood of Toulouse, was, as we know, the rule with the ancient Aymaras in Peru, whose cranium was hardly like that of other human beings. It was even, it would seem, a distinction reserved only to the chiefs. We find also the same custom was common in North America with the Chinuks, the Chactas, the Natchez, etc., who for nearly a month, by means of a very simple instrument made of two laths of wood and some string, used to subject the cranium of the newborn child to a gradual and constant pressure.

A similar custom was also practised at Tahiti, and in many of the Polynesian islands; and again, we see it in Sumatra and elsewhere. Taking everything into consideration, insane as these practices may appear to us, they nevertheless afford some mark of superiority, for they ought to be considered as one of the many attempts that man has made to change his nature and his person, according to his caprice or his want at the moment. It is by making these bold endeavours that he differs from and raises himself above the level of other mammalia.

We will briefly mention a few other kinds of mutilation that we find practised amongst different people. To describe them would lead us too far into details, and it is not our object to write a minute treatise on ethnography. Among these deformations, there are some that would seem to have sprung from a strange notion of æsthetic feelings, such for instance as the atrophy of the foot with the Chinese women—though the Chinese appear to attach some erotic signification to the custom. But amorous desires and an idea of the beautiful will often be found hand-in-hand together. Other mutilations, such as circumcision, infibulation, etc., may at first have arisen from an idea of æsthetic sentiment, and then religious notions intervened and gained the upper hand. It has been thought sometimes that these last-named mutilations were adopted from hygienic reasons; but hygienic notions do not readily enter into the brain either of the barbarian or the savage. However this may be, circumcision is found in many and in very different countries. Total circumcision is the law both with the white and black Mussulmans in Asia and Africa; it is customary also with the Kafirs and with the Hovas in Madagascar. Circumcision by simple cutting, without loss of substance, is practised amongst all the Melanesians, Australians, Papuans, New Caledonians, New Hebrideans, etc., and over the greater part of the Polynesian archipelago, as far as Easter island. According to Moerenhout, in Polynesia this operation was reserved for the chiefs, and was performed ceremoniously, as is the practice in the Mussulman country. We may mention the semi-castration in some Hottentot tribes, and then we will close this enumeration, already long enough, by remarking once more that the taste for mutilation and deformation lessens gradually in proportion as man becomes more and more civilised. The love of jewellery, of clothing with bright colours, of artistic head-dresses, seems to be the last phase of evolution in the toilet; and on these points we have now to say a few words.

III.

Jewellery, Clothing, and Head-dressing.

In saying that the taste for jewellery and bright-coloured clothing is the last stage in man's love for attirement, we do not at all mean that this phase is necessarily built upon the others and that it commences only when the others cease to exist. The most extreme state of savagery will not exclude any kind of ornament, but the practice of mutilation and of coatings of paint disappears much earlier than the love for jewellery. It is true enough that all these customs may exist at the same time—indeed we see that they do coexist—for certain mutilations have been invented for no other purpose than affording a means of increasing the number of ways of wearing jewellery: for instance, the boring of the ears, the nose, the lips, and even the cheeks. The Esquimaux who live on the side of the Mackenzie river make a hole in their cheeks to put in a sort of stone button—"a cheek-button," as they call it. But these artificial means of adorning oneself do not at all exclude the more natural means. Everywhere, since the age of cut stone, man has found pleasure in wearing necklaces and bracelets, and in putting similar ornaments around his arms and legs. From the lowest to the most refined phases in civilisation the same taste exists: the matter employed and the workmanship only vary. At first man was content with shells, animals' teeth, bony fragments worked in various forms, coloured and wrought stones, pierced and then joined together by means of a thread. As soon as metals were known, we find him decorating himself with copper, gold, silver, brass, bronze, and iron. This jewellery, coarse as it may now appear to the eye of the civilised man, still continues to delight the savage. Men and women wear it in emulation of each other; the men are often more richly adorned than the women. Hutton reckons up the ornaments worn by the king of the Ashantees: a necklace of gold and stones that in his country were considered precious; on his shoulder were sheaths of gold containing the saphis or talismans; on the fingers were a profusion of gold rings, and gold castanets were on the little finger and on the thumb; around his wrists, his knees, and his ankles, were rings and bracelets. In countries where precious metals are unknown the people make use of others. The Denka and the Bongo women on the Upper Nile overload their neck, and arms, and legs, with masses of iron, of which the total weight will sometimes be as much as fifty pounds. This heavy jewellery naturally does not exclude ornaments of less value, such as feathers, flowers, leaves, and bright-coloured berries. For we find this taste common everywhere all over the world; the poor Tasmanian women, the Veddah woman in Ceylon, wear, or used to wear, necklaces made of shells, and to put flowers in their hair. Like our great ladies in Europe, women amongst the most primitive savages have also a desire for some ornamental form of head-dress.

This is a matter of much import with many people, and especially amongst those whose hair is woolly; for the fleece will readily adapt itself to the most ingenious forms of construction. The Fiji women sometimes wear head-dresses that measure as much as five feet in circumference. And in Africa many travellers have been astonished at the wonderful variety of head-dresses, both masculine and feminine. At Jenneh, in the valley of the Niger—a country relatively civilised—the plaited hair of the women looks like the helmet of a dragoon soldier. And to set off the complicated architecture of these head-dresses, the people will put slabs and diadems on to their heads, and round them twist the different plaitings of their hair. Formerly in Abyssinia the head-dresses of the men used to serve as a means for registering their acts of valour. Each enemy killed or taken prisoner gave the right to wear a tress; so that when a man had been ten times victorious he might then dress his hair as he pleased. Not only are these people careful to adorn their heads, but everything which is capable of receiving adornment. The New Caledonians, whose only article of clothing is a belt, etc., round the loins, will dress their belt as carefully as they would dress their hair. In other places people will ornament their teeth. For instance, in the skull of a Dyak, six incisor teeth had been carefully bored, and into them was inserted brass wire, at the end of which there was a little ball. The aborigines of India, whom Marco Polo saw, used to cover their teeth with a case of gold.

As our social life becomes more refined, our taste for ornament changes; our jewellery gets to be smaller in size and more artistic, the quality is also rarer and more precious.

But it is more especially by his clothing that man has always tried to adorn himself. The Polynesians knew how to stain red and yellow the stuffs that they got from the mulberry tree. The natives of Brazil plaited for themselves real clothes from the parrot's feathers. In Maouna, in the Navigator islands, people obtained the same result by intertwining the leaves of the palm trees of different colours. As a general rule the love of bright colours is shewn in articles of clothing, in proportion as the people are in a wild or barbarous state. We all know the importance that was given to purple in the days of classical antiquity. Homer clothes Ulysses with a double woollen cloak stained with purple. Martial estimates at ten thousand sesterces the price that in Rome would be paid for a cloak of Tyrian purple. Pliny reckons that a pound weight of the best wool dyed with Tyrian purple was worth a thousand denarii.

IV.

Evolution of Taste in Ornament.

Some few fundamental notions may be gained from the facts which we have just mentioned.

The taste for ornamentation seems to be common to all members of humanity.

We see it first in the custom of man covering himself all over with paint, and the colour red is the one generally preferred. We are therefore obliged to admit that primitive man is not altogether inept in distinguishing the various shades. It is therefore difficult for us to believe, as H. Magnus would have us do, that the authors of the Vedas were deprived of the chromatic sense and capable only of distinguishing light from darkness. According to this hypothesis we should have to admit that the ancient Aryans, who were both shepherds and poets, were, as regards their sense of colour, inferior to the besotted Fuegians and Tasmanians.

In another phase of civilisation, a little less degraded, but still very savage, to the coatings of paint we may add the mutilations and deformations, and to both of these we may again add the wearing of jewellery. This seems to have been customary from the very earliest times.

By degrees the deformations and mutilations become less serious and less hideous-looking; the jewellery becomes lighter and more artistic. The taste for ornamentation is shown mainly upon the clothing, which is ever brighter in colour in proportion as man is less far removed from the state of barbarism. In Madagascar the officers bedizen themselves with embroidery to an unlimited and boundless extent.

In the primitive phases of human development the desire for ornamentation is common to both sexes; and often, as in the Azore islands, the men are decorated more richly than the women. This is also the case in the basin of the Upper Nile, and among certain tribes in North America, where the women spend a great part of their time in painting their husbands. In Yucatan, the men used always to carry about with them looking-glasses of polished pyrite, as also did the Mexicans. Then as a man grows more intelligent and less sensitive, the desire for ornamentation becomes the taste of the woman. In this respect the women now alive in Europe are nearer to the barbarous times than the men—the dressing of their hair, over which they take so much pains, their taste for bright colours, the paint which many women still continue to use, are all relics of a savage past epoch; they are small details which have not yet become extinct. The boring of the ears, even, is connected with one of the rudimentary phases of civilisation —with that of mutilation.

The more man progresses, more his reasoning power develops, more his intelligence predominates in his mental life, and more also he gives up wearing bright colours and ornaments of every sort. Perhaps, in looking at the dull costume of the European citizen of the present time, some learned man in a future age may conclude, after the example of H. Magnus, that we could not distinguish the difference in shades of colours.

Even nowadays the remnants of ancient barbarism are seen in certain costumes which we are scarcely allowed to change—in the dress of official functionaries, judges, priests, and especially in military uniforms. The red coat of the English soldier, the dyed trousers of the French soldier, the epaulets, and all the trappings of the military costume are the last manifestation of the æsthetic savage. We shall eventually give them up as we have given up boring holes in our noses, and depressing our craniums. The march of civilisation is assuredly both slow and halting. Not until after much trouble does the human kind free himself from the instincts of inferior creatures. Poor worms that we are; our existence is but for one day! In the eyes of the individual, almost as soon closed as they are opened, everything would seem to be fixed and unchangeable if the annals of humanity were not opened and laid out before us to show with unmistakable plainness that progress is not altogether a dream.

CHAPTER VII.

ON THE ARTS IN GENERAL.

Man has undoubtedly talked nonsense on every subject; but no subject has been the cause of so many of his wanderings as that which we call æsthetics—and in this respect a special privilege is given to music. Every time that an author opens this chapter, he runs a very considerable risk of losing his head; sometimes, indeed, he affects to do so—it is thought to be a sign of good taste.

We should assuredly be more moderate and more rational, if we formed for ourselves a true notion of the origin and the design of what we call the fine arts; and on this head biology may serve as a guide to sociology.

In man, and even in every conscient animal, a strong impression will always be inclined to spread itself all through the nervous system. When the impression makes itself felt in a man who is very intelligent, in whom the field of a conscient life is very wide, the nervous shock is transformed first into sentiments, into ideas, and then, if his strength is not exhausted, into a reflex power of action. In the animal, in the child, in primitive man, in woman, this strong impression will oftenest directly show itself under various forms, according as the impression has made itself felt in this or in that organ.

Ordinarily, in the creature who, intellectually, is but slightly developed, this excess of nervous shock shows itself mainly in muscular contractions, in movement of the limbs, in gestures, or in cries, which are the gestures of the larynx. And these phenomena may, to a certain extent, be reversed. If any given impression provokes ordinarily such a form of gesture, such a movement, or such a cry, it will often be sufficient to perform, or to see performed, this gesture, to utter, or to hear this cry, in order to feel more or less the impression to which they correspond. Man may therefore reproduce, or excite at will, in the cells of his own consciousness, or in those of others of his fellow-creatures, a certain number of impressions, or of sentiments.

This is the whole foundation of æsthetics. From the cry we get song and music. Gesture, performed more or less in cadence, becomes a dance.

As every strong impression is surrounded with a crowd of images and mental visions, man, in reproducing or in trying to reproduce these images, will invent drawing, painting, sculpture, in a word, the graphic and the plastic arts.

Naturally, too, the degree of perfection or of imperfection in these arts will be closely connected with the degree of development in the man who shall put them into execution—with his sort of life, his tastes, his passions, and the kind of civilisation of his tribe or of his race.

CHAPTER VIII.

DANCING.

An art becomes more dignified as it becomes more intellectual. The art of dancing is therefore, surely, of all the arts the most inferior and the most savage. We find it even among certain animals. The males of certain birds go through real dancing evolutions before their females, in order to charm them and captivate them. The measured salutations of the male turtle-dove and also of the male pigeon, when the amorous passion seizes them, are in fact choregraphical exercises.

In the human species the practice of dancing is almost universal. Among certain people dancing mixes itself up with everything, and discloses characters of very various kinds. The three principal categories are—the hunting dance, the war dance, and the dance of love.

The lowest of all is the hunt dance, for this ordinarily reduces itself to a coarse imitation of the movements and the manners of the animals which are habitual game of the people of the tribe. It was after this fashion that the poor Tasmanians and Australians tried to imitate the movements of the kangaroo and of the emu; for the hunting and the taking of these animals was with the negroes of Tasmania and of Australia the main affair and the great joy of their life. And also, for the same reason, the dance of the Kamtschadales is copied from the awkward movements of the bear. The Red Skins performed their buffalo dance before a hunting expedition, always wearing the skins of the animal. Many other instances of the same kind might be given.

And war, too, being only another kind of hunting, in which man seeks his game in man—war also had its dance, which was held both before and after an expedition. The war dance is very common, from the dance of the New Caledonians, which is accompanied with anthropophagical songs, down to the Pyrrhic dance.

The more a people are savage, the more the war dance is characteristic, and is oftener practised. For instance, the New Caledonians, before going upon an expedition, would, while they were dancing, hold the following dialogue with their chiefs: "Shall we attack our enemies?" "Yes." "Are they strong?" "No." "Are they brave?" "No." "Shall we kill them?" "Yes." "Shall we eat them?" "Yes."*

Like the hunt dance, this war dance is performed only by the men. The war dance of the New Zealanders has always made a strong impression upon European travellers. In this country the dancers used to brandish their lances and their darts, and would strike at an imaginary enemy with their patou-patou, singing the while some wild song as an accompaniment; for the primitive dance was rarely performed without song or without music. We may imagine that an exercise such as this could hardly be other than masculine. And it was the same too in the various dances of the Red Skins. These people looked upon their dances as serious occupations, characterising all the incidents in their lives: a treaty, a reception of foreigners, a war, a birth, a death, a harvest, a religious ceremony, etc.

* De Rochas' "Nouvelle Calédonie."

But as soon as women begin to dance, either before men or with them, the dance has quite another character; it is then more or less connected with ideas of an amorous nature, and often becomes indecent. It is generally the measured motions of the haunches or of the basin of the woman that give to the dance an erotic character. At Tongatabou, in the Sandwich islands, these dances, always performed by women, were a pleasure much enjoyed by the people. We find similar kind of dances everywhere, specially at Madagascar. In India the lustful dance became a religious art. Each pagoda had its bayaderes, brought up to their profession from their infancy, by means of a methodical and constant training, and these women were hired out at high prices to rich men, and they thus gained considerable sums of money. Everyone has heard of the almées in Egypt; they were a sort of unprofessional dancing women.

All over negro Africa the people dance wildly. It is an amusement of which both the sexes are passionately fond. And the dances are often very indecent, but the negroes seem to take special delight in the very quick movements, which are always performed in time. "As soon as they hear the sound of the tom-tom," Du Chaillu says, "they lose all hold over themselves." A real fit of dancing fury comes over them, and for a while they forget all their public and private misfortunes.

As we might expect, the people least given to dancing are the serious and methodical Chinese. Although in China the people are very fond of scenic representations, the idea of dancing seems to have occurred to no one; in the eye of a Chinaman dancing is a ridiculous amusement, in which man seriously compromises his dignity.

In a word, we may say that dancing is to reproduce interesting scenes by preparing different kinds of pictures of animal life; to move one's body more or less violently, singing at the same time, or following the measure of, a rhythmical air. That is the foundation of real dancing, of dancing such as we see it amongst the primitive nations, or, in our European societies, amongst the men of the people. We have now nothing to say of our opera ballets, or of our drawing-room dances, which are exactly what Mr. Tylur has called "the relics of past times."

CHAPTER IX.

VOCAL MUSIC.

As we have observed, the art of dancing is not peculiar to humanity. We may say the same also of music, and to a very much greater degree.

Certain singing animals remind us in this respect very strongly of primitive man, or of man not yet civilised. The organist of the East Indies (*euphonia musica*) can perform the seven notes in the scale. The chaffinch goes farther still; he sings real songs, some of which he has invented, others have been taught to him by man. One of his songs has as many as five long strophes; it is much more complicated than many of the songs of the savages, which never run to any length. At the time of Cook's second voyage, the women of Middleburg island, in the Fiji archipelago, could only sing from *la* to *mi*. But the song, both of birds and of savages, moves only at short intervals, and they do not adapt themselves easily either to measure or to rhythm. We may observe also that the song of the bird is quite as artistic as that of the man, for he perfects himself by practice, and also he gradually improves with the progress of time.

In large monkeys also we find both vocal and instrumental music in a rudimentary state. Darwin has seen a gibbon who knew how to modulate an octave; and Savage relates that the black chimpanzees sometimes come together in twenties or in fifties, to hold a concert by beating a hollow and sonorous piece of timber with small sticks which they hold in their hands. It is perhaps only a noise that they make; but it was only by slow degrees that music has grown to be other than noise; and all over the earth the drum seems to have been the first instrument of music known to man.

Like dancing, music is only the art of expressing, or of awakening our mental feelings, which are more or less coarse, more or less refined, sensitive, or affective. But it is an art very differently expressive from that of dancing, for it imitates or reproduces the infinitely various modulations of man's cry and of his voice.

Every strong impression in man, and especially in primitive man, is shown by reflex movements, principally by the contraction of the muscles of the larynx; and from this results either the utterance of cries, as with animals, or the special variations of the sound of the voice. These larynxated manifestations of the affective life are fatal, and are so closely bound up with the psychical phenomena, that every man can understand what they mean, and also how certain vital modulations excite the same feelings, more or less, in everyone. In making music rest upon the animal cry of passion, Diderot expressed an idea both very profound and very true.

In truth, music has the same origin as speech, and at first one was taken for the other. Then as language by degrees became more perfect, and words were gradually created, the spoken language became divided and separated from the language which was sung; this latter was mainly used for the expression of some few sentiments, very intense and very narrow in their bearing.

Among all primitive people vocal music hardly differs from a monotonous recitative in a minor key. Man is aware of an agreeable impression in perceiving a musical sound, that is, a sound which causes one or more of the terminal fibres of the acoustic nerve to vibrate thoroughly and without confusion.

The roaring sound of the waves, on the other hand, is bounding and irregular. One hears only the noise. But as regards music, the ear of the savage is not exacting. When he has once perceived the sound of a regular note, the man whose intelligence is but slightly developed is quite struck by it, and he feels a great pleasure in repeating it over indefinitely, without gradation of sound, and with little or no notion of the half tones. Certain races, very far developed, the Chinese for instance, are in regard to music very poorly endowed. The Chinese scale in music has only five notes, and no half tones; whilst the ancient treatises on Sanscrit literature divide the scale into seven intervals, with twenty two intervals of lesser dimensions. Later, the Greeks invented the diatonic system, distributing the succession of sounds into a series of intervals called tones and half tones: hence arose our modern music.

But the Grecian music was only an accentuated melopœa, marking and exaggerating the intonations of the voice. Often in primitive music song is accompanied with movements and with gestures; music then is hardly different from dancing. The savages seen by Cook on the western and northern slopes of North America had not got beyond this; they accompanied their songs with regular movements of their hands, and keeping time by striking their paddles against the planks of their canoes, and performing a thousand other expressive gestures. The women of Tongatabou used to sing and keep time with their music by making a cracking noise with their fingers. Everywhere the primitive song was only a recitative, sometimes interrupted by voices imitating the cry of certain animals. With the Bongos of Central Africa their song is still only a hurried sound of broken words, imitating, it would seem, the barking of a dog or the lowing of a cow.

If the primitive song is only a recitative, we may suppose that it was intended as an accompaniment, to set off the narration of some acts of warfare, of a love story, or sometimes of a mythological legend. In Tahiti and in New Zealand the people used to sing, to airs that were more or less monotonous, of the deeds of valour of their ancestors, of the exploits of their gods and of their heroes, of the creation of the Polynesian universe. That was one of the favourite pleasures of the voluptuous Areois, of whom we have already spoken.

As it is natural to suppose, vocal music seems to be the more primitive. It is that which still makes the most impression upon civilised man, for it must have sprung quite naturally from our feelings, from our emotions, from the human passions; and it shows that we possess a rich mental foundation which lies concealed, but which we nevertheless inherit. There are races who feel a considerable enjoyment in vocal music, and who in that way have received certain endowments—we may take the Esquimaux as an example—to whom, if we except the drum, which is not noted for the sweetness of its sound, every other instrument of music is altogether unknown.

The taste for music, especially for song, exists everywhere, and often among races also but very slightly advanced. The Esquimaux of North America, the Polynesians, the Indo-Chinese, the Malays, the Hovas of Madagascar, and the aborigines of South America —these latter, according to D'Orbigny, are born musicians—they all gladly sing their melodies, which are often dispiriting and sad. On the other hand, there are superior races, such as the grave Tuarick, the prudent and industrious Chinaman, who sing but little and seem to value slightly the effects of music. Music will please the sensual and impressionable races, whatever may be their other moral or intellectual acquirements. If we bear in mind the lessons taught us by ethnographical and social experience we may even assert that, without being absolutely incompatible with the force of our intelligence, and the rectitude of our understanding, the artistic aptitude, and especially that of music, will rarely be found to accompany them.

CHAPTER X.

INSTRUMENTAL MUSIC.

We have seen that vocal music has proceeded from a cry; in the same way instrumental music has proceeded from noise. It is not difficult to retrace the reason.

Instrumental music was first a chimpanzee music. Man felt a sort of pleasure in hearing certain noises; he therefore studied to reproduce them. As the chimpanzees used to have concerts amongst themselves by striking the hollow branches of the trees, so primitive man, rather more intelligent, invented the drum, the first of all instruments, but of which the construction was then far less perfect than it has since become. The drum was at first either a cylinder made of bamboo and closed at one end by a knot and left open at the other, or it was the trunk of a tree hollowed out, the aperture always remaining open laterally. Man played upon the first by knocking the closed end against the ground, and the other he used to beat with sticks. These two kinds of drums were still much in use in Tongatabou at the time of Cook's voyages. With a little more inventive effort, to this primitive drum were added one or two diaphragms of skin. Hence we get the tom-tom, that we find almost everywhere in the world except among the Tasmanians, the Australians, and perhaps the Fuegians, who possess no musical instrument. The Hottentots, even, have invented or adopted the tom-tom. The Tahitians used to play upon it. All over negro Africa men delight in it; it is also the favourite instrument of the islanders of New Guinea. Clavigero says that the Mexicans formerly used drums which they sounded as an alarm, and which could be heard a mile distant. Russian travellers have found the drum in the island of Oumnak, at the north-western extremity of America, in the Esquimaux country; and Parry has seen it in Esquimaux regions still farther northward. Some other percussion instruments, rather more complicated, probably sprang from the drum; for instance, the metallic slab placed on the top of the Toltec temples, and which is beaten to invite the people to come to say their prayers. The Chinese gong is also an instrument of the same species. The harmonica, the most musical of all percussion instruments, is only an improvement upon the drum. We find it even in the Marquesas islands, where it is made of slabs of wood of different sizes, and on which the women tap with a small hammer of casuarina wood. We find similar instruments in Malay, where the slabs are sometimes made of metal, and in Senegal, where the touchstones are methodically graduated and support the gourds placed there to give a richer and a fuller sound. In Java the slabs are sometimes replaced by two ranges of metallic gongs graduating in size; in form they are like saucepans overturned, and are supported by pieces of cord intertwined one with the other. The harmonica made of wood is found also in Cambogia. But we should very greatly deceive ourselves if we imagined that these similarities supposed a conclusive proof of any intercourse between the different races. In all members of humanity there is a common groundwork, an analogous impressionability, which prompts in us all similar ideas.

After the drum, which already had the great merit of producing semi-musical noises, and of serving as an instrument to mark the cadence in all his choregraphical amusements, man then invented wind instruments—trumpets, whistles, flutes—at first very rudimentary, and each giving only one note, and afterwards by degrees two, three, and four notes. The Bushmen manufacture for themselves a call whistle out of one of the bones of the leg of the antelope, and for the same purpose the New Zealanders make use of a wooden trumpet, about four feet long, which makes a sort of bellowing noise. But over all Polynesia the people had wind instruments of a somewhat less imperfect kind—coarse flutes pierced with two, or four, or even six lateral holes. Men used to blow into them with their nostrils, and they knew how to produce a few notes of different sounds. They also invented a sort of Pandean flute with six or eight reeds placed beside each other, but without thought as to the length of the reed; and this same instrument is found also among the Moxos in Central America, but made after a more complicated fashion, and sometimes as much as five or six feet in length. With these rude implements one can produce melodies of some sort, of a very simple kind; for the number of sounds that these musical toys can give is extremely limited, and the same notes must necessarily be repeated at very short intervals.

But with string instruments the field of instrumental music becomes considerably larger, and we can without much difficulty construct musical implements on a wider and more varied scale. We hardly ever find string instruments among the quite primitive races. They seem to have been unknown to the Polynesians, to the Melanesians, and to the aboriginal Americans. Even the ancient Mexicans, who enjoyed a sort of advanced civilisation, and according to the chronicler, B. Sahagun, were expert in vocal music, were acquainted only with the drums and gongs that were made with a slab, and a kind of flute. It would seem that string instruments were first invented in Asia, that is, in Indian and Semitic Asia. These ingenious constructions gradually became known farther and farther afield, first in Egypt, where the lyre and a sort of guitar were already in use; they afterwards came into Europe.

From Egypt and Nubia the art of manufacturing string instruments spread itself all over the African continent, but with very different degrees of quickness. The Mandingos make use of a harp with as many as sixteen strings. The Niam-Niams, in Central Africa, make little mandolines, correctly constructed, and very similar to the "rababa" of the Nubians. The Bongos in the valley of the Upper Nile, who are very fond of music after their own fashion, have a monochord harp, similar to the "gubo" of the Zulus, and which is known throughout all the southern part of Africa. The Hottentots have improved, or rather disfigured, this instrument: a small lath taken from the ostrich feather and fixed to one end of the bow nearly touching the catgut string, has made of the Hottentot monochord a reed instrument, producing fluted sounds. That they call the gourah, or gorah. And the same Hottentots make use of a guitar with three strings, but on which they are incapable of playing any kind of tune. In a word, the cord instruments in Africa become more simple in construction the more one travels towards the south. In the centre and in the north-west of Africa a real musical instrument does exist, and the people know more or less how to play airs upon them; but in the extreme south they are content with merely making a noise.

From a musical point of view there are no people more poorly endowed than the pure Mongolian race. The true Mongolians are not at all a musical people. For instance, the Chinese will often prefer a noise to music. They know the art of music, because they have made it conform to certain rules; they have invented a scale, they know how to mark their airs, but their music is very scanty and poor of its kind, and semitones are to them quite unknown. On the other hand, the Mongoloids of Indo-China and the Malay peninsula, a mongrel race of people, descendants doubtless of a mixture between the Mongolians and the Asiatic black men with smooth hair, are passionately fond of music. At Bankok people devote themselves to it with much eagerness. It is mostly a music of Aryan origin; the airs played and the instruments have both come from Burmah.

Like the Siamese, the Malays are a very musical people. They have composed melodies that are full of expression, and have shown themselves very clever in manufacturing, with little labour, various kinds of instruments: harmonicas, gongs (graduating according to the exigences of the gamut), coral instruments, etc. A few strings and a banana leaf bent in the form of a shell are sufficient means for them to extemporise a lyre at once, on the spur of the moment (Exposition Universelle de Paris, Colonies Hollandaises, 1879).

CHAPTER XI.

THE TASTE FOR MUSIC GENERALLY.

Man invents music for himself in nearly every part of the world; and it is only natural that this music should be more complicated and more ingenious in proportion as his race is more intellectually advanced. As soon as intelligence exists it serves for every purpose, it vivifies everything; but the energy of the taste for music would be a very bad criterion by which to appreciate the degree of intelligence in any people, or in any race. Though the inferior races possess only a very rudimentary kind of music, they are much more sensitive to its sound than our European dilettanti are to the sound of our own music. A whole army of Tahitians would dance and beat time when one of their tunes was being played, and a most lively sense of pleasure might be seen marked on their faces. The Esquimaux women, when they heard Parry's fellow-travellers sing or play any air, would eagerly put their heads forward and push their hair away from their ears so as not to lose a single note. The Hottentots seem to have felt an ecstatic pleasure in hearing the sound of the Jew's harp. The Hovas in Madagascar are madly fond of music, and every Hova of any position has his own troupe of musicians. The Niam-Niams, who are a most voracious people, forget to eat and to drink as soon as they begin to play their music, and they will go on playing it indefinitely. The sound of the tom-tom will almost produce a fit of epilepsy upon the negroes near the

Gaboon river. A sheik in Central Africa, when he heard the "Ranz des Vaches" played by a musical box, was so moved that he covered his face with his hands and listened in the most still silence. But on the other hand, the Chinese, who are so much superior to all these people, are insensible to music.

Among civilised white men, as everywhere else where music is the most perfect, it is most strongly and sincerely felt by the common people, and by a small number of men and women of an artistic, and often of a feminine nature. But for the mass of our people music is only a thing of fashion; it is much more affected than real. The voracious Niam-Niam will forget his dinner to play a tune, but there are very few French bourgeois who would do so.

We also see that music has everywhere become melodious as soon as it has freed itself from the primitive stage of barbarism. It is that the rhythmical amplification of the cry of passion is the basis and the foundation of music, and in this respect the full bloom of music has only been reached in our later generations. As time goes on, humanity inclines to value the sensitive and even the affective pleasures less highly; we are beginning to see the melodious phase in music already draw towards its close. We have now music in its old age. For the civilised races the main interest is elsewhere; the large melodies of Beethoven, of Mozart, of Verdi, of Bellini, etc., are as the song of the swan. The prevailing fashion, in which there is so much affectation, is music without expression—so called harmonious music. It is the decline; for melody is to music what imagery is to poetry: it is the very essence of it. In fact, musical art is withering, and it threatens to end as it began—in noise.

CHAPTER XII.

ON THE GRAPHIC AND PLASTIC ARTS.

In describing the arts we are obliged to isolate them, to range them in a series, according to their value as a means of expression; but that does not in any way prejudge them according to the order in which they would appear in our social life. In reality the genesis of the different arts has been at one time successive, at another time simultaneous, according to the very different aptitudes of various races. Their origin has been most often synchronical. As soon as man became sufficiently freed from the animal life to feel a desire to draw out certain mental images, and to give to them an actual existence, he had recourse indifferently to sounds, to forms, to lines, and to colours; it was all-important to him that his sentiment or his idea should be visible in the signs which he had chosen. The troglodytes of Perigord practised line drawing and sculpture both at the same time. And the Esquimaux do so now, for they have arrived at about an equal degree of artistic development. The Tasmanian would sing and at the same time he used on the bark of the trees to make rough outline drawings of fish, of quadrupeds, of men, and of women. But the Tasmanian had not commenced the art of sculpture, and the Australian knows only sculpture of a decorative kind. In the Australian section of the last Universal Exhibition one might see an Australian lance, of which the handle was cut and delicately ornamented with lines and arabesques chased in relief. In this respect the Perigord troglodytes were very superior to the Australians of the present day, because they knew how to carve figures of reindeers and of mammoths. But still, to be able to cut in relief ornamental lines, even though the lines are purely geometrical, argues a certain skill in the art of wood-cutting.

The taste for sculpture and the cleverness in executing it are very unequally divided amongst the various races. The Polynesians, for instance, who are more civilised and more intelligent than the Melanesians, care much less for sculpture and are much less practised in the art. The monstrous giganteque statues in Easter island, and others like them found in Pitcairn island, in Laivawai, and in many of the islands in the Pomotou archipelago, are the work of the ancestors of the present existing Polynesians. They remind us of statues of the same kind near the lake Titicaca, and must have been erected by emigrants from Peru, at a time when the

Peruvian civilisation was still in swaddling clothes. The Polynesians themselves are certainly incapable of performing work of the same kind; they cannot get beyond cutting and carving much smaller objects, and even in doing these they are awkward enough, excepting perhaps in New Zealand. The Tahitians, for instance, confine themselves to decorating the posts which serve as supports for their rickety dwelling-houses with figures very coarsely cut, representing men, women, dogs, and pigs. The New Zealanders, who are more clever, are masters in the art of decorative wood-carving. Their war arms, their utensils, ornamented with chasing and arabesques, are often traced with a good deal of taste, but in the art of sculpture, properly so called, they are less clever. The prows of their canoes are usually set off with the figure of a human being with his tongue drawn out, and with eyes of mother-of-pearl; but the execution is very feeble, as is also that of their little wooden statuettes, the images of their household gods, which the people used to place before the door of their huts whenever the master of the household was absent. It was the sign given for a temporary interdiction to enter the house, and any slave who disregarded this notice, and passed the threshold of the hut, underwent capital punishment.

But the taste for sculpture is much stronger with the Melanesians of New Guinea; and these people may perhaps have taught it to the New Zealanders, who, there is reason to suppose, were preceded in their islands by the Papuans. The Papuans of New Guinea cover with chasing the planks and the posts of their houses, the prows of their boats, the pestles in which they mould their clay for their pottery, the floats on their fishing-lines, their tobacco boxes, etc. With them every piece of wood serves as a pretext for ornamental carving, sometimes for cutting human figures, of which they will load the head with feathers, in imitation of their own woolly fleece. These latter subjects are often very rudely executed, but the smaller cutting, in New Guinea, is often done with much good taste. Wallace, from whom we learn these details, wonders that so much artistic feeling can be co-existent with an intelligence so blunt, and with such coarseness of manners. But artistic aptitude is *sui generis*. Intellectual and moral improvement will doubtless enliven art, but it is very different from it, and often enough refuses to ally itself with art.

Except the colossal statues in Easter island, the authorship of which is unknown, the Polynesian and Melanesian sculpture is mostly in wood. For these primitive artists have but very imperfect implements at their service. They are generally made of stone; and to realise their most artistic conceptions, the carvers must content themselves with notching their wood, for which purpose bone is their most serviceable tool, or with modelling their clay when the art of pottery is made known to them. But the Polynesians are as yet ignorant of the potter's art, and the Papuans, to whom it is familiar, seem to have kept it for their domestic uses.

In the most civilised part of Africa that is inhabited by negroes, near the bay of Benin, on the banks of the Niger, wood-carving seems to be practised by the people. At Katunga the people decorate their doors and the posts of their verandahs. They cut bas-reliefs representing sometimes a boa holding in his jaws a hog or an antelope, sometimes war scenes, cavaliers bringing home their slaves, etc.

In the same country, at Kiama, R. and T. Lander saw an ornamented stool, supported by the figures of four men, and, according to their description, the stool reminds one of somewhat similar furniture amongst the ancient Egyptians. At Jenna they found work which, if we can believe their report, is of a still more remarkable kind—drums ornamented with bas-reliefs chased on brass, and on which were the forms of men and of animals.

The civilisations in Central America, far superior to those in North Africa, had given birth in Peru, in Yucatan, in Mexico, to a sculpture of a barbarous kind, but still showing a certain degree of elevation. There the artists, furnished with instruments of copper, of brass, of bronze, or of a very hard stone, cut large statues out of stone and made bas-reliefs of a complex kind.

In its general features this statuary reminds us of the Egyptian art. Often enough the Peruvian artist did not know how to detach from the body the members of the figure he was cutting; and the Mexican statuary was hardly more skilled in his art, but he was more inventive. In fact, ancient Mexico was filled with hideous idols, overladen with ornaments, with bas-reliefs in which nearly always the personages are shown in profile, as was the case in Egypt. And the two peoples also knew how to inlay in gold and in silver the forms of animals and of plants. Shortly after the landing of Cortes in Mexico, Montezuma sent him as a present small figures of this kind, representing ducks, roebucks, dogs, tigers, and monkeys. In their sumptuous gardens, the Incas made artificial ground-work on which all the plants were imitations in gold and in silver of the real plants of the country. The Spaniards were specially struck with admiration at the maize, with large silver leaves, from which an ear of gold might be seen mounted on a tuft made of the same metal as the leaves. But it was mainly in their ceramic art that the Peruvians liked to show their dexterity. Matter is more pliable in the hands of the potter than in those of the sculptor; a finish in the workmanship, and truthfulness in the shapings and in the expression are more easily obtained in clay for the artist who is but ill provided with tools. Therefore the human figures that the potters both of Peru and of Yucatan took a pride in modelling are much superior to their statues; there are certain faces that show a lifelike expression; and we may say that art has here begun to free itself from its earliest imperfections.

In this ceramic art in Central America care was taken in the exact reproduction of nature; and even many Peruvian vases showing the imitation of human figures are now considered as precious ethnical relics. This art, still in its infancy, would have certainly developed itself by degrees, it would in time have attempted to render facial expression and individual types of men. But the Spanish Conquest brutally stifled it, not only in staying the execution, but in destroying pitilessly the works already executed. The chroniclers tell us that the foundations of the cathedral in Mexico are built with the statues of the Mexican gods.

In Central America, the idea of artistic beauty was still ugly enough; and this may be accounted for from the slightly advanced state of civilisation in Peru and in Mexico. But as we have already observed, more than once, æsthetic aptitude is not closely connected with the degree of moral and intellectual development in a race of people. The Mongolian race, specially in its ethnical groups, in China and Japan, are certainly far superior to the races in Central America; but the art that they have hitherto produced has been of a very imperfect kind. It is almost uniquely in decorative art that in these countries æsthetics have attained to any degree of perfection; and this decoration consists of little more than ingenious combinations of lines. It is art of a geometrical kind. In statuary, or speaking more generally, in the reproduction of human figures, of animals, etc., the Mongolian has not yet got beyond the lowest grades of art. In China and Japan, statuary, properly so called, hardly exists. The Chinese rarely trouble themselves but with sculpture of a small kind: statuettes or ornamentation, on which there is no elegance, no idea of correctness, but everywhere grotesque fantasies. As regards the larger statuary, it is hardly ever seen, except to reproduce in their temples the conventional types of their divinities—specially Buddha; and the greatest effort that Mongolian notions of æsthetics have accomplished in this respect may be seen in the colossal bronze statue of the Japanese Buddha of Koumakoura, which after all is only a mechanical copy of a hieratic type. But the Japanese are superior in carving of a smaller kind, and specially in the reproduction of animals and of plants, which they imitate with scrupulous exactitude, and not excluding elegance and a sort of animation. In an artistic point of view, as well as in every other, the other branches of the Mongolian race are very much below the Chinese and the Japanese. But we must mention a curious kind of sculpture seen by the Père Huc in a Buddhist lama temple in Tartary. In the temple at Kounboun is celebrated, on the 15th day of the first moon, the

Feast of Flowers, and the principal attractions of this feast consists of bas-reliefs, which twenty lamas take several months to model. These are religious scenes, in each of which Buddha is supposed to be present, represented by a person of Caucasian type, whose face is white and rose-coloured, and he is surrounded by a crowd of people of the Mongolian type. The most curious thing is, that in all these works of art the personages, the clothing, the landscape, and everything else are all worked up in fresh butter, which is carefully modelled under water, so that the heat of the fingers should not disfigure the workmanship. These strange designs, of a most perishable nature, are not without merit in the eyes of the traveller who looks at them carefully and can appreciate them. In short, for sculpture, as well as for music, and, as we shall see later on, for painting also, the æsthetic bloom has been luxuriant and full only amongst the white race of people. Among them we will now continue to follow the evolution.

CHAPTER XIII.

GREEK SCULPTURE.

It is only in Greece that sculpture has yet attained extreme perfection; but even in Greece the art did not come all at once to its ideal state. There, as everywhere else, where human work is seen, the evolution has been slow, resulting from early efforts, feeling its way gradually as time has progressed. Greece, like other countries, borrowed considerably from foreign sources, and this has very materially shortened the period of her æsthetic infancy. Her earliest masters—those which she copied first—were the Egyptians and the Assyrians, who enjoyed an anterior state of civilisation. But, as the Assyrians and the Egyptians also took their lessons more or less directly from ancient India, we may consider as the distant teachers of Grecian art the unknown artists who have cut and engraved thousands of statues, on mythological subjects, shown upon the walls of the hypogeums in Ellora, and in other places.

In Assyria and in Egypt the general characteristics of sculpture are nearly the same. They are either statues more or less of colossal size, representing their divinities, in which the animal and the human form are often capriciously united, or else bas-reliefs, in which are pictured mythological or warlike scenes. The statues are heavy, stiff, without expression, and without animation. Generally speaking, the artist has not been able to detach the limbs from the body. The whole is a rough-hewn block, showing a conventional figure. In the bas-reliefs there is a little more animation and variety; but even that is very infantine. Nearly everywhere, as we see in the bas-reliefs at Palenque in Mexico, the personages are all shown in profile, and are placed all in a string, one after the other, without the slightest attempt at any sort of perspective. The execution is of the simplest kind; and we can determine very exactly what was the process from the sketch of a bas-relief taken by Lepsius to the museum in Berlin. The artists began by drawing squares on the wall which they wished to paint; then they marked the points where the features or part of the body of the individuals ought to come. They then drew the figures, drawing a red line along the tracing of these points. The rude sketch thus made was corrected, as far as was found necessary, and it was afterwards definitely drawn with a brush. And not until then did the sculptor begin to cut away the stone and allow the bas-relief of his drawing to show itself. It was principally the Phœnicians who so widely disseminated—and specially in Greece—the products of the Assyrian, and even of the Egyptian arts. These were generally vases with paintings and ornamental designs, frequently also statuettes, which were copied sometimes upon a larger and sometimes upon a smaller scale. The Phœnician artists were also employed to adorn the temples and the public buildings. The Bible tells us of their adorning Solomon's temple: "And on the borders that were between the ledges were lions, oxen, and cherubims." In cases of this kind they were real originators. They seem to have been specially expert in inventing the bas-reliefs and the kinds of ornamentation; for their statues were mostly in wood, and simply covered over with leaves of metal beaten out with a hammer. It was, doubtless, in this way that they taught to the ancient Greeks the art of taking castes, which the Greeks practised from a very early period, and for a considerable length of time.

For many centuries we knew nothing of Greek art but the masterpieces, which showed us the full bloom of the Hellenic genius. A few erudite men only concerned themselves with the more primitive attempts; and their archaical studies did not go very far back into past times. It has seemed to us that most of the finest productions of Greek statuary sprang up, so to say, without preparation, without gestation, by some sudden and spontaneous generation. One fine morning a Greek sculptor awoke, conceived in his brain, and then worked out in marble, the Venus of the Capitol, this most perfect model of beautiful form and serene grace. Hellenic mythology has thus made Venus rise up out of the waves, already clothed with irresistible beauty. But we know in fact that the growth of Greek art was slow and laborious. Its beginnings were very humble, as has been shown and proved to us by the shapeless forms lately excavated at Mycenæ. The plastic art of the Greeks of those days was perhaps inferior to that of the Papuans of to-day. It is shown by objects in terra-cotta, similar to those which the children now make for their own amusement in the Frœbel gardens: figures of cows, of women, in impossible and unrecognisable forms. The mythological conceptions of the Mycenian savages of those days often associated together the form of animals and of human creatures; their rude statuettes of women often show the head to be mounted with two horns. For a long time their posts served them to represent their godlike images, and they supposed them to be invested with the necessary attributes. If it was a question of a goddess, they put round the post a woman's dress. In the temples at first they erected only wooden statues, painted with raw colours, and dressed very much like dolls.

It is probable that these models of Assyrian art assisted the Greek in freeing himself from the period of æsthetic infancy. In a tomb in Attica we find Assyrian objects: a head of a beardless man, wearing a mitre, with slabs of bone cut in the form of the female sphinx. All these objects were much superior to what the Greeks could have then produced, for the idol of Juno, so much worshipped by the people in Argos, was but a coarsely-cut plank of timber. Pausanias also saw in the town of Argos a statue of Jupiter in wood, found, it is said, at Troy, in the palace of Priam, and which had three eyes, and one of the eyes was placed in the middle of the forehead. The early Greek statues were stiff square figures, the arms fastened to the body, the two legs and the two feet joined together, the whole showing no animation and no expression. And Homer puts into the palace of Alcinous golden statues serving as lamp-posts, intended to be young people bearing torches; and dogs of gold and of silver, he tells us, were watching the gate of the same palace. We must, of course, suppose that these statues were in cast-metal, similar to the golden masks covering the faces of the bodies in the tomb of Mycenæ, and to many other objects of the same kind, dug out of the same place: flying griffins, women holding a dove, imitations of leaves, cuttle-fish, etc. The Greeks, too, were probably disciples of the Phœnicians, for in a tomb in Phœnicia has been found a mask in cast gold, covering the skeleton of the face of a newborn child. All these objects are proof of art in its most rudimentary state. But the people were beginning to trace bas-reliefs, and thus gave an outward and ill-drawn picture of a battle or of a hunting scene. Dating from this humble commencement, archæology enables us to follow each step made in Hellenic statuary; it is ever in a slow state of progress towards the reproduction of true and of idealised nature. After the early primitive statues which were painted, clothed, ornamented, the hair combed, laden with necklaces, diadems, and earrings, purely hierarchical for the most part—after these we see the figures of athletes. A statue was made of every athlete who had been three times victorious, and the artist naturally endeavoured to copy exactly the shape and form of the body, and to bring all the muscles into relief. After they had succeeded in giving a true picture of the forms, in observing the proportions, which indeed became canonical, the artists then tried to put some life into their figures by giving them animation and natural attitudes. The more we approach the Macedonian period, the more we shall see the statuary endeavouring to vivify itself by the expression of the human face. After Alexander, Greek sculpture ceases to reproduce the figures of athletes, and begins to execute the portraits of well-known personages. And when liberty was dead the supple Greek genius languished and withered away; the unhappy period of decadence set in, and the true art was then forever lost in Greece.

To trace the history of the development of sculpture would lead us into endless details; we are, therefore, compelled to confine ourselves to noting only the principal phases. But the facts already quoted will be enough to show us the art, first in its embryo state, producing only infantine and shapeless sketches, very clumsy copies of the objects and the creatures that man saw everywhere around him. Afterwards imagination finds its place; sculpture becomes mythological; in the bas-reliefs, in the first place, it attempts complicated subjects—war scenes, or hunting scenes—anything that impresses itself strongly upon the mind of man. Stiffness and conventional types are the characteristics of this period. At last the artist frees himself. By degrees he discovers exact and scientific aids to his art; he learns how to fashion stone and marble, to make them give not only the likeness of form, but even the peculiarities, the individual features, and the passing habits of the time. This is the last phase of sculpture, and we shall also see that it is the last degree of progress that has been reached in the art of painting.

CHAPTER XIV.

ON PAINTING.

PAINTING, like everything else, has had a very humble beginning. Such is nature's law, with which the great doctrine of evolution has now made us familiar. Did the embryo period of painting precede, or did it follow, the not less modest early days of sculpture! If we were to look at the plastic and graphic arts only when they are at their apogee, in what we may call their adult age, we should be inclined to say that the sculptor's or the modeller's art was more ancient than that of the painter. It is surely easier to copy in relief, more or less awkwardly, the forms of men's bodies than to realise upon a plane surface the looming of perspective, or to give light and shade or any notion of colouring. Nevertheless, excepting any particular case of special aptitude for this or for that art, it would seem that the commencements of both were simultaneous. The man who, in his state of savagery, tried to hew roughly with his stone knife a piece of wood or bone, to extemporise thus a living image out of his own imagination—the same man would generally attempt to produce a like result by means of lines either traced or drawn. These first drawings are never more than outline; and if, as was usually the case, the artist wished to represent the forms of animals or of man, it is always the profile that he shows to us.

The Tasmanian, who was ignorant of sculpture, the Australian, who did not get beyond a little sculpture of an ornamental kind, have both invented this primitive method. On the stone rocks in Sydney, and on the rocks in Tasmania, have been found rude images of fishes, of quadrupeds, of men, of birds, and of kangaroos. At Port Jackson there is still existing a more complicated sketch, in which the designer wished to represent one of the great dances of the country, called the "corroboria." The Hottentots, and even the besotted Bushmen, have left upon the rocks in their country sketches of a similar kind. The Papuans of New Guinea, relatively clever at sculpture, have shown less ability for drawing; still, they can trace well enough, on the sand or elsewhere, sketches of canoes or of men, often obscene phallic figures, which their sculpture has again reproduced.

For the art of drawing, like many other arts, there are special aptitudes, which are very unequally divided among the different races of men. The Polynesian does not know how to draw, but the Esquimaux, in many ways inferior to him, is a good draughtsman. In this respect, this human type, of the age of the reindeer of our own times, resembles man of the age of the prehistoric reindeer, of which MM. Lartet and Christy have dug out the remains and some samples of their works, in Périgord and elsewhere. Very often the weapons and the utensils of the Esquimaux are ornamented with outline drawings. We may see cut upon them flocks of reindeers, which a hunter is pursuing as he crouches along, or else a man lying down with his harpoon in his hand, and the skin of a stuffed seal close by him serving as a bait to attract the animal which it is clumsily intended to represent. Or again we may see a drawing of men fishing for whales, a different scene in Esquimaux life. From the manner of execution, and often from the subject represented, these Esquimaux drawings remind us singularly of those left to us by man of the age of reindeer in Périgord; and it is also the same with regard to sculptured objects. It is curious that neither the Melanesian nor the Esquimaux seems to have had any idea of bringing any parts of the bodies into prominence, by means of light and shade, whilst the prehistoric artist had already tried to realise this sign of progress. Upon an ornament found by M. Lartet in the grotto of Bas-Massat, we see the outline drawing of the profile of a bear, cut with great firmness, and there are notches made, meaning to represent the shadow. On the other hand, the intelligence of certain savages of our own times is so little awakened to any perception of the graphic arts, that to them a drawing is quite unintelligible. The fact has often been already mentioned, notably as regards certain aborigines of Central Africa; and also some of the Australian tribes.

Hitherto we have only spoken of drawing; for it is probable that drawing preceded painting. Nevertheless man began very early to endeavour to produce certain tints, some of which he observed more strongly than others, and specially the strongest of all, that which is seen in the central portion of the retina—the colour red. It is a certain fact, and we have already quoted instances to prove it, that nowadays there are many primitive races, and also many of the poorer people in civilised countries, who have a very marked taste for this strong colour.

Painting, properly so called, began probably by the colouring of sculptured objects, of statues, of designs, and bas-reliefs, as soon as men knew how to execute them. The New Caledonians will colour everything that they can with red; the roughly-cut statues dug out at Mycenæ were painted mostly in red, sometimes in yellow; it was the same with posteries found at the same time. We know from other sources that the statues of primitive Greece were coloured with raw tints. Each divinity had its special colour; Bacchus, for instance, was always painted red. Our sculptors have now discontinued to use colours, except in polychrome statuary; but the popular taste is very different, as the Catholic and the Buddhist religious images prove to us clearly enough. It is in this way that the lamas in Tartary are careful to colour their curious bas-reliefs, made in butter, of which the Père Huc has given us a description. This custom of covering with bright colours the products of the plastic arts is very general, for we find it in *New Caledonia*, in Tartary, and in Greece. In the mind of the primitive artist, this colouring gives greater life to the statuary, and also he thinks that the brightness of the tints gives an additional pleasure to the eyes. It is likely enough that the ancient Greeks adopted this mode, perhaps taught to them, like many others, by the Assyrians, for it was their constant practice.

The colouring of the statues and bas-reliefs and the colouring of drawings are so closely connected that the two methods must have been conceived and followed simultaneously. Among people who have never got beyond the lowest grades of art, painting has always been the simple colouring of outline drawings, designed without any thought of perspective. The Mexican artists before the days of Cortez, those of China and Japan, those of ancient Egypt, did not get any further. Soon after the landing of Cortez in Mexico, Montezuma sent out artists as "special correspondents" to draw with their pencil and colours, on some cotton stuff, the faces and the war arms of the foreigners. These very elementary designs were only outline drawings; and as far as possible the artist sketched the men and the animals only in profile. In simplifying them a little they were easily turned into hieroglyphics, and words were attached, so to say, to the mouth of each individual. We know that the Mexicans made constant use of this drawn or painted form of writing, and the Egyptians also did very much the same. The same device was also practised by the Chinese, who attained to a greater degree of ingenuity, for their painters knew how to represent men and animals in the most varied attitudes; whilst their hieroglyphics have become a regular system of writing, and the characters hardly continue to represent the objects symbolised.

What is always more or less wanting in these primitive paintings is mainly the art of giving to the drawing, to the fresco, or to the picture, any appearance of relief or of background, so as to produce any illusory effect upon the imagination. A notion of light and shade or of perspective is quite unknown; we see hardly a hint of it. In the Mexican drawings, in the Egyptian frescoes, the personages are all put in file one after the other, and all set on the same alignment. In the Chinese and Japanese pictures it would seem as if the artist had taken his point of view from a balloon, with the sun vertically over his head. The idea of showing sunshine and shadow, the gradations of tints, the distant objects, can hardly have been present to the artist's mind. In the choice and the grouping of colours more skill is shown. The Egyptian artist would, from preference, choose the brightest colours, and he knew how to combine them tastefully. The Mexicans excited the admiration of the Chinese by their skill in painting and arranging feathers. The Chinese are masters in the art of choosing and grouping their colours. It affords a curious contrast to watch their skill in harmonising the tones, while they are so absolutely ignorant in the art of drawing. But still it seems that the Chinese artist is conscious of his own ignorance, for often he will purposely deform real life to hide his own inability in expressing it. Caricature is to him, in drawing, what buffoonery is in the conversation of a fool. In the miniatures which adorn the Arab and Hindoo manuscripts we admire the same variety of different shades of colour, and the same delicate taste shown in their arrangement by the side of some drawing much less fantastical, and certainly very much more simple. Sometimes on the Arab manuscripts we see rose-leaves painted in arabesques, and these are masterpieces of their kind, in which the artist has shown his power of harmonising the tones with infinite art. Do not these ancient and backward civilisations contradict absolutely the theory of H. Magnus, as to the gradual development of the colour-sense? When we see an Egyptian fresco, or a Chinese vase, perhaps some thousand years old, are we not then convinced that these cannot be the works of men afflicted with Daltonism?

Painting, like sculpture, was only fully developed amongst the Aryan races, so called; but even with them it rose from very humble beginnings. In the excavations in Attica, and specially at Mycenæ, Mr. Schliemann has found remains of vases, generally painted in red, and ornamented sometimes with circular or irregular bands, sometimes with black bands, or sometimes with bands of a deeper red colour than the red shown in the groundwork. That is no doubt the first state of ornamental painting, since these people have tried to picture, by lines of the same kind, birds and quadrupeds. The work is so rudely done that it is often difficult to recognise the animal that the artist has endeavoured to paint. The handwork shown in these drawings reminds us nearly of the totems in use among the American Red Skins. On other fragments we see represented six warriors completely armed in coat-of-mail, and painted in red upon a bright-yellow grounding; for the primitive Greeks, like other people, showed a strong preference for the colour red. At a later date the Phœnician cups served as models to the Greeks, and aided them in raising their art from its barbarous condition. For a long while archaic art in Greece was confined to copying the drawings in these Phœnician cups; at first the imitations were very servile, but afterwards they grew bolder. Certain passages in Homer—the description of Achilles' shield, for instance—may be seen illustrated accurately enough in the Phœnician vases which have survived down to our times. To express the variety and the succession of events, the actors concerned were painted a second or a third time, and in this way whole legends might be read written upon these vases. On this Greek pottery, as on the Etruscan pottery, the personages are almost invariably represented in profile.

On all these vases, and later upon the frescoes or upon the pictures, the artists were satisfied with clothing the men and the animals represented in one uniform colour. Then dawned upon them the idea of indicating the shadows, but very simply, by means of brown or black notchings. That was a great discovery; it soon grew to be more perfect. The first man who expressed in the shadows as well as in the light part of his pictures the same tints as might be seen in the model itself, became a celebrated artist. His name has been handed down to us—it was Apollodorus. Henceforward men knew how "to colour the shadow," as the Greeks used to say; the art of painting was then established. But it was still very elementary. For a long while the only kind of painting was ornamental, either on the vases or in the temples. The personages were very simply ranged along in a file, and their figures were always seen in one bright shade of colour. But by degrees painting freed itself, and became quite an independent art; it was then highly appreciated. We are told that the king Attalus offered to the painter Nicias, who belonged to the Athenian school, a sum equivalent to £10,800 in our money for a picture representing Ulysses evoking the shades of the dead.

Nevertheless, the ancients, both the Greeks and Romans, in painting, made only limited progress; their painters were much inferior to their sculptors: we may judge of this by the ancient frescoes which have been preserved. In this respect the Pompeian museum at Naples will dispel any doubts, not to speak of the frescoes to be seen in Pompeii itself. It will be sufficient to mention "Diana and Actæon," "Orpheus," "Bacchante borne by the sea-panther." No doubt the best of these works are remarkable for the elegance and the beauty of the personages represented. The forms are sometimes superb, and remind us of those of statuary, but the modelling is far from perfect, and the art of perspective, still in its infancy, often resembles that of the Chinese pictures.

During the time of the Lower Empire and the Byzantine period, painting retrograded more than it advanced. In the way of stiffness and absence of animation, certain frescoes in the Catacombs—for instance, the one of "Death"—are like, and are even inferior to, the frescoes at Pompeii, to that of "Diana and Actæon" among others, and the designs and the modellings are much more imperfect. Every one knows the rude state of art in the Byzantine period; the traditions of the antique art are barely mechanically preserved; the Catholic mythology has replaced the Græco-Latin mythology, to the great detriment of grace and beauty. The personages are stiff and the drawing is incorrect; light and shade is barely shown in the colouring; the landscape, which is never used but to ornament the background of the pictures, is hieroglyphical, and shows an utter disregard for any laws of perspective. M. E. Véron rightly compares this barbarous painting to that which decorates the Etruscan vases at Corneto; we see the same simplicity and the same ignorance. It was not until the fifteenth and sixteenth centuries that the Renaissance, the spring of European intelligence, began to show itself; it was not until then that art began slowly to throw off its swaddling clothes. Raphael in his first manner distantly reminds us of the poor Byzantine period; but yet we ought not to think of this when we admire the Farnesian frescoes, and the admirable drawing, colouring, expression, and truth to life, shown in the "Santa Cecilia" of Bologna.

It is not our object in this short notice to give a history of the beaux-arts, but merely to indicate very generally the principal phases of their development. We do not speak of Raphael's competitors, of Titian's colouring, nor of the bold frescoes of Giulio Romana, or of Michael Angelo, nor of the modern painters who have been able to follow with more or less success the way shown to them by their glorious predecessors. With regard to the work, to the mechanical portion of the art, we shall never surpass the great Italian painters of the Renaissance. Unrivalled for perfection in colouring, and for the beauty and the fusion of their tints, for the exactness of design and of modelling, they have made of painting an art able to rival the objects represented; to seize and mark the most delicate fleeting shadows, the most fugitive passions and sentiments of man. The artists of the present day, or of the future, can only surpass their predecessors in giving ideas which were wanting to the masters of the Renaissance.

But there is one branch of this art which the Italian artists did not bring to perfection, and that is landscape. They had doubtless learned the principal laws of perspective; their paintings show much depth, and light seems to circulate in them; but with a few more or less happy exceptions, certain pictures of Salvator Rosa for instance, landscape barely interested them, and, like the artists of antiquity, they used it only as a background to their work. It was with them no more than an accessory, and was always neglected even by their greatest masters. As it is natural, the most scientific part of the art of painting was perfected last; and in this respect the palm belongs no longer to Italy, but in some measure to France, if we wish to value rightly the pictures of Claude Lorraine, and especially those of the Flemish school, in which Paul Potter, Ruysdael, Hobbema, and others, have done for landscape what the Italian masters did before them for the grand school of painting: they have brought it to a degree of perfection which will be always difficult to surpass.

We have now come to the end of our description, very short as it has been, of the origin and development of the fine arts. The real side of our sensitive life lies in that direction. There is there a sort of language, much more confined than our language, properly so called, but more picturesque and more expressive. It is almost unnecessary to remark that in the fine arts, as in every other direction, where human activity is shown, the white or Aryan races very far excel all the others, though they have had to pass through similar stages. Like that of all other human races, the Aryan intelligence has sprung from the most humble sources; like others, she had to stammer before she learned to speak. But her development is now less incomplete. She has given more light to the world than all the other races; her wisdom has spread itself more widely, and she has left deeper marks of a durable nature. It is no doubt very far from being the case that in any given race the most intelligent will make the best artists. The perfecting and refinement of our senses plays a most important part in our artistic aptitudes; but the progress of art will ever depend upon the general development of a race, for in everything intelligence will form the basis.

CHAPTER XV.

THE EVOLUTION OF SENSITIVE LIFE.

A FEW general ideas may be gathered from what we have just said.

The study of the different races on the surface of the globe, in whom we see human intelligence so variously developed, and the history of those who have become gradually civilised, will show us the different phases through which the sensitive life in humanity has had to unfold itself.

A certain acuity of sense may be often found in the primitive man, but the object is confined; for with him the mental register of the sensations is poor and ill kept. His power of fixing his attention is weak, and his memory is short. Either the gradations escape him, or they are not sufficiently remarked.

Man begins by showing a preference for strong colours, especially red and yellow; he chooses them to adorn his works of art or industry, or to ornament his person. By degrees this taste grows weaker; he asks for milder and more varied colours, for darker shades, and he gives less thought to ornamentation.

All our artistic tastes have modified themselves in the same way. The dance has gradually changed its character. At first it was a rhythmical mimicry of hunting or of war; it afterwards became that of love, and often joined itself to the most important acts of our social life. It gradually ceased to be customary, until it degenerated, in civilised societies, into an amusement, in some sort archaical, and until it was hardly more than a reminiscence of past times. Music, at first extremely simple, purely melodious, and mainly vocal, a sort of modulated cry, has gradually acquired a greater fulness. Instruments, more and more varied in their kind, ingeniously constructed, are used to accompany the recitatives or the songs. The science of instrumental music is attained. Quite at first, harmony sets off melody and makes it sound richer, but by degrees melody loses its freshness and its power of expression. Music tries to produce complex effects, it becomes less attractive; but as intelligence is denied to it, it languishes and loses flavour.

The graphic and plastic arts, on their side, at first very rudimentary, slowly improve and grow to be more varied; their power of expression increases; their representations by degrees acquire a more life-like character, and a greater exactness in detail. They succeed sometimes in making us believe that they are in fact the objects intended; they are often more beautiful than the image represented.

The genesic instinct undergoes a metamorphosis of the same kind. It is there that the sensitive life is joined most closely to the affective life; it began by being absolutely bestial, similar to that which we may observe among the coarsest animals. From relationship, at first purely physiological, between the sexes, a few affective sentiments are born. By degrees shame arises to stimulate and also to guard against the gratification of the sexual needs. Love is no longer an instinctive brutal pleasure. Individual sympathies and antipathies show themselves, jealousy is awakened, and the genesic need begins to be transformed into love; for there are tendencies more and more strongly marked towards exclusiveness. It is at first in woman that the gradual ennoblement of the genesic instincts at first arise, for in her this instinct is nearly always the pivot upon which her whole moral life is made to turn.

In a word, this evolution in our sensitive life unfolds itself in the cerebral activity, which is always expanding in our conscient life. As regards æsthetics, properly so called, the fact is evident. The memory takes note of an ever-increasing number of sensitive impressions, the imagination revivifies them and brings them together, the intelligence endeavours with more and more good effect to make them appear in works of art.

The amorous need, which, to satisfy itself, claims the assistance of another individual, seeks mainly to enhance our affective and even our social life; but it has strong indirect influence upon the æsthetic taste, which it often stimulates. It was undoubtedly one of the principal agents in human societies and one of the great sources of our affective life, which we shall now endeavour to describe.

BOOK III.

AFFECTIVE LIFE.

CHAPTER I.

THE REFLEX ACTION ACCORDING TO RACE AND CIVILISATION.

In mammalia and in man, who is only the first of the order of mammalia, the nervous centres, the spinal chord and the brain, receive through the sensitive nerves the incitements of the outside world, and reflect it all along the principal nerves, and they command the movement of the muscles. We have elsewhere* described at length these actions and reactions, which in physiology are called *reflex actions*. Of these some are called inconscient and others conscient. The inconscient actions operate mainly in the spinal chord, the conscient are seated only in the cerebral hemispheres. In our ordinary life both are mixed up and joined together, but according to certain laws.

If we were to let fall upon the hind leg of a frog, whose members were all perfect, a drop of nitric acid, the frog would pull back that leg, move his other members, shut his eyes, etc. We should thus have provoked a general reflex action, conscient and irradiated. If the same experiment were tried upon a frog whose spinal chord had been divided, all that portion of his body anterior to the spinal chord would remain immovable, but the frog would draw back more rapidly and with greater energy the leg that had been cauterised. The reflex action becomes both simplified and made more perfect. The reason of this is quite clear.

In the vertebrated animal, especially in those of a superior kind, conscient life intervenes in most of the acts of daily life.

* "Biologie," 120, 156.

Every sensation, every impression, is caused by a molecular shock transmitted all along the nervous sensitive fibres to one or more of the conscient cells—that is to the cerebral cells. When it has reached the heart of these cells the molecular vibration there becomes changed and gives rise to subjective phenomena, to impressions, to ideas, etc. In a nervous centre so complex as the human encephalon, the molecular current, stimulated by the shock of a sensitive nerve, must surely divide itself into many secondary currents, going from one cell to the other by means of the fibres which join them together. The fractioning of the molecular wave may be compared to the subdivision of a river, of which the bed, cut into many islands, is thus divided into several secondary arms.

If the conscient cells are numerous, perfected, susceptible of collecting and combining a great number of impressions and ideas, the molecular movement stimulated by the sensitive nerve may extinguish itself in the brain, there wholly transforming itself into conscient phenomena. In the contrary case the conscient cells do not suffice to absorb the wave stirred up, a portion of this wave jumps over them, and in spite of the will of the patient, reflects itself upon this or that nervous impelling branch; hence the involuntary or reflex movements. We can therefore easily understand why the reflex action is achieved more easily and more rapidly in the decapitated frog where no conscient inhibition can arise to hinder the reaction.

The greater or less degrees of energy in reflex irrepressible actions may therefore give a fairly good idea of the cerebral development. We may say that the superior cerebral nervous centres are the more perfect as they preserve and turn to better use, in transforming into conscient acts, the nervous stimulation, come from outside—as they oppose it more completely to the reflex action.

In this respect monkeys are very inferior. With them the reflex action is excessive. Always excited, always grinning, always in motion, they show an extremely mobile disposition. Similar characteristics may be noticed in our children, and in many women; we may, therefore, expect to see them also in the inferior human races, who, relatively to the superior races, are the infantine races.

Like the monkey, like the infant, the savage, and even, more generally, the human being who is but slightly developed, to whatever race or sex he may belong, is incapable of governing himself; he is the plaything of outside circumstances. He will laugh or cry for the most futile reasons. For laughter is common to man and to many monkeys. But Darwin says that he has not observed laughter in idiots. The least intelligent of men, the Veddahs in Ceylon, laugh very little, or not at all. Tears are not man's exclusive privilege, nor is laughter. According to Humboldt, the saïmiris monkeys in Peru—monkeys of a very inferior kind—would laugh for the very slightest cause.

Among the inferior races, laughter, tears, or the wildest movements break forth at every instant. Sturt relates that a young Australian woman, when she first saw him and his companions, threw herself down on the ground exclaiming loudly. In Tasmania the women were always in motion, ever gesticulating; and, according to the Rev. Mr. Bonwick, were like so many monkeys. The Papuans also are ever in a state of perpetual movement, always singing, crying, gesticulating, laughing, and jumping.

Nothing can be more changeable than the humour of the Polynesians. All travellers liken them in this respect to children. A chief in New Zealand burst into a violent fit of tears because some sailors had covered one of his smart cloaks with flour. The Tahitians, ever ready to laugh or cry, would change all suddenly from one state to the other. Explosions of joy or of sulkiness would disappear with them in an instant. A woman who was crying bitterly because her child had just died broke out into laughter when she saw Captain Bligh. In the same way the Noukahivan is fantastic, insensible, subject to feverish over-excitement; he becomes depressed very suddenly, he is uneasy, and little capable of gratitude. His mental instability is excessive.

These characteristics are not peculiar to the Polynesians only; we see them in all races who are poorly developed. The African negro women burst into tears upon the slightest cause, or without any cause. Du Chaillu has seen some who were crying in torrents and laughing at the same time.

The Chiquitos in South America are exceedingly gay, madly fond of dancing and of music. The Guaranis constantly pay visits one to the other, and each visit opens with a fit of crying in memory of dead relations; they afterwards dance, drink, and enjoy themselves.

The Cochin-Chinaman cannot fix his attention; it jumps about from one thing to the other. He changes suddenly from gladness to sorrow, and to fits of violent passion.

The Aryan races, too, when only slightly civilised, closely resemble the inferior races by their moral instability. The Afghans quarrel amongst themselves, they fight, and become reconciled from one instant to another for mere trifles; to conceal their thoughts is altogether beyond their power. A Persian grand seigneur, Mirza-Selim, broke into tears on hearing the sound of music.

Nothing would be more easy than to accumulate facts of this kind; but those already mentioned will suffice to give an idea of the affective mobility of the savage, or of the man whose intelligence is slightly developed. In some civilised societies we find also the same results. The untaught white man, the infant, the majority of women, in short, every organisation that has not been modified by a long moral and intellectual culture, have cerebral springs always ready to make themselves elastic. The outside physical and moral causes annoy and disturb them repeatedly; their mental equilibrium is at the mercy of a thousand daily incidents; they have never complete mastery over their own actions.

CHAPTER II.

ON POLITENESS AND CEREMONIAL BEARING.

It is not our purpose to speak at length on this subject, which necessarily forms an accessory part of sociology, but it may be advisable nevertheless to say a few words.

It appears that gestures, and consequently the very different ceremonial formula customary in all human societies, have their origin in reflex action.

In sociable animals the excess of nervous shock caused by a strong sentiment often strikes back upon the nervous peripheric system and stimulates involuntary movements, which will vary according to the kind. The restless horse will point his ears towards the object which causes him to feel emotion; the dog will wag his tail to show that he is pleased; under similar circumstances the cat will pur and raise her back.

The turtle-doves show their love by kissing each other. And the dog, too, kisses his master, after his fashion, by licking him, to show his affection. But if on the other hand he is afraid of his master he will put down his head, or if need be crawl, and lie down on his back.

As man is but the highest of earthly animals he expresses his actions, as they do, by a reflex mimicry, which is the more instinctive as he is less civilised, less master of himself. Certain of these expressive acts, maintained in some societies from the time of their commencement, have been adopted as symbolising this or that sentiment. Prostemation, a custom in some eastern despotic monarchies—in Siam, for instance—is evidently similar to the cowering of a dog when he is frightened. The kissing of feet, and even of hands, closely resembles the act of the same dog in licking the feet or the hands of his master. But in proportion as man acquires the knowledge of his own dignity, of his liberty, ceremonious mimicry becomes less servile and less animal. An individual is no longer, as he is in Siam, "the animal of the king," he no longer consents willingly to humble himself; and therefore the gestures intended to convey marks of respect become more simple. Our slight bowings of the head are no more than a recognised and shorter form of the more ancient custom of prostration. Nearly all over the earth, if a man wishes to salute respectfully, he bows his head, and places at least his right hand against his chest; this is a movement opposed to the attitude of defence. And again, to put one's right hand into the hand of another, is, in a certain measure, giving oneself up to him.

Man being an intelligent animal he has naturally learnt how to perform this comedy of ceremonial form in various forms—not taking into account that his movements of expression will differ widely according to his race. The kiss, which seems to the Europeans the most tender and the most natural mark of affection, is unknown to the Australians, to the Papuans, to the Esquimaux, to the Fuegians, to the people in the west of Africa, to the Lapps, and to the greater part of the Polynesians—though not to all of the Polynesians, for Cook tells us that the natives of Tonga island kissed his hands. But nearly everywhere in Polynesia the friendly kiss is replaced by a more or less complicated ceremony, of which the most important part consists in rubbing one's nose against the nose of the person whom one wishes to welcome. In the Gambia islands people used to utter violent exclamations, or else they growled between their teeth. In Malay and in China the people used to and still have the same custom; but they sniff at the same time, so as to inhale the perfume of the individual whom they like, or with whom they pretend to be on terms of friendship. Here we see an instance of direct mimicry of the animal.

Polite usages, when once adopted, have been refined from whatever may have been their former state. They have grown complicated and diversified, until many of them have become peculiar only to human beings, varying according to the degree and kind of civilisation.

The salutation shown in uncovering the head indicates a very advanced social condition, for the head is the last part of his person that man thinks of covering. To touch this or that part of one's body, or of the body of someone else, is sometimes equivalent to a mark of gratitude, or to the making of our engagements. The inhabitants of the Tonga islands carry on their head everything that is given to them, and also everything that they get by way of exchange. In the latter case this is a token that the terms are accepted.

In countries where human life is but little thought of it is the rule to walk before anyone whom one respects, or to go first into the market. This is the case, for instance, in New Caledonia. Also in

Malayala, civility exacts that one should turn one's back and keep one's eyes closed and one's head covered; the reason for this form of politeness may be easily guessed. In Tartary, at the time of William de Rubruquis, it was held to be a grave offence to the proprietor of a tent to touch its cords. It was thought that the slight wall of a tent, not protecting one like the wall of a house from the enemy or any rash-minded person, ought to be guaranteed by signs of mutual respect.

In many countries also people have not stopped at gestures; they have adopted emblems of peace, as having the value of a formal promise. And on this point, too, it is curious to see how races of the most different kinds will resemble each other. White staffs and green branches were used as announcing pacific intentions both in Europe and in Polynesia.

At Noukahiva, politeness went in reality as far as it did in appearance, judging by the formula of ancient societies. A man identified himself with his host, and this perfect intimacy was symbolised by an exchange of names. When the pact of friendship was concluded, the stranger absolutely took the place of his friend in everything, and for everything, making such use as he pleased of his house and of his wife; and she would often feel herself slighted if her husband's *theo* did not assert his rights.

Verbal formula were very early introduced into the usages of politeness. The caprice and the notions peculiar to each race of people will here strongly show themselves. The least civilised, such as the New Caledonians, do what they please without any phrase-making. Like other islanders in the Pacific ocean, and other savage people, notably the Kalmouks, they have no word expressing "thank you." But in the majority of races, however little civilised, there are sacramental forms of politeness. We find them practised in the steppes in the south part of Asia. Two Kirghises who are well-mannered ought to accost each other by asking "Who are your seven ancestors?" This is the genealogical salutation. In the presence of a foreigner of distinction the Kalkhas Mongolians throw themselves on their knees, exclaiming "Love, Peace!" Then they ask the foreigner: "How do you like the Mongolian waters?" The ceremonial politeness of the Chinese is justly celebrated, for they have precepts and rites ruling every formality in their social intercourse. According to Confucius, ceremonies are typical of virtues; they preserve them, they recall them to one's mind, and may even replace them. This substitution of ceremonies, and even of ceremonial grimaces, for generous sentiments, is the law among people who have fallen to a low moral state. When two Arabs from Zemen meet they assault each other with compliments. It becomes a question as to who will kiss the hand of the other, but it is understood that the elder, or the more distinguished man, shall finally accept the offered civility. All travellers are agreed in accusing the Persians of theft, falseness, hypocrisy, want of moral sense, and yet they of all people are the most polite. It is a matter of contest with them who shall give way to the other. If they receive a visit, their visitor is for the while "their master;" their pipes, their horse, their clothes, are "presents for the master;" their house and all it contains, still more their villa, and still more again their fields, are at the visitor's disposition. We may easily understand that these magnificent offers are not made under the idea that they will be accepted.

This brief anthology of ceremonial observances may suffice to indicate the principal phrases of politeness and of other things connected with it. Like all human actions, those of which we are now speaking change their character in proportion as man himself changes and develops, morally as well as intellectually. In the earliest stages man servilely copies the reflex animal action; afterwards he simplifies it, he abridges the mimicry which often becomes repugnant to him; at last he replaces this mimicry by formula, and as words cost little, these formula become all the more exaggerated and the more debasing as his normal condition is less sincere. Of all the courtiers who kissed the feet of Heliogabalus and of other emperors during the period of the decadence, there was not one who would of his own free-will have given to this arch-sycophant a single drop of his blood. And the formula of the Lower Empire, of which the traditions have come down to us, were not, and are not now, taken in earnest by anybody.

As a general rule, among individuals, and among races of people, extreme politeness is shown in inverse ratio to their moral worth. Our European politeness, grinning and insincere as it is, comes to us in a great measure from the Lower Roman Empire. But it is destined to grow more simple, if, as we may hope, we can raise ourselves to show dignity and fairness in our dealings.

CHAPTER III.

LOVE FOR THE YOUNG IN ANIMALS.

When we try to seek out the causes of the phenomena of the affective life, so varied in the animal and in man, we soon find that we are obliged to stop suddenly short in our investigation. Our knowledge on this subject is still very limited, and a high wall, as yet but slightly broken, separates it from the unknown. No doubt, in this as in other things, the great doctrine of transformation may guide us and light our path; but in order that our instruction be perfect we shall have first to become acquainted with the genealogy of the animal kinds through its long and innumerable vicissitudes. And at the outset we can scarcely distinguish and mark the principal facts.

Up to the present time the apparition, among living creatures, of the conscient phenomena is a fact as inexplicable in its essence as gravitation. How have the affective sentiments, the most interesting of all these phenomena from a social point of view, been developed in us? To this question we can only answer by conjectures.

Let us take, for instance, the most constant and the strongest of all affective sentiments—the love of parents for their offspring. The origin of this is still very obscurely hidden.

The modes of animal generation may be reduced to a small number: fission, gemmation, and endogenous division; and these would seem to have sprung one from the other. In the first of these two modes we may easily understand the interest that the progenitor would feel for his descendant if this progenitor were impressionable and intelligent. The young creature produced by a divisioning or a parcelling out of the elder, it would be very natural that the parent should interest himself in that portion of his own being which frees and emancipates itself. But in this inferior phase of civilisation the psychological me does not yet exist; the animal has surely only a vegetative life, for nothing authorises us to admit a conscient life in the absence of a nervous system, however slightly developed it be. We do not even understand the existence of the cares of progeniture among the inferior invertebræ already provided with ovulation. Why, for instance, do the ascidiæ keep their eggs under their cloak? How can we explain that certain female insects who have not known their parents, and who will never see their little ones, should seem to interest themselves in the fate of their eggs? Why do we see them sometimes prepare for their carnivorous larvæ a food which is unsuitable to themselves?

It is not much easier to resolve the same problem with regard to certain fish who appear to care for their progeniture. In fact, with fish the part of the male is often confined to milting the eggs laid by the female. How does the male stickleback acquire the instinct or the desire to build a nest for his little ones, to protect his female, to bring his little ones home, in case of danger? *

How has the toad midwife learnt to roll his hind legs around the viscous thread of the eggs laid by his female? Why does he carry those eggs? and why does he think of plunging into the water the moment they are hatched?

No doubt many of these acts are automatical. They are hereditary habits, and are instinctive; but how were these instincts first acquired? In order to explain these curious facts we should have to make ourselves acquainted with the details of the numberless adventures through which the animal kinds have passed; we should require to know the whole of the long history of zoogeny.

* These facts and others of the same kind have been enumerated and discussed by M. A. Espinas in his book "Les Sociétés Animales."

We may abstain from making idle conjectures, and confine ourselves to stating that love for their young, one of the most powerful sentiments of which man is capable, is also seen in the lower stages of the animal kingdom.

In the superior classes of vertebræ, in birds and in mammalia, the hereditary instinct is not the only motive which causes the parents, and specially the female, to care for their young. The foundation of psychical life is the same in them as in man; the conscious will intervenes and brings intelligence to its aid. Animals, too, have received from their parents the lessons and the cares which they give to their offspring.

Maternal love is certainly the strongest sentiment of which many birds and mammalia are capable, and this sentiment sometimes inspires them with a devotedness which would do honour to the human kind. A female wren, observed by Montagu, spent sixteen hours a-day in looking for food for her little ones. At Delft, when there was a fire raging, a female white stork, not being able to carry away her young ones, allowed herself to be burnt with them. In 1870, in Paris, during the German bombardment, a shell bursting in a granary did not drive away a female pigeon who was sitting upon her eggs.

In the mammalia the maternal love shows characteristics that we may call human. J. J. Hayes tells us of a female white bear forgetting the Esquimaux dogs, the huntsmen, and her own wounds, in order to hide her own little bear with her body, to lick her and to protect her. In Central Africa a female elephant, all covered and pierced with javelins hurled at her by the escort of black men attending on Livingstone, was all the while protecting her young one with her trunk, which her own large body enabled her to cover. In monkeys the intelligence is mixed up more strongly still with maternal love, sometimes to thwart and stifle it, sometimes to exalt it and to prompt it with ingenious resources. Occasionally the females of the ouistiti (*hapale*) will commit infanticide. If they grow tired of their children they eat their heads, or crush them against a tree. Often, in the midst of the greatest danger, the female anthropomorphs will give proof of the most touching acts of devotion. In

Sumatra, a female ourang-outang pursued with her little one by Captain Hall, and wounded by a gun-shot, threw her infant on to the highest branches of the tree on to which she had climbed, and continued until she died exhorting her young one by her gestures to escape. In Brazil Spix saw a female of the *stentor niger*, who, wounded by a gun-shot, collected her last remaining strength to throw her young one on to one of the branches close by: when she had performed this last act of duty she fell from the tree and died.

The main characteristic in the love of animals for their young, strong as the love may be, is that it is short, and is strictly confined to the time necessary to allow the offspring to provide for themselves. Parents and children then become complete strangers to each other. As an instance, it is curious to see the turtle-dove maltreat and drive out, as soon as they have grown up, her young ones, which only a few weeks previously she had watched over with so much care.

From this point of view, man, and man of the inferior races especially, differs only slightly from animals. We may affirm that among certain savage people the tenderness shown to their progeniture is less than that which we see in many animals who are highly endowed. In these belated races, the purely animal instincts are held in check by an intelligence that relatively is more fully developed. Man's insight, even when it is most confined, will enable him to see farther ahead than we find to be the case in the majority of animals. He foresees from a distance troubles, annoyances, the cares of a family; and as his morality is still but poorly developed, he often sacrifices his posterity for the sake of his own personal welfare. We can hardly doubt of this fact, so little flattering for the human kind, if we study abortion and infanticide among the different races.

CHAPTER IV.

ABORTION.

In the primitive forms of human societies morality is of the most rudimentary kind; the public opinion of the horde or of the tribe bears very slightly upon individual actions. Parents can dispose of their children as they please; *a fortiori* they have the power of preventing their birth. Even the greater number of the written codes of law in use amongst societies where civilisation is progressing are silent as to abortion; and we must go back to the Zend-Avesta to find legal decrees upon this subject. Among races that are still quite savage it is as lawful for a pregnant woman to disburden herself of her fruit as it is to her to cut her hair. We shall here follow our usual custom in citing facts in support of our opinion. Amongst all races we shall find instances of the same custom.

The Tasmanian woman, anthropomorphous as she is, used to practise abortion very frequently. She was unwilling to become a mother until after several years of married life, in order, as Mr. Bonwick says, to preserve the freshness of her charms. The means used to procure abortion were primitive, as was the intelligence of the race: to obtain the desired effect an old woman would beat the belly of her who was pregnant. A similar custom used to exist also in Australia at the time of the arrival of the first European colonists.

Mr. Bonwick assigns only a sentimental reason for the existence of the practice. There is often another cause, and a powerful one—the difficulty of finding food for the children.

In New Caledonia the women, whether married or not, procure abortion very constantly. They employ different means: the most simple is "the banana process," which consists in swallowing green bananas cooked and made boiling hot. This usage has become proverbial; it is said of a woman in the island who has procured abortion: "There is another who has eaten bananas." According to the morality of the country, there is not the slightest blame attaching to any woman doing such a simple thing.

In Formosa island, which is inhabited by a more developed race, it is not allowed to women to have children before they are thirty-six years of age; and the priestesses perform an act of social duty in stamping upon the belly of every woman who becomes pregnant before the lawful time. This is not done from caprice or from individual egotism; the State ordains it, to prevent the population from becoming too great for the resources of the island.

In America also we find the practice of abortion very widely spread. It was customary with the inhabitants on the shores of Hudson's bay. And nowadays in La Plata the Payaguas make their wives procure abortion as soon as they have given to them two live sons. Their neighbours the Mbayas do the same. According to Humboldt the aborigines of the basin of the Orinoco river have similar customs. The women there use abortive drugs very largely, and they usually postpone the labours of maternity until they have come to an age relatively advanced.

Among the civilised people belonging to the white race, with whom material life is more easy, with whom morality is more developed, whose codes of law endeavour with more or less success to preserve the public welfare, abortion is stigmatised by public opinion and is severely punished by law. We know that it is still practised amongst ourselves upon a large scale, as it was formerly practised by the Greeks and the Romans. Our law reports give us abundant information on the subject; and there is not a doctor to whom women have not come to ask for abortive drugs. On this matter, as on so many others, we must, if we wish to learn the truth, look below the moral varnish of our modern civilisations. In the heart of those of our societies apparently the most refined, there still subsists, and for a long time will subsist, an old remnant of barbarism.

CHAPTER V.

INFANTICIDE.

ABORTION is a learned form, and at the same time a dangerous one, of the system of Malthusian prevention. As infanticide is both more simple and less dangerous, it is for this twofold reason more largely practised. Even among animals instances of it are not uncommon. We find this foreseeing cruelty among wasps, who, as they do not make for themselves a winter storehouse, kill their young who are hatched too late in the autumn. In primitive societies life is rough and food is scarce. Man lives from hand to mouth, eating when he can, and a numerous family is an intolerable burden. He therefore remedies this by putting to death a great number of children, especially girls. Such acts are thought to be natural enough, and no one finds fault. For in the heart of these rudimentary societies the instinctive feeling of affection for the young is soon overcome by the desire to leave the future unfettered. Examples of this are numerous enough.

All through Melanesia infanticide used to be practised, and perhaps is so still to a very considerable extent. The Tasmanians, often famished with hunger, thought little of the lives of their own offspring. There as elsewhere, it was mainly the children of the feminine sex who were sacrificed. In the case of the death of their parents their children would be buried with them alive. No one had leisure to trouble himself about orphan children. Maternal love, thwarted by the inexorable necessities of a miserable life, showed itself in another direction; the same women who would kill their children without a flush on their cheeks would bring up and fondle little dogs.

The Australians, so similar to the Tasmanians, are also quite as pitiless. They rid themselves of a great many of their newborn, especially of the girls.

This is the principal reason of the numerical inferiority of women amongst them, and surely one of the causes of their bestial promiscuity, of which we shall again have to speak. And male children, too, in Australia, were spared only relatively. Sturt relates that an inhabitant of the interior of Australia made use of his sick child by breaking his head against a stone, and then eating him after he had roasted him.

Among the other Melanesian races the children are not much better cared for; but we must curtail our enumeration, as we have many facts to quote of the same kind observed among the greater part of half-civilised races.

In certain tribes in Southern Africa the aborigines lay great stone traps for lions who disturb them, and they bait these traps with their own children. The inhabitants of Fullindochie, in the valley of the Niger, will gladly barter their children for the smallest trifle. And according to M. Raffenel, so do the Zulas in Senegambia.

But it is principally in the islands where food is not plentiful that infanticide is general. It was very generally practised all over Polynesia. In the Sandwich islands more than two or three children were never kept in one family; the others were strangled or buried alive. In Tahiti there was hardly a woman who had not put to death at least one of her children. We know that in the Tahitian archipelago the association of the Areois—of whom we have already spoken—held infanticide to be obligatory on all its members, except in the cases of some few exceptions, which were clearly specified. For instance, of the chiefs, the firstborn child was spared; but the most distinguished chiefs were held bound to put to death only their eldest sons and all their girls. We must remark that the brotherhood of the Areois was composed of the flower of the population, that its practices were authorised and consecrated by the religion of the country. But the human plant, as Alfieri expresses it, grew in this happy climate, and the Society islands were overflowing with their population. We see there a most practical instance of the utilitarian origin of morality.

Also the natives of Tikopia, an island only seven miles in circumference, imposed upon themselves the law not to spare more than two of their male children; the others were strangled. In some exceptional cases the girls were respected, hence the absolute necessity for polygamy. And for the same reason the natives of the Radik islands (Aur) put to death the third, or at least the fourth child of every wife.

Salus populi suprema lex.

At the risk of being monotonous we must continue our disagreeable but instructive task; for it is important to show how weak in the poorly-cultivated man are the sentiments which our moralists and our philosophers commonly regard as the glorious appanage belonging to our kind.

Among many American tribes the people used to hold the lives of their children quite as cheaply as in Polynesia. The Yuracares in South America made a sport of abandoning and of burying their children. The Moxos of the same country also did the same; and specially, as is customary among savage races, they did not spare the twins. The Peruvian aborigines, too, who are more or less christianised, do not baptise their twins, they bring them up always regretting their birth. Charlevoix has observed similar facts among the Red Skins; he saw a foster-child buried alive with the corpse of the mother who had nursed him.

Socially speaking, the Esquimaux and the Red Skins have nothing in common, for the Red Skins of North America exterminate the Esquimaux as venomous beasts wherever they find them. But still the one race of people hold the lives of their children in as little estimation as do the other. The American and the Kamtschadalian Esquimaux do not hesitate to put their children to death if they are in the slightest way weak or deformed. In these arctic regions the struggle for life is hard to bear, and a poorly-organised person cannot be tolerated. Two Esquimaux women volunteered to Captain Parry to barter their children for some trifles, and thinking the bargain concluded, began to strip the children of their clothes, as not forming part of the sale. In Greenland, or more generally among all the Esquimaux, if a mother were to die, the children would be buried with her. Their religion justified this custom, as it everywhere justifies social necessities, and our uncontrollable wants. The Esquimaux believed that from *Khillo*, or from the sojourn of the dead, the mother would call for her child, and the father therefore took care to bury with him the straps which she had used to enable her to carry him.

In China, a country so enlightened in many respects, a country in which the established religion is little more than a code of morals, abandonment of their children, and also infanticide—specially of the girls—has long been customary. Marco Polo observed it even in his time, in spite of the foundation of hospitals for foundlings, and in spite of edicts. The Chinese morality had to give way to the over abundance of their population. Besides abandoning their children the Chinese allow the parents to sell them, and the sales are openly and publicly made.

We find the same customs in many districts of India, from Ceylon up to the Himalaya. It is chiefly among the black Indian natives that infanticide is most openly practised; and indeed it is customary with every inferior and poorly civilised race. Here too, as elsewhere, it is the girls who are mostly sacrificed. The Katodis used to keep only one or two in each family. The Ghauts in the Vindhya mountains act in the same way, and they also have sanctified the custom by making it one of their religious precepts. From manners such as these, polyandry, as we find it practised in Ceylon and in the Himalaya, will follow as a necessary consequence. But the custom of female infanticide is in no wise peculiar to the remains of the inferior aborigines of India. Among the highest born of the Rajpoot it is a common practice. They consider it dishonourable to have an unmarried daughter; it is degrading to marry her below her rank, it is ruinous to marry her to a man of higher rank who asks for a dowry; as a last resource, therefore—and this clears away every objection—by the sacrifice of a girl "the bad powers" are appeased.

It is instructive to note that these people who do not hold infanticide to be even a peccadillo, would undergo anything rather than hurt a cow. In many places, girls who have been spared are considered as merchandise. The Indians at Tullen, near the source of the Jumna river, will barter them for the merest trifle, or will sell them for a few rupees. But, on the other hand, it is a matter of great difficulty to get them to sell one of their sheep; for they say: "The sheep gives us wherewithal to clothe ourselves, but what can we do with a girl?"

Among the Semitic and European races, even those who are the least civilised, the moral sense has made some way. The historians tell us that the inhabitants of Mecca, in times of famine, have been known to sell their children for a measure of corn. But these are cases in which a superior force overrules everything; and, as a general rule, among the Semitic races, and with Europeans, the abandonment or the murdering of their children are individual acts, and are relatively rare. In Europe it will be found that as time has progressed the occurrence has become gradually less and less frequent. For if we may believe Saint Vincent de Paul, the abandonment of children in the seventeenth century was then common. And here is one of the thousand arguments, and not one of the weakest, which we may adduce in proof of the progressive movement of humanity. Man has sprung from a very low state; but he can, and will, ever continue to rise to a higher and better condition.

CHAPTER VI.

LOVE FOR THE YOUNG IN HUMANITY.

ACCORDING to European notions the facts which we have just mentioned are atrocious. There is still taught in our schools the obsolete theory that our moral ideas—the ideas that we Europeans of the nineteenth century now possess—are innate in all the human kind. How can those who defend the commonplaces of university teaching explain the fearful state of manners which we have just described? The doctrine of advancement, of slow and progressive evolution, is the only argument that can rest unshaken by facts such as these. That doctrine accepts them because they are found to exist, and classes them, as do naturalists, considering them characteristic features in the inferior phases of civilisation.

Are we to understand, then, that love for their children is unknown to primitive man? By no means. How should we feel, we the descendants, still rough and uncouth after the efforts of so many ages, if these ancestors, coarse as they appear now to us, had not bequeathed to us the rudiments of love? Can it be that a sentiment, so primordial that we find traces of it even in the inferior animals, should be foreign only to man?

Even amongst the wildest hordes, parental love, especially that of the mother for her children, exists, but in an instinctive state, and is overcome by the inexorable necessities of existence. *Primo vivere.* This sentiment has not been exalted as is the case with us, by education, by literature, and by tradition. Sentimentality is unknown in these embryo societies; animal ferocity is not kept in check by moral feeling, by human respect, or by the severity of the laws. In truth, during these inferior phases of joint evolution an idea of morality hardly exists; laws that are purely traditional scarcely influence individual actions; children are the absolute property of their parents. And when hunger does not cry too loudly primitive man will love and fondle his children, perhaps as tenderly as he who is civilised. This may be proved by numberless facts.

Even the poor Fuegians on the Magellan straits caress their children and play with them. Wallis saw the parents make their young ones jump about in their canoes, lift them up into the air, and hold them over the water to amuse them by frightening them.

In South America the Yuracares do not allow themselves to scold their children; they consider that it would be very wrong to vex them. The Esquimaux make dolls for their little girls, and small bows for their boys; they never eat until their children have had their share. An Esquimau father and mother, passing a spot where the summer previous a child whom they adopted had died, knelt down and began to cry and to mourn. An Esquimau father advised that the dead body of his child should be buried in the snow; for he said that the mother, who had died before, would cry out in her grave if stones or blocks of ice were to weigh heavily upon her offspring.

The Polynesians, prodigal as they were of the blood of their newborn children, were very fond of those whom they had spared. In the Marquesas islands the women used to nurse their foster children most tenderly, they heaped upon them the most delicate attentions; the men pressed between their arms the children whom they thought were their own; it was not otherwise in the Sandwich islands.

The coarse Hottentots are as fond of their children as people of other races. As soon as he is born the little Hottentot is fastened with straps on to his mother's back, and he does not leave her. On the banks of the Niger the maternal affection is so strong that after the death of their children the mothers will carry upon their heads small wooden images in commemoration of their little dead ones, and they will not allow these emblems to be taken from them. They seem to consider them as living images, and before eating themselves they always offer food to these little wooden children.

Nevertheless, among savages, love for their children does not last very long. We can understand that it should be stronger in the woman than in the man; in every race, civilised or not, woman is in this respect more instinctive than her companion. Maternal affection increases very considerably with the constant intimacy which the early cares towards the child render necessary; for nursing among savages lasts sometimes for five or six years, to such an extent that a foster-child in the Marquesas islands has been known to take a cigar out of his mouth before beginning to suck the breast. In these same islands, where life is easy, the boy, when he has grown up, builds for himself an ajoupa of branches and of leaves, and troubles himself no longer about his family. The parents at first appear still to be fond of him, but they resign themselves to the life he has adopted, and soon they think no more about him. In much the same way many birds and mammals drive out their little ones as soon as they have arrived at an age when they can provide for themselves.

It is not until a long process of culture has developed the mental life, enlarged the intellectual horizon, and created affective sentiments, that the instinctive love of the progenitors for their offspring can ennoble itself and fully conquer physiological wants. Lasting ties then unite the parents and their children, for the sentiments of the children model themselves more or less upon the sentiments of the parents by reason of a long and intimate daily intercourse. Parents and children have both acquired their share of the mental qualities of their race and of their nation—without speaking of money or the care of social position, two very powerful factors in societies which are called civilised. But all that with the savage exists only in an inchoative state.

CHAPTER VII.

FILIAL LOVE, ASSISTANCE TO THE OLD, TO THE SICK, ETC.

In humanity, parental love for children is assuredly the foundation-stone of the affective life. In the animal kingdom it is also the most fully developed of all the benevolent sentiments. It has certainly been the object of selection, for it is indispensable to the keeping together of all the superior kinds. On the other hand, the love of the young for their progenitors is less necessary, and is therefore rarer and much more feeble. The majority of animals do not know it, the savage feels it only slightly; even amongst civilised people filial affection is much less energetic than the paternal, and specially than the maternal love.

The affection shown towards parents and assistance given to old men are very closely allied, especially in primitive societies, where all the members of a horde or of a tribe are more or less closely related; or again, when promiscuity is common, and the children do not know their own father. We shall therefore study simultaneously filial love and assistance, given or not, to old men, and more generally to the infirm and to the sick.

In order that they may be more developed in man, the noble sentiments of which we are going to speak do not appertain to him exclusively; we may mention some undoubted instances that have been observed in animals. Latreille had cut off the feelers of an ant; she was assisted by her sister ants, who, when they had examined her wounds, covered them with a mucus taken from their mouth. A queen bee, nearly drowned, was surrounded by the working bees, who took care of her and licked her untill her strength was restored.

It is not wonderful that some feeling of joint responsibility should exist in ants and in bees, for they are both very sociable animals; but this feeling can develop itself in animals whom the caprice of man has compelled to live together, as may be seen from the following fact. A friend of mine, M. Frère, a chemist in Paris, brought up a couple of canary birds, who had come direct from the Canary islands, and he put them both together into a garret in his country house at Nanterre. This couple, well fed and almost free in their actions, increased and multiplied. Fifteen or sixteen years afterwards the garret was inhabited with a tribe of canaries, sixty or seventy in number; and among them there was some mixture of the green canary, for strangers had been intruduced into the family. The mother bird, then seventeen or eighteen years old, was so enfeebled by her great age that she could hardly flutter. She could barely drag herself to join in the common meal. Two of her descendants—two only, and they were both of the pure breed—perceived this and came to her assistance. They took care of her until her death, as much as nearly two years afterwards. They fed her from their own beaks, as they would a little one; and what is equally singular, the old grandmother welcomed them by beating her wings, as the young ones do. This was not an instance of filial love, for the two charitable birds were distant descendants of the mother ancestor; it was an act of what we too proudly call "humanity," though this noble sentiment is far from existing in the hearts of all men.

In fact, in all primitive humanity the lot of the aged and the infirm is generally pitiful. Over all Melanesia it is customary to put to death the aged and infirm, as they are so many useless mouths. The New Caledonians, who regard the head of their father as a sacred object, consign their parents to some secluded spot, and leave them there to die. They sometimes bury them alive, and the sufferers take the whole thing as a matter of course. The old people will often ask for death, and will walk towards the ditch into which they are felled by a blow on the head. The same custom was generally practised in one of the Fiji islands. There it had been established by the religion of the place; for religious ideas are most frequently first prompted by the requirements of a people or of a men. The Fijians believed that a man goes into the future life exactly in the same condition as that in which he has left the present. There was therefore a very strong argument to prevent him from allowing himself from falling into decline. Hence arose the duty of the children to warn their parents in time, and to kill them was their last act of earthly gratitude. The children did not fail to obey. A mortuary feast was held, to which friends and relations were invited; then the victims walked quietly towards their ditch, and after a tender farewell the sons would, with their own hands, strangle their parents.

Similar facts have been noticed nearly all over the world. Campbell relates that among the Matchappi Kafirs, the old men are held to be contemptible, and are forsaken; they die of hunger and their bodies are left as prey for the wild beasts. In Polynesia the fate of the aged and of the sick is not less hard. They were often driven out of the house, and sometimes were buried alive. According to Robertson, to put one's old relations to death, was a general custom from Hudson's bay down to the La Plata river, and we may almost say down to the island of Terra del Fuego. The Esquimaux either buried them after they had strangled them (H. Ellis) or they shut them up in an *iglou* of ice. The Itonamos of South America used to stifle their sick. In times of famine the Fuegians asphyxiate and eat the old women, in preference to their dogs, who they say eat otters.

The Kamtschadales used to kill their parents to be rid of them, and would then throw their bodies to the dogs. The Kamtschadalian conscience justified itself on religious principles: to be eaten by the dogs was a sure way of being taken to the next world, for the dogs themselves were so good. The Koriaks and the Tchuktchi were like the Fiji islanders, who wished to leave this present world in a good condition, so as to be happy in the next; they therefore required of their children to kill them before they got old.

The Thibetans, respectful towards their parents, give very little assistance to the sick, especially to those afflicted with any contagious illness. As soon as a case of smallpox declared itself in a house the inhabitants would all leave it: they would go away from the town, and the sick person would die uncared for and alone.

These animal customs are not peculiar only to the inferior races, properly so called; they will always be found where civilisation is in a backward state. The Massagetæ used to bury their old men. A tribe at Sardis used always to fell their old men by hitting them with a stick. In ancient Bactriana dogs were kept for the special purpose of devouring the aged and sick; they were called the "dog buriers" (Strabo). It is not long since that the Abasians would willingly sell their fathers or their relations.

We have only to make a choice of facts related by historians and travellers in order to prove how little susceptible of affection is man, in the primitive stages of his existence, for his aged parents, and for the infirm of every sort.

Are we to understand therefore that the poorly developed man does not feel filial love? By no means. But the capacity of altruism, or the feeling for others, is with him very slight, and is easily quenched by contrary sentiments. During this period of his mental evolution man's moral sense is very unsteady; he may be charitable or unforgiving, according as circumstances will sway him.

The New Zealanders used to show a great respect for their old men; they used to place them in the best seats at their feasts; and often the chief would feed the common people for the simple reason that they were old. In Senegambia the phrase: "strike me, but do not curse my mother," is familiar even among the slaves. At Kaarta the Bambarras call all the old men "baba," or "papa;" they pay a sort of respect to their white hairs.

The Dayours on the Upper Nile venerate the old men, and in all their hamlets some gray heads may be seen. This is a very fertile region; life here is easy, the struggle to live is not cruel, benevolent sentiments can therefore afford to show themselves. Also in California, among some of the Catholic missions, when hunger did not make itself too severely felt, the old men lived at the expense of the community, and were fairly well cared for; and here the aborigines belonged to a very inferior race.

The Tartars, much superior to the Californians in the hierarchy of the human types, and also much more civilised, whose pastoral life guarantees them against hunger with tolerable certainty, are kind and hospitable, and they show a great respect for paternal authority even after they are married. In China these sentiments have grown with the march of civilisation, and respect for their parents and for the aged has become with them an imperative moral duty. After the death of their parents their sons continue to celebrate each decade of their existence as though they were alive. To forsake one's old father is a crime that they rarely commit. The emperor will sometimes give to the old men a yellow dress as a mark of homage. Asylums for the aged, for widows, for the infirm, hospitals for foundlings, where they are well cared for, benevolent societies for affording assistance, houses for the education of the poor, are numerous, and in some cases date from a very early period. As long back as the days of Marco Polo the emperor decreed that the homeless children should be brought together and be educated. We cannot but see in all this an indication of a high degree of moral elevation. Let us add that in China the respect for intellectual accomplishments is also not less noteworthy. A Chinese society is formed with the object of collecting every old piece of paper on which there is any writing or printing, so as to save it from the rubbish heap. The tramps, the literary scavengers employed in collecting these scraps, carry placards on which may be read: "Carefully respect the paper on which any letters are written." This is in truth a degree of educational respect both touching and piteous. One must go to China to see characteristics that are so pathetic.

The Chinese laws have pushed the case of joint responsibility so far that some of them defeat their own object; for instance, the law holding amenable to be punished with death the man who was the last to see any other person, for this will often prevent a man from affording assistance to one who is drowning, or to one on his deathbed.

CHAPTER VIII.

THE FEROCIOUS INSTINCTS IN HUMANITY.

THE organisation of benevolent institutions can only be possible in a society that is itself well organised. But still true humane sympathy for one who is suffering is not unknown among many primitive people. From this point of view, what are the characteristics that we find in different races? Or, in a more general manner, how is human life thought of in the different human groups? From an ethnico-psychological point of view these questions are all-important.

The altruist sentiments, as the positivists say (the power of being able to put oneself in another's place and feeling for him), are certainly the result of a high degree of culture. Doubtless they are not altogether wanting to the inferior races, and we shall have to mention some instances of this; but they are uncommon and inconstant. In the early phases of civilisation this regard for the feelings of others has only dawned upon the human conscience.

The Australians do not think any more of the life of a man than they do of that of a butterfly. But they feel with great violence the desire for vengeance, and will satisfy it upon any member of the tribe to which one who has offended them belongs. In speaking of the genesis of the moral sense we shall have to return to this curious fact. The same contempt for human life may be observed all through Melanesia. We shall see later on that in New Caledonia the chiefs will gladly, in their own family circle, eat one of their subjects. Similar manners used to prevail in the Fiji islands. A Fijian named Loti ate his own wife, after he had roasted her over a fire which she, by his own orders, had prepared. He committed this atrocity simply to attract attention, to acquire some degree of notoriety. To kill anyone in this country is an act that entails no consequences. A man may bring himself into good repute by so doing; consequently the natives take care always to be well armed.

The negroes in Africa are hardly more humane, in the European sense of the word, than are those in Melanesia. The contempt with which the Ashantis regard human life is well known; in fact it goes beyond all belief. At the death of one of their princes, these people are sacrificed by hundreds, and even by thousands. Other butcheries take place at stated periods: at the beginning of the season for gathering the ignames, or of the harvest, etc. Sometimes a young virgin girl is impaled in order to remedy the slack condition of their commerce; and worse still, the people are not content with killing, they often exercise their ingenuity in torturing their victims before sacrificing them. Bowdich saw a man whose hands were tied behind his back, and who had been tortured as follows: a stake hanging down in front of him was fastened to one of his ears; the other ear, nearly altogether detached from his head, was hanging only by a shred; a blade of a knife ran through his two cheeks; he had several large gashes cut into his back; a knife was passed through the skin under each shoulder-blade; and he was led about as a beast of burden by a cord pulled through a hole made in his nose.

The negroes in Senegambia, who are in some slight measure civilised, and have in them a mixture of Moorish blood, are excessively violent and cruel. Murder is very frequent there, and undying vengeance is with them an act of duty. According to Mungo Park, they make a study of doing harm to others, and take pleasure in watching their sufferings.

The vast American continent used to contain, and does still

contain, nations or tribes whose manners are very widely different; but except perhaps some of the tribes in Central America, ferocity of disposition is everywhere the predominant character. In speaking of cannibalism, of the treatment inflicted upon prisoners taken in war, of the different religions in America, we shall have to mention many very atrocious facts, among others the punishment of prisoners by the Red Skins and the Brazilian Indians, and of the human sacrifices of the Mexicans.

The Indians whom La Pérouse saw at Port-des-Français, in South America, were very irritable, subject to fits of anger, perpetually quarrelling amongst themselves and threatening each other, and extremely vindictive; they were much more savage in their manner than many wild animals. The Malays, who appear to be cold, taciturn, cunning, masters of their own actions, bold, and despising peaceful occupations, are, we are told by travellers, singularly cruel, and at the same time they have always a noble bearing and an extreme politeness of manner. If we may believe Niccolo Conti, an old traveller who wrote in the year 1430, the Malays looked upon homicide as a joke. He says: "If one of them bought a sword he would try it at once by plunging it into the chest of the first man who passed him." Public opinion was in nowise shocked; the skill of the murderer would rather be praised if the blow had been artistically dealt. We see another instance of this ancient ferocity in the well-known Javanese custom "to run a muck." In Malay, as elsewhere, for one reason or another, a man would be tired of life. Then instead of killing himself, as a man might do in Europe, he clutches his kriss (a sort of dagger)—he often first makes himself drunk with opium—and wreaks his wild fury upon any one he meets. Ten, fifteen, or twenty persons may be killed by a man in this state of madness, before he is himself either killed or even arrested. The Malays do not consider that "running a muck" is at all dishonourable. A man will adopt it from any or from the slightest cause; the "muck-runner" may consider himself unfairly dealt with, or he may have lost at play, or any other reason will prevail with him. In the time of war, a whole body of men may sometimes be seen concerting their plans to "run a muck" through the enemy's ranks. They will rush with most furious energy, and allow no obstacle to stay them.

The Polynesians too, relatively a soft-mannered people, as gay and fickle as children, held human life quite as cheaply. The chiefs would maltreat, wound, or kill the common people, as their caprice prompted them. With the consent of a priest, any Polynesian, especially one of the lower classes, might be seized and sacrificed to the gods. We shall again revert to this in speaking of the religion of these people.

Humanity and philanthropy are the smallest among the good qualities of the different branches of the Asiatic Mongolians. The nomad Turkomans, who wander about through the Khorassan, are literally beasts of prey. They run madly through the agricultural districts, and make a sport of killing the people or of making slaves of them. For the slightest fault, for any caprice of their own, they will put to death their wives, their children, and their servants. If we may believe Father Huc, the Mongolians, those properly so called, have lost the old ferocity which at one time characterised their invasions into Europe and into Asia. They would now seem to be soft mannered, peaceful, and hospitable. According to a more recent traveller, this softness has grown into apathy and treachery among all the tribes in which there is any trace of Chinese customs or of Chinese blood.

The Chinese, so estimable in many ways, and even so admirable, seem to be able to reconcile their extreme contempt for death with a degree of treachery that is also equally excessive. In their history we read of a series of frightful civil wars, of bloody rebellions, followed by most cruel tyranny. On the other hand, we all know with what placidity the Chinaman will look upon capital punishment. Quite recently, a man who was condemned to die readily found a substitute, who, for a small consideration, at once consented to undergo the punishment. At the same time, these people, who hold in such slight estimation the lives of others and also their own, will submit to any burden without a murmur; they will, in their own country, let themselves be governed by a handful of Europeans; they will servilely obey the Malays, who are so inferior to them. It is difficult not to trace this enervation of character among the Chinese to their own civilisation, which during the course of some thousands of years has ever tended to govern all the acts in their daily life, to paralyse individual enterprise, to hold implicit obedience to the State officials as absolutely imperative, and to despise every warlike profession. The complication of these results has at last ended in stifling all the primitive natural energy of their race. According to Father Huc, it is always the Chinaman's endeavour not to compromise himself; and this sentiment has grown into a saying which the Chinese have ever ready in difficult situations: "Lessen your desires." May this maxim be taken to heart by certain European nations, in which the governing classes are evidently trying to become Chinamen.

This enfeeblement of the Chinese is not a characteristic originally natural to them; it has been acquired. It is the result of institutions, otherwise beneficial, but which have taken thought only of intellectual development, and given none to the strengthening of the character, or to the training of the moral system. The Japanese, on the other hand, so similar to the Chinese, and to whom they owe the brightest spots in their civilisation, have always maintained the primitive energy of their ancestors. With them, to forget a wrong done is stigmatised as an act of cowardice; they hold military courage to be no more than ordinary virtue.

Their suicidal custom of embowelling themselves, which is practised even now, must no doubt be called an act of madness; but it is certainly incompatible with an enfeeblement of the national character. Now, among any people, no quality is more primordial than freshness of mind; without that, intellectual development, even to a considerable extent, will benefit little. But alone, it can assuredly bring *forth* no fruit. To extend the frontiers of human knowledge man must generally brave the prejudices of his time, forget himself, disdain his own personal welfare: to think strongly, he must first have willed strongly.

It would be equally rash to deny or to exaggerate the influence of the institutions with which we are surrounded upon the character of a man or upon a race of people. Every man when he is born inherits a moral system, and this system will form the basis of his nature during the whole course of his existence. Education is not disarmed by the force of instincts transmitted to him by his ancestors, but its power is greatly limited. The influence of social surroundings upon each single individual is very small, but it goes on increasing mathematically from one generation to another; it is the untiring effort of the drop of water ever falling upon the granite which bores the hole and at last splits the rock. The slight mental changes produced in each man by the social atmosphere in which he lives add themselves together and make a sum-total, and in a given time they may entirely metamorphose the character of a people or of a race. In China the Mongolian has become cowardly and servile, but in Japan his primitive energy has been more fully preserved.

Similar effects may be observed in the different ethnical groups of the white race. The real Hindoo, who is the Brahmin offspring of the ancient emigrants come from the steppes of Central Asia, is now enervated and effeminated; but at the same time he is humanised. In him all trace of his ancestral ferocity has disappeared. With certain Hindoo Buddhists the humanising sentiments have been so exaggerated as to become animalised, for they have founded hospitable asylums even for their dumb creatures. The mainspring of his character has become weakened, and hundreds of millions of Hindoos now submit themselves to a handful of English conquerors. In India if we wish to find some little courage or energy, we must go up to the northern regions, where the stamp of native wildness still exists. The Sikhs are fearless; the Rajpoots are drunkards, sensual, and cruel, but they are still courageous.

In Persia the moral degeneracy of the white race has, we are told by all travellers, grown to excess. The people there will tolerate with abject servility the most capricious despotism. The Persian himself is hypocritical, cowardly, and ferocious. The only white men not yet crushed by European civilisation, and who, during the primitive epochs, gave to our ancestors their ethnical superiority, are the clans in the Caucasus, whom the Russians have had so much trouble in bringing under their subjection.

Among these mountaineers, as among the Arabs, the human sentiment is shown by their religious feeling of hospitality; and the sanguinary instincts of past ages is still seen in their desire for vengeance, which they hold to be a sacred duty. The obligation is always handed down from father to son. We know that it is almost the same with the Arabs, both in Asia and in Africa, who have come to an equal stage of civilisation.

Among the so-called Indo-European nations, now the most civilised, the sanguinary instincts of the animal, though very much deadened, often awake and still show themselves in a thousand different ways. It may be that abandonment is no longer practised, that we do not eat our old men, as the Thracians used to do in the days of classic antiquity; but, nevertheless, in the heart of nations apparently the most civilised, acts of cold inhumanity, of actual savagery, are of daily occurrence. As has been recently shown by Dr. Bordier, in examining the skulls of assassins, atavism reproduces even now in Europe a certain number of savages who belong to the age of polished stone. This ferocity is not yet quite extinct, and shows itself only too plainly in a great social crisis, when all legal restraint is thrown aside, and especially when, in a cause more or less well generally understood, the fluctuations in public morality appeal to the sanguinary instincts against a foreign or domestic enemy.

Our modern humanity is much oftener upon our lips than in our hearts. It is indeed narrow enough, and hardly held to be obligatory except among people of the same race, or more strictly, people of the same country. Christopher Columbus, who passes with us for a type of noble heroism, thought that he was not doing wrong in ordering that the natives of the Antilles should be extirpated, that they should be devoured by bloodhounds; and until the end of the eighteenth century this practice, worse than savage, was continued in Cuba and in the island of St. Domingo by the French and Spanish colonists against the maroon coloured negroes. In Cape Colony the Dutch drove out the Bushmen, and even the Hottentots, as though they were wild beasts; and we know only too well that in the same savage way the English colonists have exterminated the Tasmanians.

Does all this mean that, as regards benevolent and humanising sentiments, man has not progressed since the early primitive ages? It would be ridiculously absurd to maintain such a theory. At first man was but little different from other superior mammalia. His kindly feelings were weak, intermittent, easily mastered by his instincts and personal wants; but as hunger, the all-powerful incentive, gradually slackened its hold, man's egoism has by degrees become less violent. At first people were fond only of their children, and were fond of them only for a short time, after the fashion of animals. Then man began to show care for the aged and infirm. For a long while humanity was not extended beyond the members of one's own family or one's own tribe. In modern times, among civilised nations, except when war is raging, the fact of being a man has asserted a claim to certain rights. We need not be accused of over optimism if we believe that humanising sentiments are destined to extend themselves still much more largely. But this noble side of the moral man has grown very slowly in the human conscience, as we may see clearly by some examples among the inferior human types; for they, in this way, as in many others, show us the successive stages through which the superior branches of humanity have passed. It will therefore not be uninteresting to study in the different races the manifestations of altruist sentiments, to see the gradual passage from the bestial to the human state.

CHAPTER IX.

BENEVOLENT SENTIMENTS.

WE may easily understand how the feelings of pity, of compassion, etc., arise and have their existence. If any organised creature is touched by the suffering of one of his kind, it is painful to him to know of the suffering. The exterior signs of the pains of others strike back upon the individual who watches them; they call to his mind the recollection of the same pains which he himself has undergone, they reproduce a more or less weakened image of what he himself has felt. From such a condition to that of aiding one who is in pain, the step is but a short one; it is a generous way of relieving and aiding oneself. According as man's imagination is more or less strong, the reflection of grief will be more or less strongly pictured, and the feeling of pity will also be strong in proportion. In all this, intelligence, properly so called, can have no place. We need not, therefore, be astonished to see the feeling of pity very strongly developed in certain animals, and only rudimentary in certain men.

If, in a flight of parrots, some are killed by a sportsman, the others will for five or six minutes flutter round about the dead bodies of their lost companions; they will cry plaintively, and will also allow themselves to be killed in their turn. Bullfinches, cardinals, and siserins (a sort of linnet) will also do the same. J. Franklin tells us a touching story of two little parrots, called "the inseparables." The female had become gouty, and the male fed her for four months. He assisted her in climbing on to her perch, and when she was dying he increased his marks of care and tenderness, mourning all the while most pitifully. After her death, he languished and died himself at the end of a few weeks.

The ouistiti, when they are taken prisoners, hasten to their sick and nurse them. This surely is a case in which humane sentiments intervene, for there is ordinarily no relationship between these animals, brought together merely by chance.

It would seem that, as regards sympathy for the feelings of others, the lowest of the human races are inferior to some kinds of animals of which we have just made mention. It is certain that the *faculty* of being moved by the sight of suffering is both a sign and a cause of social progress; we must, therefore, expect to find it very weak in those human types still in a primitive state. The Fuegians, the Tasmanians, and the Australians are celebrated for their total moral insensibility. The New Caledonians are almost incapable of gratitude.

According to A. Bourgarel, the New Caledonian woman is the most vicious of all animals. The aborigines of South Africa, the Hottentots and the Kafirs, are also extremely hard. It is only in the equatorial regions of the African continent that travellers have observed acts that are really human, and there they would seem to be specially the privilege of the woman. In the Gaboon country, where women are treated just as beasts of burden, they are nevertheless capable of pity and of compassion. Du Chaillu relates that he himself fell sick, and they overwhelmed him with care and attention. And we must here notice that humanity for another race is not common even among the white races.

In Senegambia, where the negro blood has become more or less mixed with Moorish blood, these altruist sentiments are considerably developed. After the burning of the town of Bali, Clapperton saw the inhabitants of Koulfani, a neighbouring town, send to the houseless creatures everything except that of which they themselves were in urgent need. The women in this region will sometimes show their compassion by acts of the greatest delicacy. An old woman met Mungo Park in a famished state, and completely stripped by a black chieftain; she gave him something to eat, and went away without waiting to be thanked. On another occasion, the same traveller, possessing no other worldly goods but the saddle of a horse which had been stolen from him, was taken in and lodged by the women, whom he heard before he went to sleep singing as follows: "The winds are roaring, the rains are falling; the poor white man came and sat under our tree; he had no mother to give him his milk, no woman to grind his corn. Let us have pity on the white man. He has no mother," etc. In the same country, the French traveller Raffenel was treated in the same way. "They knelt," he says, "upon the mat on which I lay down; some were fanning me, others were rubbing me, others were giving me milk and baked pistachio nuts. Their song, which became gradually more and more melancholy, ran something in this wise: 'The white man who has come from a great way over the sea has stopped here. He was tired, tired because he had walked in the heat of the sun. He was very hot, and water was running down his cheeks from his white forehead. He was very hungry and very thirsty. Take the finest mats and spread them under the white man, that he may repose his weary limbs. And our master said to us: "'Take your big fans, and fasten your scarf-belts, and wave them over the head of the white man to dry up the drops of water that stand in beads upon his forehead. Take your gourds and fill them to the brim with the best milk from my cows, that the white man may slake the thirst which is devouring him.'" *

These facts, and others of the same kind, would lead us to think that compassion was especially a female virtue. But when we are speaking of a creature so widely and wonderfully various as man, it behoves us to draw our conclusions prudently. In sociology, more than in most other matters, we shall find exceptions to the rule. For instance one would think that there was no virtue more essentially feminine than shame; but nevertheless in many countries the men only wear any sort of clothing. We find this to be the case on the banks of the Orinoco, and among many African tribes. In some cases clothing is the privilege given to the married woman, in others to the young girl. As regards the feeling of pity, the diversity is equally great. The negro women in Senegambia are susceptible of acts of delicate charity, but the Bechuana Kafir women would, without blushing, watch their husbands, after a victory, decapitate the wives of their prisoners simply to possess themselves of their necklaces, which bore too

* Raffenel, "Nouveau Voyage aux pays des Nègres," i. 275.

heavily round their necks. Thevet relates that the Brazilian aborigines used to keep their prisoners of war for a certain time, they would bring them up, and provide them with wives; then at a given moment they would eat them amidst pomp and ceremony. Then, says Léry, the wife of the dead man, after some feigned show of grief, would always be the first to eat a piece of her late husband. The aborigines of Central America were always, and are still, generally soft mannered and sociable. When they were once conquered they submitted themselves docilely to a few Spaniards; they would often show themselves to be devoted to their masters, in spite of the masters' cruelty towards them. D'Orbigny has seen them weeping for grief because they were obliged to kill one of their high-priests. The islanders of Cuba showed to Christopher Columbus the heartiest and the warmest welcome. A chief whose wife had been taken away from him went with tears in his eyes to entreat Barthélémy to give her back to him; and after his prayer was granted he returned with four or five hundred of his subjects to clear away a piece of ground for the Spaniards.

It may astonish us to learn that the poor Esquimaux, coarse mannered as they are, have sometimes shown instances of great generosity. They gave to Ross and his companions a handsome present of fish, and looked for nothing in return. With delicate politeness they thanked the English, who had allowed themselves to be lodged by them. A woman, whom the doctor upon the expedition had attended, brought to him the most precious object she possessed: a stone from which fire could be struck.

In the primitive man, whose morality is still in the course of formation, whose animal instincts speak so loudly, and whose disposition is so extremely changeable, there is no steadiness of character. Kindness and ferocity may be found co-existent. During this early phase there are in man, so to say, several psychical beings; mental life is fragmentary, and man's actions will depend upon the impression of the moment. The Polynesians, who show a most infantine versatility, are both soft mannered and cruel. Porter extols the courtesy, the bravery, and the affability of the Noukahivans; he is astonished at their brotherly dealings. Cook has seen the Tahitians divide into equal portions one single piece of bread-fruit, mutually exchange their clothes, and show themselves willing to oblige each other. The inhabitants of Easter island, during a time of famine, offered to Cook some potatoes from their own scanty meal. Bligh speaks with admiration of the gaiety, the good-humour, and the sociability of the Tahitians, whose life is ordinarily a long period of amusement; and in Bashee island, some of Dampier's sailors, who had deserted, were kindly received by the natives, who gave to each of them a wife, a field, and all the implements necessary to cultivate it.

Let us repeat, the most striking moral contrasts may be seen in the ill-developed man. The Malay tribes, for instance, have justly earned their reputation for being ferocious; but Wallace speaks in the highest terms of the morality of the Borneo Dyaks. They appear to be honest, scrupulous, never committing acts of violence among their own tribe; but the very same men are, tribe against tribe, intrepid in pursuing each other.

We know that among the Chinese humane sentiments may be seen in their large number of hospitals, of institutions giving assistance to the weak, to the infirm, to widows, by the establishment of pawnshops, etc. We need not be astonished at all this, for in China we find an ancient and ingeniously contrived system of civilisation. We may be more startled to see the refined notions of chivalry existing among the Mongolians in Tartary. For instance, in Tartary, if one wishes to pass unmolested through an enemy's village, one may do so by putting women in front of the caravan, and giving to them the care of driving the animals. It is held to be a point of honour in the country not to attack women, or to steal from them the animals they are driving. And yet these people, who have such generous manners, are the descendants of the terrible Mongolians, whose bloody invasions cast so much terror upon the whole of the old continent.

In the same way, too, the Turkomans, in whom the Mongolian blood predominates, and whose life is but a long series of thieving and murder, practise hospitality in a most generous way. The stranger, unless he be a declared enemy, is welcomed into their tents with the greatest courtesy; his only danger is if he should wish to exchange the hospitality of one host for that of another.

This virtue of showing hospitality is not, as we are aware, the peculiar privilege of the nomad Turkomans. It is also in some way a moral characteristic in the generosity of the Arabs. In Arabia, everyone, both rich and poor, chiefs and simple Bedouins, are strictly enjoined to exercise it. The force of public opinion has so willed it, being prompted thereto from the simple feeling of neighbourly duty. The most cutting reproach that can be made against an Arab tribe is that "the men have not the heart to give everything, and the women can refuse nothing."

Later on we shall have to inquire the reason for the feeling of duty. It is beyond all doubt that this feeling becomes increased by inheritance; but, as with everything appertaining to the human conscience, its manifestations will be found to differ very widely. In the sixth century of our era Antar practised hospitality upon a truly grand scale. It is therefore considered a most imperious moral obligation among all the present existing Arabs who have preserved the civilisation peculiar to their race.

The sentiment of a joint responsibility, of a joint feeling for the sufferings of others, has in the moral system of the Indians shown itself in a form that we may well call *excessive*. If a man has lost a lawsuit, or undergone an injustice, he will kill himself in order that his blood may fall upon the head of his offender. Heber relates that in the district of Ghazeepore (Hindostan) a man who had been nonsuited in a trial as to the possession of a field, brought his wife there and burnt her alive, in order that her spirit should come back after her death, and that the ground might be cursed. Before the English conquest the last resource of the people against the tyranny of the rajahs was to assemble silently before the palace of the master, and, if need be, to let themselves die of hunger if the rajah did not yield to them. The quite primitive man—the Australian, for instance—will kill and eat his own child without any feeling of regret or of remorse; the Hindoo, on the other hand, has such confidence in the altruist sentiment among the men of his race that he will base his vengeance upon it. The two men rush to the opposite extremes of the scale of humanity. The Hindoo has assuredly once been in the present position of the Australian, and his example, along with many others, proves to us how indefinite is the field of our moral evolution.

Nothing, however, could be further from the truth than to consider the humanising sentiments as virtues invented by the Aryan races. We have seen instances even among the inferior human types; they will be found to generalise themselves in the hearts of every people who have arrived to a certain degree of civilisation. European Christianity has taught us charity and neighbourly love; but the Chinese have built and dedicated temples to Pity.

We all know very well how slowly has sympathy, the power of feeling pain ourselves because of the suffering of others, developed itself from the earliest times of the Greco-Roman antiquity down to our own days. Between Ormuzd and Ahrimane the struggle has been long, and it is yet far from terminated. In the beginning of their history the Grecian heart was hard; the foreigner was to them more or less an enemy; the slave, a sort of domestic animal, whom we might, according to Aristotle, hunt out as though he were game. But the Greek religion soon instituted places of refuge for the outlaws, the conquered, the slaves, and even for the guilty. Euripides even went so far as to define the truly good man: "he who lives for his neighbour." * In Rome nearly the whole population used to take their delight in watching the bloody sports in the circus; the vestal virgins (oh feminine sensibility!) by lowering their thumb would thereby command the deaths of the gladiators. But by degrees the Greco-Roman philosophy encouraged humanising influences, which Christianity claims for itself to have invented. Have we yet reached the ultimate stage of this evolution? In order to think so we must both ignore the past and blind ourselves as to the present.

We do not need to go back to the primitive phases in our social evolution to show irrefragable traces of the progress of humanity.

* This humane evolution in the Greek mind has been admirably described by M. E. Havet, in his "Origines du Christianisme."

Our own history will supply us abundantly. Let us take one instance: During the first centuries of the dark ages, the right of shipwreck was a right belonging to the monarch, and a prince of Léon, in Brittany, calmly contemplating the terrible rocks off the coast of Finisterre, said, without feeling the slightest twinge of conscience, that they were worth to him more than all the precious stones of the most opulent monarch. As a matter of fact, the confiscation of waifs of all sorts coming from shipwrecks, then so frequent upon this coast, brought him, one year with the other, an income of ten thousand golden crowns. We nowadays spend a much larger sum in constructing with great labour and at great risk ingenious watch-towers upon this same dangerous coast.

Our own European history, bloody as it may be, is but a long effort, not always conscious of the past, towards humane progress. During the early ages treatment of the slaves was gradually softened; then the slave himself at last became politically equal with his master. The distinctions of caste and of class wore themselves away; the vassals began to encroach upon the fiefs of their lord. The nobles, who were dominant only by the right of conquest, slowly lost the warlike virtues of their ancestors, without always acquiring those belonging to a more humane epoch. This slow work of equalisation is, doubtless, far from complete; but even now in civilised countries, the fact of being a man carries with it its "rights." Instruction has become, and will tend to do so more and more, a common source of wealth; consequently real inequality, moral and intellectual inequality, will always fatally diminish. Now, this gradual elevation of the mind and of the conscience is necessary, for it is the result of ethnical competition; and the people, who, in this salutary state of rivalry allow themselves to be outrun, are destined to disappear from off the face of the world.

CHAPTER X.

THE CONDITION OF WOMEN.

THE ideas implanted in the human brain of right and of justice, and also of the feeling of respect for the weak, are the fruits of a high degree of culture, little recognised, perhaps unknown, in primitive civilisations, in which man, realising certain conceptions of the Greek mythology, has not yet emancipated himself from the animal state. It is woman's misfortune all over the earth to be weaker than her companion; we may therefore expect to find that her lot is harder in proportion as the society to which she belongs is more rudimentary. The condition of women may even furnish us with a fairly good criterion of the development of a people, as we shall see when we examine the principal human races from this point of view.

In Australia woman is a domestic animal, useful for the purpose of genesic pleasure, for reproduction, and, in case of famine, for food. Hunting and fighting are man's occupations. The woman has to follow him on his excursions, carrying her children and a flaming brand to light the fire—this being the stock of furniture belonging to the family. On the seashore it is she who has to go into the water in search for shell-fish, which form the staple article of food. She does not eat until after her master has first filled himself; she then feeds upon the remains which he has thrown to her as though she were a dog.

As we have already seen, the life of the Australian woman is a long series of prostitution; her savage owner does not seem to have for her the slightest feeling of affection. He regards her as a thing, and as a thing of little value. The Rev. Father Salvado says: "One evening, as I was repeating my breviary, I heard outside a noise resembling a constant repetition of blows, and also the cries of a woman. . . . I immediately went out, and saw eight savage women mercilessly belabouring each other with their cuanas, or their sticks. I rushed into the middle of the fray to try and separate them, but the wind rendered my words inaudible. One can hardly say they were women; they were rather wild beasts. I then took a stick and laid it well over the shoulders of the most savage of them, and so put an end to their squabble. Some of them had their heads cut open, others their shoulder-blades broken, and blood running down in torrents. There was not one whose black skin was not all stained with blood from her head to her feet. Seeing their husbands sitting round the fire looking at them and laughing all the while, I scolded them roundly. 'What!' I cried, 'your wives are killing each other, and you stay there quietly and do not think of separating them!' They answered me: 'Who would wish to interfere when women are quarrelling?' 'You, of course, who are their husbands.' 'We? It is no matter to us.' 'What do you mean, it is no matter to you? Suppose one of them were to die, wouldn't that be of consequence to you?' 'Not in the least; if one were to die we still have a thousand others.'"

The placid contempt for women in Australia is shown in the most bestial way: three days after a man's death his wife becomes the property of his brother-in-law. It is rarely enough that an Australian woman dies a natural death. "They generally despatch them before they grow old and thin, for fear of losing so much good food. . . . She is so little thought of, either before or after her death, that we may ask ourselves if a man does not put his dog, when alive, exactly upon the same footing as his wife, and if he thinks more often and more tenderly of one than of the other after he has eaten them both!"

In the other Melanesian islands the lot of the woman is hardly softer; for among the Oceanian negroes humanity is not the predominant quality. In Fiji a man has the right to sell his wife, or to kill her if he so please. They often fasten their wives to a tree or to a post and then whip them. In New Caledonia the wife is not allowed to eat with her husband; she lives in a detached part of the habitation. The hardest part of the work is imposed upon her, she is even subject to bad treatment, and she often puts an end to her trouble by suicide.

We have already remarked in the course of this book that

primitive races of the most different kinds will often resemble each other in many ways. And in no respect do they differ less than in the slavery to which they subject their wives. It is a great defect to be weak, even in our most civilised societies, but in the early stage of human development it is an unpardonable wrong. In the negro districts of Africa woman is not treated much better than in Australia, but in Africa she is rarely eaten—certainly much less often than in Australia. The reason is that on the African continent game may be found, and the pursuits of the African negro are pastoral, or more generally agricultural. Baker has nevertheless related, according to the testimony of an eye-witness, the history of an anthropophagical feast, which was held at Gondokoro on the Upper Nile, and for which the female slaves and the children furnished the dishes. As a general rule, however, the female negro is not eaten in Africa by the stronger race, though she is forced to do the hardest and most laborious work.

Among the Hottentots and the Kafirs man hunts or is engaged in warfare—occupations which have all over the world ever been considered as the most noble. He also looks after the cattle, the enclosing of his fields; he tans the skins of his oxen, with which he clothes himself more or less. Among the Kafirs, to look after the cattle is considered as a superior occupation; and in Kaffraria the cow is called "the hairy pearl." With primitive man the domestication of the bovine race has always marked a stage of progress. But all the other less distinguished labours have devolved upon the woman. In the Hottentot country and in Kaffraria she builds the habitations, she plaits the mats, she bakes and moulds the pottery. Among the Kafirs who are agricultural she digs the earth, sows it, and gathers in the harvest. Man has no idea of helping her, and all over negro Africa woman's condition is very similar. With the exception of the Hottentots, nearly all the African negroes practise agriculture in a certain measure; but the care of cultivating the land is specially incumbent upon the women and the slaves. As everywhere else, she cultivates the soil; but when she has sufficiently fed her husband she may then dispose as she pleases of the surplus produce of the harvest. As in Kaffraria, these poor creatures have to build the habitations. In some districts half negro and half Moorish, for instance among the Solimans, the women are the barbers and the surgeons. On the other hand, the men sow and wash the stuffs. The women have everywhere to share with the oxen, the mules, and the asses the labour of carrying the burdens; the men do not degrade themselves by giving any assistance. Women also grind the corn with stones in round holes dug in the rocks. In this region, as in Southern Africa, man reserves to himself the care of the cattle, he milks the cows and leads them out to grass, etc. He is everywhere the blacksmith. To beat and fashion iron in Africa is considered a noble profession, which sometimes confers special political rights. It is also a scientific trade; in Kaffraria the blacksmith is called "the iron doctor."

We find the same custom in the basin of the Upper Nile; all over Africa man is a huntsman, a soldier. During his many leisure hours he will lie down lazily in the shade, smoking or chattering while his wife is digging and doing the hardest part of the work.

In the middle part of Africa, as in Australia, the woman never shares the repast with the man; her children show their disdain for her and do not listen to her; the head of the family will often knock her down upon the most frivolous pretext. And everywhere the poor creature will submit humbly to her sad lot, bearing it all without a murmur. It appears to her natural that she should be so oppressed. And in countries where the Moorish influence is predominant the lot of women is not much better than among the negroes. In Senegambia they cultivate the soil, bear the burdens, and even watch the cattle; they have not the privilege of eating with their husbands. When he mounts his horse they must hold his stirrup for him. They are beaten and cast off at pleasure. In Darfur they are treated in the same way; one may often see them, laden with baggage and provisions, following on foot their lord and master, who is comfortably seated on an ass. Their husbands will gladly lend them to strangers—it is their right to do so—for a fitting retribution. In this country, only the sultan's daughters have the right to exercise any will of their own without caring for that of their husband. The power of their father suffices to overrule every other authority.

Among the nomad races the absolute subjection of women becomes more difficult. We need not, therefore, be astonished to find that the Tuarick women are much more free than the Moorish women. They enjoy an amount of liberty which is relatively large; they move about without restraint, and will gladly talk with the men. They are not even always bought by their husbands, as is usually the custom in Africa; the purchase-money will often take the form of a handsome present made to the father.

It seems that everywhere the primitive lot of women has been, or is still, a state of servitude more or less hard, more or less capricious, according to the race and the country. In Polynesia, among this infantine race of people whom travellers, and especially those of the last century, have delighted to paint to us in too bright colours, woman was, as everywhere else, considered as a thing belonging to the man. No doubt, nearly everywhere, before her marriage she enjoyed the fullest degree of amorous liberty; but once married, she became as a field let out to hire by her owner. Adultery was strictly forbidden to her unless in the case of previous authorisation; but it was her duty to prostitute herself if her husband told her to do so—it always being for his advantage—whenever he pleased to give her to a friend or to a stranger. In most of these islands the women were obliged to find and prepare the food for their husbands, either in laboriously breaking the fruit of the pandanus, so as to extract the nut, or in spending whole hours, barefooted, on sharp coral reefs, up to their waist in the water, exposed to the heat of the sun, endeavouring to catch shell-fish or other sea produce. When the feast was made ready the men would gluttonously devour the best parts, and the women had to be content with what was left, or what was thrown at them. Nearly everywhere they were forbidden to sit down to eat with the men. In Tahiti, they were even obliged to cook their food upon different fires, and to eat in different huts. They were obliged to respect all the places frequented by men, also men's war tools and fishing implements. The head of their husband and of their father was to them a holy thing. They were forbidden to touch any object which had touched those tabooed heads, or to pass in front of the men when they were lying down. In the Marquesas islands, the women were not allowed to go into the boats; it was said that their presence frightened away the fish. The best and most choice food, that which was called divine, because it might be offered to the gods, was altogether forbidden to women. At Tahiti, this comprised all winged creatures, cocoa-nuts, and plantains. The pig was everywhere reserved for men and for gods. The power of being able to eat as much pork as they pleased was, for the women in the Sandwich islands, one of the greatest attractions of the Christian religion. In the little island of Rapa, all the men were considered as holy beings by the weaker sex; and all through the year the women were obliged to put the food into their mouths, but in the other islands this was only done during the periods of taboo.

In some of the archipelagoes, at Noukahiva, in the Friendly islands, the men used to cultivate the land, used to construct the houses and the canoes, but nearly everywhere all the hard labour was the lot of the women. They had to plough, to manufacture stuffs from the mulberry tree, and to carry the burdens. In New Zealand it was considered dishonourable for one of the masculine sex to carry a burden. Man there confined himself to fishing and to fighting.

Over all Polynesia we find a general sameness; but that does not prevent certain local differences, specially as regards the condition of the women, which became harder in proportion as the natural resources of the island were less meagre. In New Zealand for instance, feminine servitude was more severe than at Noukahiva, where the alimentary resources were more abundant. In this last island the women were obliged only to take care of the children, to fabricate stuffs from the mulberry tree, to prepare the *popoi*, the *kokai*, the *koku*, which were pastes or thick soups made of the bread-fruit. The men would cultivate the ground and they would fish. As these occupations employed only a few hours in the week, the rest of their time was passed in sleeping, singing, bathing, plaiting wreaths of flowers, etc. In the Tonga archipelago the condition of the feminine sex was, exceptionally, very much softer; they were there considered, not as beasts of burden, but as companions. Rare instance, but still giving another proof of how different is the condition of man.

In the savage parts of America, from the island of Terra del Fuego up to the Arctic regions, the condition of woman is almost everywhere that of a beast of burden. The Fuegian woman has in all weathers and at all seasons, notwithstanding the severity of the climate, to go into the water to look for shell-fish, or to empty the water out of the canoes. It is she who builds the rudimentary wigwam; it is often she who has to punt the canoe. She acquits herself of all these labours, even when she is acting as nurse, and is carrying on her back, in a skin, the child to whom she is giving suck. As her recompense, when she is old and can no longer work, man will generally eat her, if he is slack of food, after having suffocated her by holding her head in the smoke of a fire made of green wood. If one asks them why they do not rather sacrifice their dogs, they will answer: "The dog eats the jappo"—that is, the otter.

On the continent, where life is somewhat less difficult, and somewhat less bestial, woman is not ordinarily kept as reserve food in case of need, but she is always condemned to do the hardest work. In Patagonia man sleeps when he is not hunting, and when he is not hunting he is sleeping. The woman skins the animals that the man has killed, she prepares the skins, she softens them, joins them together with the sinews of the animal, either to make large cloaks, ornamented afterwards with paint, or to serve as walls for their tents. When the game in any district has grown scarce, the woman rolls up the skins and the posts which support the tent, she runs in the horses and loads them, or, when there are no horses, she loads herself—and then the tribe decamp from their habitation in search of a more propitious dwelling-place.

Farther north we find the same manners, in spite of the difference in the climate and in the kind of life. While on their journeys the man thinks that he ought to have no other care than to ward off the jaguar; he therefore carries only his bow and his arrows; but the woman is burdened with the baggage, the provisions, and the children. When they halt she has to collect together the wood, and do the cookery, while the man is lying down in his hammock. He will put an end to his wife without the slightest scruple. A Moxos Indian does not hesitate to kill his wife if she miscarries, an accident that cannot be infrequent in such a kind of life. In Paraguay it is considered criminal in a woman to be intimately acquainted with a man of another tribe.

This harsh treatment of women does not depend upon the race, but upon the degree of civilisation. The ancient Peruvians had rid themselves of their barbarous manners, and among them the hardest labour was always performed by the men. But with the Red Skins of the north it is very different, for they are still in a state of savagery. Except the fabrication of their hunting and war instruments, all the labour there is incumbent upon the woman. Upon her devolves the care of the household and the kitchen; she prepares the skins and the furs; she gathers in the wild rice; she digs, she sows, and plants the maize and the vegetables; she dries the meat and the roots for the winter stock; she makes the clothes, the neck collars, etc. The man will disdain to aid the woman even in the construction of their canoes when it becomes necessary to make one; fighting, hunting, smoking, eating, drinking, and sleeping are his only occupations. He considers it dishonourable to be obliged to work.

The Nootka Columbia woman collects the mussels and other shell-fish, as does the Fuegian woman. Like the man, she rows and paddles the boats; she makes the flaxen and the woollen clothing; she brings the sardines into the habitations and there she is obliged to prepare them. In Sitka island, in New Archangel, the women act as night-watchers. In this country the natives build their temporary dwelling-houses on high places, in some way protected by nature, and at night, while the men are sleeping in their huts, the women watch outside, collected together round the fire, and they pass the hours relating their own domestic quarrels.

On the other side of Behring's straits we find in Kamtschatka a somewhat improved division of labour; this is the case in all the Asiatic Mongolian or Mongoloid races, among whom the most important and probably also the most ancient features in our civilisation first took their rise. In Kamtschatka the man undertakes to build the *iourtes*, to fabricate the utensils for household use, to manufacture his implements of war and the chase, to prepare the food, to skin the wild animals and the dogs, out of which skin he afterwards makes his clothes. The women tan the skins by scraping them with a stone knife, then they rub them with fishes' eggs more or less fresh; they afterwards stretch them, cut them, and sew them together again, and so make their clothes and their shoes. In the nomad districts of Mongolia, where the women enjoy a considerable amount of liberty, they ride, straddling as they do so, from one tent to another; but they are very far from being idle. They have to fetch the water, often from a great distance, to collect for fuel the *argols* (the manure of their flocks), the only combustible matter in the country, to milk the cows, to make the butter, to tread the wool, to tan the skins, and to make the clothes. The work of the man consists in putting his cattle out to graze; the rest of the day is spent in galloping from one *iourte* to another, where he stays awhile chattering and drinking either tea or koumiss, or else he is shooting as well as he can with his stone gun, or with his bow and arrows. He is ever ready for a gallop over the steppe, and his horse rarely remains long unsaddled. In Thibet, where people are civilised, but in a theocratical way, the condition of the women is perhaps less favourable. All the hard work falls upon them. They have to dig, to buy the provisions, and to weave their stuffs. Each woman has generally three or four brothers for her husbands, and as a dutiful wife she is bound to make herself equally agreeable to them all. In an amorous point of view she has great liberty, for in Thibet adultery is an unknown offence; the husbands have long been accustomed to share their wives, and consequently they ignore every sort of formality.

In China, where the Mongolian race has acquired, after its own fashion, a high degree of civilisation, the highest attained by men of the yellow race, woman, without being made absolutely servile, is considered as a minor, and kept in a state of perpetual subjection. When a Chinaman has girls only born to him he does not consider that he has got any children. The Chinese girl is merely a creature of traffic; she will sell herself to the man who offers her most. "The newly-married woman," a Chinese author says, "ought to be merely as a shadow and as an echo in the house." She does not eat with her husband nor with her male children. She serves at table in silence; she lights the pipes; she has to be satisfied with coarse food, and is not allowed to eat even what her son leaves on his plate.

The women, still in a state of slavery in Cochin-China, where they have to perform more than their proper share of hard work, especially the paddling of the boats, are much more free in Burmah, where the white and Mongolian races have become mixed; but even here their legal status is very inferior. As in Rome, where women had not the right to profane by their presence certain sanctuaries, so in Burmah it is forbidden to them to enter a court of justice. Their legal deposition is held to be of slight value; they are obliged to give it while standing at the door. They are held responsible to answer with their person for the debts of their father or of their husband. The Burmese woman is merely a thing possessed; she is quite a venal creature. Chastity is altogether disregarded, the girls are allowed to prostitute themselves just as they please.

In India, the lot of the woman is little better; and yet there the Aryan blood and Aryan influence is predominant. The Manu Code will give us ample instruction on this point. Woman, it says, during her childhood is dependent upon her father, in her youth upon her husband, in her widowhood upon her sons or her paternal relations, or in default of them, upon the sovereign (*lib.* v., v. 148). She must always appear good-humoured (*id.* v. 150), to worship her husband, even though he be unfaithful, as though he were a god (*id.* v. 154). As a widow she ought not to pronounce any other name but that of her deceased husband (*id.* v. 157). The Hindoo laws and manners, until modern times, were modelled upon these sacred precepts. At the time of Sonnerat's travels it was held shameful in an honest woman to be able to read or to dance; these useless sources of pleasure were reserved to the courtesan and to the bayadeer. The husband habitually spoke of his wife as "servant" or "slave," and she was always bound to call him her "master lord," sometimes "my god;" but she was never allowed to call him by his name. During a certain period every month the Hindoo woman was considered to be unclean, defiling everything that she touched; she was obliged to undergo legal purifications. We find indeed that there still exists in Europe a popular prejudice to the same effect. The Breton sailors declare that a woman in a state of menstruation if standing near a compass will always affect it and cause it to point untruly.

Among the Afghans the woman's condition of slavery is still the same. She is a venal thing which man buys and sends away when he pleases, which he hires for money to his friends. A widow, when she re-marries, ought to be paid for by the relatives of the second husband to those of the first, unless she marries her brother-in-law, which indeed is always considered to be her duty.

The people in the eastern part of Afghan are still more brutal. With them the woman, or rather the girl, is a pecuniary coin. Among many African tribes the cow is considered to be the coin by which every object is valued. We know that it was the same with the ancient Romans, and that one of the early Roman coins was called "the cow," and bore an impress of the animal. Among the Afghan tribes of whom we are now speaking woman is judged after the same manner. To them, a girl represents sixty rupees; and as they hold all crimes to be redeemable offences, as did also the ancient Germans, the ransom to be paid is counted by the number of girls. Twelve are required to redeem a murder, six for the mutilation of a hand, an ear, or a nose, and one for a tooth.

With the Arabs, too, the condition of the woman is still very low. Among certain wandering tribes it was held to be a strict act of duty to give to the guest for his night companion one of the women of the family, generally the wife of the master of the tent.

The young girls only were not necessarily bound by this law. Among other tribes, those of Asyr for instance, when a father wished to marry his daughters he would lead them, decked out, into the market, and would cry: "Who wants to buy a virgin?" It was the recognised law in the tribe.

There does not seem to be more care for women and her interests among the stationary Arabs than among these nomad people. At Mecca men will not give to their wives any religious instruction, for they think it would put them upon too great an equality with their masters; and also, in spite of the Koran, certain Arab theologians refuse to allow to woman a place in Paradise. Buckingham has seen in Bagdad women dressing and undressing their husbands, waiting upon them and kissing their hands.

In short, whether the race is superior or inferior, the subjection of women, more or less absolute, seems to have been the custom all over the world; and in the ancient written laws we find many traces of this state of feminine slavery. Nearly everywhere girls have been excluded from the rights of inheritance. Where landed property belonged not to the individual but to the family, it was difficult to allow the right of property to girls, who, as they married, went into and became part of another family, and would thus have caused a dismemberment of the common patrimony. In certain Basque cantons the difficulty was eluded by making of the heiress a head of the family, who stood in place of her husband. The Manu Code, the ancient laws of Athens and of Rome, simply disinherited girls; the Voconian law, which Cato caused to be passed, forbade leaving to girls more than one-quarter of the patrimony. It was only in default of male inheritance that, according to the old Germanic code, women were allowed to inherit land. And nowadays, in Russia, girls inherit only a small fraction, perhaps about a seventh. The French Civil Code everywhere treats of women as minors; and no civilised country has yet thought advisable to allow to women any political rights.

We may say that the lot of woman in humanity is harder as man is more bestial, and that her gradual emancipation increases concurrently with the progress of civilisation. We shall probably, sooner or later, arrive at a system of equalisation of rights; but it is prudent to advance slowly and with caution. It is an undoubted fact that in the human kind woman has been the oppressed half. As we have already remarked, the feminine portion of humanity have the unpardonable fault of being weak, and unjust as it has been, this tyranny, much more than millenary, has left its impress upon the feminine nature. That woman is organically inferior to her companion, and specially as regards cerebral development, is a point which anthropology has decided beyond all possible doubt. But feminine inferiority is not only physical, it is also moral and intellectual; debility has its action upon the mind as well as upon the muscles. Muscular inequality will no doubt last as long as the human race, it is only of secondary importance; but cerebral inequality is a very different matter. If the political emancipation of women in our democratic countries were to come prematurely, it would certainly be followed by a general backward movement. After having first been used as a beast of burden, as a domestic animal, woman became a slave, then a servant, then a subject, then a minor. We have yet to promote her to the major state, to strengthen her brain by instruction that is befitting to her, to prepare her by degrees to bear her share in political equality, in order that she may exercise it for the common welfare.

CHAPTER XI.

WARLIKE MANNERS.

I.

TAKEN in the aggregate, such as he has been and still is, due regard being paid to time and place, man is, as we have already seen, a singularly mischievous animal. But we are yet a long way from having finished the enumeration of the misdeeds of the human kind, and a glance cast over the warlike manners of the different races will furnish us with much unpleasant and disheartening information.

Man is not a creature standing apart in the universe. By his origin, by his organisation, he belongs to animality; and, more or less, he has the same wants and the same instincts. To kill in order to live is an imperious law, which the animal world cannot disobey. This is an inflexible law, as hard as brass, to which every creature is more subject in proportion as he is less clever in ingeniously supplying the parsimony of novercal nature. For the creature, man or animal, without invention, without foresight, the furthermost limits of subsistence are soon reached, and he must, under pain of death, drive out his rival competitors. This is the case with the herb-eating animals, and also with men who are omnivorous. As regards the flesh-eating animals, the murder of the weak is the whole essence of their existence; many delight in it, and find a real mental pleasure in causing suffering to their prey. Audubon has given us a graphic description of the cruel pleasure of the white-headed American eagle when he has captured a swan: "He drives his sharp-edged beak into the bottom of the heart and into the bowels of the swan; he roars with delight in gloating over the last convulsions of his dying victim, while he is using all his efforts to make him endure the greatest possible horrors of agony." The anthropomorphous monkeys, too, treat the inferior quadrumanes very harshly; they beat them, they bully them, and they kill them. The so-called Aryan races conduct themselves nowadays much in the same way with other human races inferior to themselves. Gorillas fight and kill each other as men do, shrieking and bellowing forth their hideous war-cries.

This is slaughter after a primitive fashion, showing no intelligence, or no powers of strategy; the vertebrated animals hardly know of any other. They have their foreign wars and their civil wars, they have their regular pitched battles, they spread out their columns, they perform their manœuvres, they make their charges, retreats, reverse attacks, and they have their reserve forces; but they have not yet invented instruments of slaughter.

In this respect we are superior to them. In their struggles they always fight hand-to-hand; they often tear each other fearfully, though they frequently give each other quarter, and the prisoners are captured by their conquerors, who drag them off, as they also take away their dead. We may add that among the termites there is a regular class of warriors, who form a permanent army. P. Huber has been the historiographer of a great and glorious war between two powerful republics of our native ants, living at a hundred paces one from the other. While the two armies were raging, were tearing each other to pieces with their pincers, were pouring out upon each other bitter poison, and acquiring renown by their prodigies of valour, the civil population was continuing its way in the forest, occupied, as is right in human societies, in doing its peaceful work necessary for the maintenance of the two states. The thirst for glory is not easily appeased. Like many great wars between human beings, that of which we are now speaking had no other real results than enormous bloodshed and numerous prisoners captured upon both sides. The victorious army barely succeeded in driving back the enemy ten feet. With ants a retreat of ten feet means a conquest of considerable importance; but how many lives were sacrificed to obtain this result! Admitting that the fortunes of war are capricious, and that the defeated side does not always succeed in recovering its lost territory, it is still evident that, in order to keep themselves in good fighting condition, the ants must prepare in the time of peace, and they are careful to cultivate all the gymnastic and military exercises. *Si vis pacem*, etc. Poor humanity! Even in the kind of exploits of which she thought she had best reason to be proud she cannot lay a claim to originality; for as regards warlike ideas and manners, man has done little more than imitate those of the ant.

We have said that the main reason of animal and of human wars was the necessity of living, and consequently of finding for oneself nourishment. In order to live we must satisfy our needs, whatever they may be. The motives of war are multiple, and if hunger takes the foremost place, love will certainly claim the second in primitive societies. We know that many tribes are exogamic, and that in the times of historical antiquity many wars have been occasioned by the carrying away of women. In this respect also, man closely resembles the animal; but how should it be otherwise, when he is but the crowning-stone of a long line of animals? From these humble ancestors man has inherited a great number of ferocious instincts, still only half extinguished even among the most civilised races. A poet has said in speaking of mankind:

> Le vieux sang de la bête est resté dans son corps.

As we may naturally suppose, these inferior sentiments are the more imperious, and show themselves in a less disguised form in proportion as man himself is more primitive; but everywhere they exist in a latent condition. Humiliating as it may appear, this side of human nature is not less instructive; we must now notice the principal features.

II.

Warlike Manners in Melanesia.

As is our custom, we shall begin with the most inferior type of human beings; but in this instance we come across a singular exception quite at the commencement of our enumeration of sinister characteristics. Man is such a many-sided creature, he is so variously fashioned after the innumerable wanderings through which each race has been obliged to pass, that in ethnical psychology it is very difficult to separate and distinguish the different bases of his existence. The rule is often violated by the most flagrant exceptions. It will generally be found that the absence of scruple and the cowardly ferocity of cunning in the manner of waging war, is in proportion to the savage condition of man; and yet we find a sort of chivalrous loyalty in the wars of the Australians and of the Tasmanians, who are such poor and inferior types of humanity. Here, as everywhere, the rule is often broken. Sometimes they would fight without having the slightest regard for the commonest notions of courtesy; sometimes they would have recourse to cunning, as did, for instance, a body of armed Australians, who were hidden in the long grass and tried to entrap Stuart and his companions by sending them their wives to tempt them with their amorous devices. But in Tasmania and in Australia loyal warfare is, or used to be, a sort of duel. The two enemies drew themselves up face to face, and from each side the combatants stepped forth one by one, throwing their javelins at each other. When the whole series of duels was exhausted the tribe most maltreated declared itself conquered, or else they recommenced the struggle in the same way, but the second time fighting with their clubs. Each warrior was held bound to give and to receive a blow aimed at his head, and which he was not allowed to parry. The victory remained with him whose skull was the toughest. Their desire to equalise the chances was so strong that Australians have been known to give arms to Europeans who were unarmed, before proceeding to attack them.

The scruples that the Australians and the Tasmanians might feel are altogether foreign to the New Caledonians. During these last twenty years we have seen more than once in France, and especially in Germany, learned historians bring the whole battery of their erudition to prove that success is always lawful. The New Caledonians, too, are of the same opinion. According to them, all means are good to destroy the enemy. This is no more than the doctrine of the sportsman with regard to his game. In New Caledonia wars are incessant, for the country is divided into many small nations, who are always jealous and doing their utmost to injure each other. Public opinion here agrees with that in Europe, in believing that nothing is so admirable as success; but the great grievance of the New Caledonians against the French authority and against the missionaries is, that they are restrained from fighting. "We are no longer men," they say, "since we do not fight." Endless cunning practices are the basis upon which they build all their strategy. To engage themselves in a combat in which the chances are equal, and so run the risk of being killed, would, in their opinion, be mere madness. The real glory is in being able to spy out their rivals, to lay traps for them, to attack them suddenly, and even to rush in upon them at a banquet to which they have been invited. The main object is to massacre the enemy, and to eat them. Warfare as practised by the New Caledonians is therefore merely a battue of human beings.

III.

Warlike Manners in Africa.

In Africa, where anthropophagy is rare, war is nevertheless not more humane. The Kafirs will exterminate the Bushmen without distinction of age or sex, when and where they may find them. The Bechuana Kafirs will sometimes cross the Kalahari desert simply for the purpose of killing the Damara Hottentots. Being totally unconscious of what humanity is, they will rob the dead and the dying, they will kill the wounded and the helpless women, even when these imploring for mercy show to them their breasts, crying: "I am a woman, I am a woman."

In Equatorial Africa, as in all savage countries, rivalry amongst the tribes is the cause of everlasting war; and the manner of fighting as practised by the white people, attacking the enemy always in front, is considered by them as ridiculous. To lay ambushes, to rush upon the enemy unawares, to kill him while he is sleeping, to assassinate a woman carrying her water, these and such like exploits are considered worthy of praise. Farther north, in the African zone, where the Moorish and the Mahomedan blood have become somewhat mixed, in those regions where barbarism has followed savagery, warfare is still the normal condition of the people. War will be carried on between two towns instead of between two tribes, and the two rival towns are perhaps only a few hundred yards apart. It is with the men as with the ants whom we have lately described. Man spends his whole life in setting traps, in fighting, in taking his revenge, in taking others prisoners, or in being sold.

In the human kind warfare is ordinarily a masculine necessity; the woman will doubtless feel the effects very strongly, but she herself does not take active part. In the Australian wars the women and the children place themselves behind the warriors and wait patiently for the result. That is the rule, but there are exceptions. In the Kawon archipelago ([illegible]) the women form the rearguard, and laden with sacks of stones they hurl them over the heads of the warriors fighting upon their side. And near to Darfur, in Africa, there are a race who also utilise their women in their wars. The women do not take a direct part in the battle, but they stand behind the combatants and hand to them iron lances heated in a brazier. Here is a lesson for our modern statesmen whom we Europeans now hold in such high honour. The military service, obligatory upon men, has already produced all the fruit of which it is capable; but what a glorious perspective is open before the nation who shall first decree military service to be binding also upon women!

In Abyssinia we still find barbarism, but it is gradually diminishing or is modernising itself. The social condition there will bear a close resemblance to the ancient European feudal system. There are many princes who are nearly independent; they govern their subjects and maintain themselves by warfare. They have feudatories, vassals forming the cadres of their armies, with which professional braves and freebooters will enroll themselves, having, or claiming, everywhere the right to be billeted upon the peaceful population. The people are less ferocious, they kill each other less frequently, but they still practise the fearful custom of emasculating the enemy lying on the ground. The victorious warriors in this way bring home their trophies; they are very proud of them, and they go to present them to their chiefs. The Abyssinians have also another warlike custom, altogether modern,—that of insignia or decorations, awarded to them by their chiefs. These decorations are of various kinds. They are sometimes war-tippets made of lion-skin, panther-skin, or of velvet, or of blue or scarlet cloth; or they are armlets in argent or vermilion, or they may be half crowns. We see, therefore, the Abyssinians are on their way towards civilisation.

IV.

Warlike Manners in Polynesia.

A verse of La Fontaine's, which has become proverbial, tells us that childhood is ignorant of the feeling of pity. The observation may also be applied ethnographically. The most sensitive races, those whose changeableness of humour is excessive, are cruel almost unconsciously, as children are, whom they often resemble very strongly.

Among all the inferior races there are moral characteristics of a most infantine nature, and in this respect no race is more curious than the Polynesian. Its mental condition may be well compared with that of our European children of ten years of age. The Polynesian tribes were constantly in a state of warfare, often merely for the pleasure of fighting, sometimes for more serious motives—either to feast themselves upon human flesh, if they were an anthropophagous people, as in New Zealand, in the Marquesas islands, etc., or for the sake of capturing prisoners and making slaves of them. In the Marquesas islands there were those who lived on the mountains and those who lived on the plains, and both would wage war, the one against the other, for the sake of robbing the produce. The mountaineers would covet the fruit of the bread-tree from the inhabitants on the plain, and these latter used to climb the mountains to steal the fehi (*musa fehi*) which grew on the high places.

Whatever may have been the motive of the war, the conquered were always treated with unmerciful cruelty. They were generally all massacred, men, women, and children, no distinction being made. At Tahiti the people practised a sort of scalping of the beard, which they cut off together with the skin round about the chin; and this they would wear as a sort of trophy. At Noukahiva they killed their wounded, and the conquerors stained their swords with the blood of their prisoners, which henceforth acquired considerable value, and were often called by the name of the deceased warrior.

In New Zealand they would knock down the prisoners, as did

the Red Skins, first seizing them by their hair on the top of their head.

But in Polynesia warfare had lost the degrading character which specially marked it in New Caledonia.

No doubt the defeated side were horribly slaughtered; sometimes the whole tribe was completely exterminated. In New Zealand the savages have been known to cut up into pieces the enemy as they were lying on the ground, without waiting even till they were dead; or else they tortured them as did the Red Skins. The laying of ambushes was not to them the most important consideration in their strategical plans. They usually attacked each other in front; the issue was often decided in naval battles, in which they would fight hand-to-hand on planks of wood laid as platforms across their boats. There were often chivalrous challenges made from one man to another in the presence of two witnesses. Although custom in Polynesia had decreed that all the conquered people should be exterminated, sensuality sometimes took the place of rage, for young women occasionally obtained their pardon by tearing open their spare clothing and offering themselves naked to their conquerors. But the inhabitants of the Sandwich islands, though they did not abandon the cruel practices natural to their race, nevertheless attempted to lessen some of the hardships of war by adopting humanising institutions. They established places of refuge, and inside a certain boundary women, children, and conquered prisoners always found a sure protection. Flags were hoisted on to these asylums, and from a long distance the place of refuge might be seen. The belligerents might also announce or demand that the fighting should cease, in holding out green branches, particularly those of the *piper kava*, which was the symbol both of peace and of drunkenness.

But there was in all this, as in every other phase of Polynesian life, something very infantine. Before going into action the New Zealanders would excite themselves by singing a war song, and they would accompany it with dancing and contortions of various kinds. They therefore threw themselves into a fury before they began to fight. On the other hand, as we find to be common among savage people, their patriotism was very narrow; they took no regard of people out of their own tribe; the inhabitants of one village enjoined Cook to destroy all those in the neighbouring village. Their light-headedness, the main reason of their homicidal manners, often drew with it its own consequences; not counting that in many islands, specially in the Marquesas and in New Zealand, each tribe took care to fortify for themselves certain places to which the access was considered difficult. A refuge might be found there in case of defeat. They would sometimes hold a siege, which, indeed, never lasted very long; for after a few days' resistance the besieged party, from the volatile nature of their disposition, grew tired of their imprisonment, and threw down their cards, as having lost the game.

V.

Warlike Manners in America.

The human races have in the course of ages so mixed themselves up together, and we are so ignorant as regards their origins, that it is now extremely difficult for us to draw the moral portrait of any human type; we find on every side exceptions which impertinently come before us to violate the rule. Among the Australians and the Tasmanians we have seen a vague chivalrous instinct, among the Polynesians a sort of loyalty and feeble humanising desires. In America, except among the ancient Peruvians, the warlike spirit nearly invariably discloses a most atrocious character. Cunning, without any kind of scruple, forms the whole art of their strategy; and with them a victory is always followed by most merciless cruelty. Instead of the puerile light-headedness of the Polynesian, we often find in the American a long system of calculation. The aborigines of Peru have been known to plan for twenty years an insurrection against the Spaniards, and the Indians of the Grand Chaco to watch for two or for three years certain Spanish establishments before beginning to attack them.

Inhumanity towards the conquered was everywhere excessive; men would gloat over their cruelty, deriving from it evident pleasure.

The Indians of the Pampas used to fight as did the nomadic Mongolians; they would prepare a campaign merely for the pleasure of exterminating the Gauchos, and the Spanish American colonists. Their troops of horses would provide them both with animals to ride upon and also with food in case of need. They would gallop over the vast plains without stopping to draw breath, halting occasionally for a moment to change their steeds, and keeping always the best chargers for the battle. The Guaranis, or rather the natives of Brazil in general, were ever in a state of perpetual war amongst themselves; they were "hereditary enemies," according to the phrase lately brought into use by a great European nation. Every sort of truce was unknown to them. Following the American habit, they would for a long while spy out their enemy before rushing upon them as suddenly as they could; and they would often tear each other to pieces with their teeth. When surprise was impossible, the two armies would bellow, and cry, and threaten for two or three days before the attack, reminding each other in Homeric speeches that formerly on this or on that occasion they had mutually eaten each other, their parents, their relations, and their friends. When they captured prisoners they would treat them very well for a certain time; they gave them wives and abundance of food; and then at a given moment they massacred them ruthlessly and ate them afterwards in a ceremonious manner, taking care first to smear with their blood their own male children so as to make them stronger. The victim would appear brave and boastful before his torturer; he would sing his war song, and remind his enemy how many of their prisoners his own tribe had eaten when they had been victorious.

Similar customs were practised over the greater part of the two American continents, and Mollien found them still in vogue in Columbia at a recent epoch. As we shall see, the ancient Mexican had usages of the same kind. As regards the Red Skins of North America, they did not eat their prisoners—at least, not at the time of the European conquest; but, on the other hand, they exercised their ingenuity in torturing them. Their victims, tied to a post, sang their war songs, while the men, women, and children of the victorious tribe would tear them with knives, pull out their nails, burn them with hot coals, and do other acts of hideous cruelty.

The Red Skins thought that to allow themselves to be captured was as bad as letting themselves be killed; it was a sign of abominable clumsiness, which greatly tarnished the reputation of a warrior. The prisoner was for ever dishonoured in the eyes of his own people, and was considered by them as though he were a dead man. On the other hand, if the conquerors had sustained grave losses, instead of putting their captives to death they would lead them to the huts of their own warriors whom they had lost, and if the widows consented to accept them, they simply took the places of the men who had been killed. The deserters then, without a thought of hesitation, would fight against their whilom friends, who, in their turn, had turned their backs equally upon them. If we may believe Charlevoix, it would seem that the social tie between the members of a tribe was of the weakest kind. Every member of a tribe, even when he was master of his own actions, was free to accompany his companions on to the field of battle, or to refrain from doing so, as he pleased. Among the Red Skins war was carried on without the slightest feeling of chivalrous honour. To kill, capture, or scalp as many as possible of the enemy, and lose as few as possible on their own side, was to them the whole end and aim of their glory. A victory which had cost them many lives brought with it the degradation, sometimes the condemnation of their chief, who had been triumphant, but a man wanting in address.

The same absence of loyalty in battle, the same manners that we may well call savage, are practised also in Sitka island, in the extreme northern part of America. In this region the tribes, perpetually in a state of warfare, never attack each other openly; their wars are one long series of assassinations, simply for the purpose of thieving or revenge. In the latter case the desire for vengeance will satisfy itself in the Australian fashion, without the slightest care for justice. They wish simply to kill somebody belonging to the rival tribe: man for man, woman for woman.

Before bringing to a close this short study of the warlike manners of the native Americans, we must say something of Mexico and Peru, the only states in which the American man arrived at a tolerable degree of civilisation; and even with them their civilisation was more mechanical than it was moral.

As regards the treatment of their captives, the practices of the Mexicans were horrible. With them, as everywhere else, religion had authorised the sanguinary instincts of the people. The Mexican gods grew thirsty for human blood, and to appease their thirst the people undertook never-ending wars. The half-civilised Mexican did not act differently from the savage Brazilian. Like the Brazilian, he would fatten his prisoner before eating him; his only care was that the sacrifice should be performed by a priest, and that it should be offered to the gods. The priest would ordinarily begin the ceremony by cutting open the chest of the victim with a knife kept specially for the purpose; he would then take out the heart, which was immediately offered up to the idol. The owner of the prisoner then took away the body, to feast upon it at his leisure. In every village there were strongly-made cages in which men, women, or children would be kept shut up in order to fatten them. The ceremonies of the sacrifice would vary according to the divinity to whom it was offered. Sometimes the prisoners had the skin torn off their bodies. At the feast of Tezcatlipoca, a penitential festival, they would sacrifice and afterwards cut into pieces a fine young man whom for twelve months previously they had carefully fed up with delicacies of every description. May we not understand by this sanguinary custom a barbarous allegory teaching us the vanity of human pleasures?

All this cruelty, we may remark, was coexistent with an ingenious system of military organisation. The Mexican armies were hierarchical, divided into corps-d'armées of eight thousand men, and these were again subdivided into companies of three and four hundred men. They possessed a national flag, military orders were distributed, severe discipline was exacted, and they had hospitals for the sick and for the wounded.

The military organisation of the Quichuas, or ancient Peruvians, was also very remarkable. It has been reckoned that their foot regiments numbered two hundred thousand men strong; and these troops were organised into corps-d'armées, battalions, and companies, commanded by a whole hierarchy of chiefs, from the corporal up to the Inca, the commander-in-chief. The combatants were armed with bows and arrows, lances, darts, axes, and short swords, all of which were usually made of copper; but the javelins and the arrows were sometimes tipped with pointed bones, for the age of stone was not then very far distant. When they were on the march, magazines followed at regular intervals, well furnished with provisions to supply the wants of the soldiers. If we except the use of powder, we may compare the organisation of the ancient Mexican and Peruvian armies with that of the armies of modern Europe. The Quichuas erected forts upon their mountains, they had also telegraphic fires, and they instituted a system of couriers. Like all the inferior arts, that of slaughter was very soon brought to perfection.

The Peruvians differed from all half-civilised races, and even from many modern nations, in the motives which induced them to commence their wars, and in their manner of treating their prisoners. They did not go to war, as did the Mexicans, in order to offer up to the gods human victims, and then to eat them afterwards; their wars were crusades, wars of proselytism. It is surely more ennobling to make war for ideas, even though they be false, than for the sake of conquering territory and subduing one's neighbours. The Quichuas endeavoured to propagate the worship of the sun, which, of all such sorts of worship, is certainly the most excusable. They began first by persuasion; if that was not accepted they declared war upon their neighbours rebellious to their doctrine, but giving them notice beforehand of their attack, and not exacting from the conquered more than submission to their will. One of the Peruvian princes said: "We ought to spare our enemies, otherwise we shall be doing injustice to ourselves; for they and everything that belongs to them will soon be our property." The gods of the defeated party were not treated with disrespect; they were taken possession of, and were transported to

Cuzco, into a sort of pantheon. These facts may serve to show that the Peruvians were the first of the American races, and we shall find still stronger proofs when we come to study its social organisation.

VI.

Warlike Manners in the Mongolian Race

It is a difficult matter to form a true idea of the tone of mind in any race of people, but a few leading ideas may be gathered from the observations collected by different travellers, noting some and ignoring others. And from many and widely collected sources of information, it would appear that in spite of the barbarous inroads made by the Mongolians, of which Europe has not yet altogether lost all recollection, the Mongolian race, taken as a whole, is of all human races the least warlike. We do not mean that all the yellow populations are easy tempered. There are Mongolians of very various natures. The majority of the human kind are yellow men, and their different ethnical groups are still far from having risen to the same degree of civilisation. Even nowadays, the Turkomans are a ferocious people; they treat the Persians, their neighbours, as though they were wild beasts, and they religiously sacrifice their old men, under the idea that it is agreeable to their god. Certain tribes in Bhotan, or in Thibet, delight in making night attacks, in preparing ambushes, etc. They eat the liver of the men they have killed; from their fat they make wax tapers, which they burn in front of their idols; they make use of their bones for flutes; of their skulls they make cups, which they bind round with silver. The nomad Mongolians still slaughter their prisoners without any distinction as to age or sex. These people are yet in a phase of barbarism, through which all human beings must pass; but, on the other hand, there are many facts attesting the general peaceful humour of the race.

The Lapps are fond of quarrelling amongst each other, but in their quarrels they never make use of the knife which they always carry with them. The inhabitants of the Loo-Choo archipelago, the possession of which islands is the cause of a dispute between the Chinese and the Japanese, seem to be wanting in every warlike sentiment. They have no arms, either offensive or defensive. They declared to the traveller Hall that they did not know what war was like, either by experience or by tradition, and that it was with the greatest astonishment that they looked at the kriss used by the Malays.

The most civilised group of all the Mongolian race hold the art of war in poor esteem. In China the profession of a soldier is hereditarily the lot of the Mantchu Tartars, encamped here and there in different parts of the country, who mix but little with the laborious population; and these people in their turn disdain the others as useless men. But still these troops are commanded by the mandarins, themselves Chinese for the most part, and named in the lists as civil mandarins, to whom they are always subservient.

The imperial government prescribes to its soldiers precautions which would astonish most of our European generals; for instance, the bearers of calverins in the Chinese army are enjoined to stuff their ears with cotton. Certain details of Chinese tactics appear to us now to be singularly simple-minded. At Ning-Po, in 1842, when the Chinese soldiers made an attack at night upon the English troops, they were careful to carry lighted lanterns upon their heads. In 1857 they remained in their trenches, quite uncovered from the fire of balls and bullets, or else in their junks, until they were forced into hand-to-hand combat, or until a large battery of fire was turned upon them. Then they surrendered, indignantly angry at the disloyalty of their enemies.

The Chinese have in general quite lost the animal faculty which urges man to rush furiously upon his enemy; and also, on the other hand, they have not acquired the feelings of a superior kind which impels upon man the duty of sacrificing himself, in case of need, for the sake of the general welfare. Their moral evolution is incomplete, or it has miscarried. No doubt with them certain animal instincts are dead, and they are not yet replaced by instincts of a higher and better nature.

Even when they are facing the enemy the Chinese officers and Chinese soldiers will stultify themselves by smoking opium. They care very little for the interest of their country. They are totally devoid of moral impulse. They have good qualities: a great power of passive resistance, much perseverance, industry, and extreme docility; they are excessively patient, and this enables them to bear hunger, thirst, and the inclemency of the weather. There are, in fact, different sorts of courage, among others that of the wild beast, which is only the effect of reflex action; there is also the courage which is truly humane, that of a thinking creature who voluntarily sacrifices himself for a superior interest. The Chinese have lost the first of these moral forces, and they do not seem to have yet acquired the second; and in this respect many of our Europeans resemble the Chinese more closely than they perhaps imagine.

VII.

Warlike Manners in the White Race.

The different branches of the white race of man have, like all other human groups, passed through very many changes, very many admixtures; and by going back to past times we shall have the greater chance in being able to trace out, more or less intact, the primitive natural qualities of the people. Now, in the warlike manners of ancient India we find the impress of a real moral elevation, but which most modern Europeans would certainly call by no other name than folly. The Manu Code prohibits the use of poisoned, bearded, or incendiary arrows. It is ordained that the disarmed, wounded, or those who surrender themselves shall be spared. The man on horseback, or in a war-chariot, must not kill the soldier on foot. It is forbidden to attack anyone overcome by fatigue, him who is asleep or lying down, the soldier in flight, or one who is already engaged against an adversary. The conquered country ought to be respected. Security ought to be guaranteed to all its inhabitants; neither the laws nor the religion of the conquered people ought to be altered.

We see here very many precepts which may, perhaps, be adopted in Europe some thousand years hence, but which nowadays would only provoke a smile or disdainful shrug of the shoulders from any of our celebrated statesmen.

Primitive man, to whatever race he may belong, is a savage animal. And in this respect the white man is no better than the others. He has been, and is capable of again becoming, as ferocious as an Indian Guarani. We have just seen to what degree of humanity the ancient Indians had risen; but yet nothing can be more bloody than the history of all the Aryan nations. Nothing can be more atrocious than the warlike manners of the Hebrews, or, in general, those of the Semites. After victory the people of God slaughtered and massacred whole nations; they made a sport of crushing to powder the heads of children. Ninus, the conqueror of the Medes, crucified their king, his wife, and their seven children. At this time slavery was the mildest treatment which a prisoner could expect.

The Romans were not more humane, and their history affords us abundant examples of unmerciful ferocity. We will only mention the massacre of the Jews by the virtuous Titus, the Gallic hecatombs accomplished by Julius Cæsar. In this respect the annals of modern Europe, from the fall of the Roman Empire down to our own times, show us some awful instances. Even if we pass over the darkest parts of the early Middle Ages it will be sufficient to mention the Hundred Years War in France, the Thirty Years War and the sacking of Magdeburg in Germany, the horrors committed by the Spaniards in the War of Independence in the Netherlands: and during all this fearful sacking of towns, murder, theft, violation, were acts lawful, or perhaps held worthy of praise.

Even nowadays, or at least at a very recent date, the most civilised of the European people make a sport of exterminating the inferior races. In Tasmania, the English, holding their Bibles in their hands, have, with full determination and in cold blood, destroyed the aborigines of the country, no doubt following the example of the savages themselves who were swarming in the country. The American government has more than once put a price upon the heads of the Indian Red Skins. On the 2nd of

October, 1749, Cornwallis, then Governor of Halifax, offered ten guineas for every Indian Micmac killed, scalped, or taken prisoner. On the 10th of August, 1763, Amherst gave orders not to make prisoners of the Indians but to exterminate them.

We regard as atrocious the customs of certain belated groups of the white race: the Khivites, for instance, quite lately used to pay each of their soldiers so much for every one of the enemy whose head was cut off; and this, of course, without prejudice to honorary rewards, prizes for valour, etc. Within the last few years a French general, after having massacred a handful of Italian patriots, telegraphed to Paris saying that "the Chassepots had done wonders;" and in 1870 the army of the "nation of thinkers" shelled the French towns and shot down all the francs-tireurs they could find. It is not now our purpose to speak of civil wars.

In short, the Europeans who are so proud of what they call their civilisation are still only in a state of mitigated and disguised barbarism; they have yet to make very great progress before they have advanced in morality, in kindness, in humanity, and in justice, as far as the results show them to have advanced in the mechanical arts during the last fifty years of this century.

CHAPTER XII.

ANTHROPOPHAGY.

I.

EVEN though we still hear of the most civilised nations taking pleasure in the slaughter effected either in civil or in foreign wars, we cannot doubt that the moral sense is a fruit of mental maturity ever ripening in the human brain. Our ancient ancestors and our poorly-developed contemporaries used to or still feel scruples on this account to a very much smaller extent. But anthropophagy often becomes confounded with homicide, either in war or other wise, amongst the inferior degrees in the evolution of mankind. Neither public opinion nor individual conscience will trouble itself for so small a matter, especially if the man eaten be an enemy—that is, if he belongs to another tribe. In this case, to eat one's neighbour is an honourable and glorious action. Primitive man in this respect is absolutely animal; he resembles many fish, or reptiles, even mammalia, who will gladly eat animals of their own kind. There are certain human tribes, as we shall see presently, who fatten up human beings to put upon their tables, exactly as do the Mexican honey ants (*myrmecocystus Mexicanus*), who kill in the winter individuals of their own caste for the sake of the honey to be found preserved in their abdomen.

Our pharisaical morality considers such acts to be atrocious. In this we may be right; but we forget to protest when, in time of war, flaming shells are thrown into a town filled with women and children, with the aged and the sick; or when in a war, civil or other, by the help of engines as ingeniously constructed as they are diabolical in their effect, man will in a few hours crush into powder thousands of his fellow creatures. But nevertheless in the eyes of humanity and of common sense, to kill a human being is certainly more reprehensible than to eat him after he is dead.

In an interesting sketch presented to the Anthropological Society in Florence in the year 1878, M. Herzen showed plainly from observations made in the laboratory of physiology in the same town that cynophagy is repugnant to certain kinds of the canine race, and not at all to others. We may say as much of the existing human races; but there is not now one race extant who has not been anthropophagous at some past epoch. There are different kinds of cannibalism, and we have elsewhere endeavoured to classify them. The principal are cannibalism from actual want, cannibalism from love of gluttony, cannibalism from warlike or revengeful anger, cannibalism from religious motives, cannibalism from filial piety, and lastly, the most exalted of all is judicial cannibalism. We will quote a few examples of these different kinds of anthropophagy; and we shall be obliged to pick out our instances, for the stock to choose from is amply abundant.

II.

Cannibalism in Melanesia.

The most common kind of cannibalism is certainly that which is caused by actual want. It is practised almost everywhere among savage races, but specially in countries where the eatable mammalia do not exist, or where they are scarce. In the islands of the Pacific ocean, for instance, on the Australian continent, the native fauna are very parsimoniously supplied to the human beings. As we have already seen, the Australians, when hard pressed by hunger, will kill their wives and eat them; they have even dug up corpses recently buried. Cunningham found the throat of a woman in the sack of the Australians who accompanied him. Certain tribes in South Australia made use of human skulls for their drinking cups; but we must consider that these poor people did not know the art of pottery. We can understand that cannibalism should be most common among the least civilised of the Australian tribes, amongst those who were still living in hordes, governed by hereditary chiefs. The Melanesians are all more or less stained with cannibalism. In proof of this assertion, we will quote a few examples, but shortly, not to be fastidious. Cannibalism is frequent in New Guinea. O. Beccari, as well as many others, has told us that it is so. In the Fiji islands it has become celebrated. Though the Fijians were more or less christianised, they used to pull to pieces and roast their enemies killed upon the field of battle; they would sometimes devour even their own wives. With them, a dish of man was served at every official banquet; they used to call human flesh "long pork." An anthropophagous repast graced every Fijian solemnity; for instance, the inauguration of a temple. Cook has declared the same passion for human food to exist at Tanna, where, as indeed at Fiji, pigs, fowls, roots, and fruits were to be found in abundance.

As to the cannibalism practised by the New Caledonians, we have the most precise and the most authentic information. Before the arrival of the Europeans the islanders knew of no other mammalia than the vampire-bat and the dog-fish. They therefore had recourse to cannibalism from actual want of food. The desire to eat human flesh was the cause of frequent wars among the different tribes. The chiefs sometimes said to the people: "It is a long while since we have eaten any meat; go out and get some." The struggle was over when the desired object had been effected, as soon as a few men had been killed. Human flesh was considered by the New Caledonians to be a delicacy; they ate it because they liked it. There were chiefs who allowed themselves to taste one of their subjects as they were sitting in their family circle, and they would occasionally have the pieces preserved in salt. Public opinion was not severe upon these princely repasts; they were thought rather to be dignified. After a friendly meeting, those of the chiefs who had taken the lion's share in the feast would send a few pieces to doubtful friends, so as to insure their alliance and good-will.

We may now see that the New Caledonian morality was far from throwing discredit upon anthropophagy. These people had a peculiar tool specially adapted for cutting human bodies into pieces; and very generally before sitting down to a feast the natives would begin by a dance, holding their knife in one hand, and a lance in the other.

III.

Cannibalism in Africa.

The African negroes are not more scrupulous than the Oceanian negroes as to eating human flesh, and anthropophagy, at least accidental, has been found to exist almost everywhere in Africa where the black men prevail. The Kafirs, who are a pastoral and agricultural people, and who are relatively intelligent and civilised, are not ordinarily cannibal; but they will become so in times of famine, as did the Mantati Kafirs seen by Thompson. It has been remarked that some of their tribes, after having become anthropophagous from necessity, kept up the habit from sheer gluttony. Gardiner gives an instance of this among the Zulus. Not many years ago some tribes of Basuto Kafirs lived altogether upon cannibalism in the middle of a fertile country in which game was plentiful. Like the European troglodytes, our ancestors, they used to live in caverns, where they would take and devour the human meat. A misfortune which befell them constrained them in the first instance to have recourse to such an extremity, but they long kept up the practice. In the year 1868, they had not corrected themselves of it, for at this time an English traveller saw in their caverns some human bones from which the flesh had been quite lately dragged. The corpse, he relates, had been pulled to pieces in a skilful way. The lower jaw had been detached by blows from an axe; a hole had been bored in the top of the skull, so as to extract the cerebral substance; the long bones had been split open longitudinally, so as to extract the marrow, as had formerly been the practice with prehistoric man. No doubt accidental cannibalism, proceeding from actual want, is not very rare even among modern Europeans, as we learn from the stories related by many travellers; but if they ever acquired the practice of cannibalism, as did the Kafirs, they must also have lost it again quite lately.

Among the Fantis in Central Africa, one of the most intelligent people of the negro race, cannibalism is an habitual custom. They adopt the practice peacefully and commercially. All the Fantis, except the chiefs, the kings, and those who have been exceptionally distinguished in the eyes of their own tribe, are eaten after death, instead of being buried. But the Fantis have certain scruples in their notions of anthropophagy. As far as it is possible they do not eat those of their own tribe, but they procure the corpses of the neighbouring tribes by giving them their own dead bodies in exchange; and up to the present moment no Fanti moralist has thought it expedient to censure this custom of making use of the dead.

We find very similar practices among the Niam-Niams on the Upper Nile, who, besides eating their prisoners of war, also eat the bodies of those of their own tribes who had died from abandonment—bodies which may be compared to those amongst ourselves who have to undergo the scalping knife of the anatomist.

Schweinfurth was once present at a sort of idyllic anthropophagical feast among the Niam-Niams. Between two huts, the doors of which were opposite to each other, a newly-born child, on the point of death, was laid out on a mat. At the door of one of the huts a man was calmly playing upon a mandoline; at the door of the other, an old woman, surrounded by a group of young boys and girls, was cutting up and preparing gourds for the supper. A caldron full of boiling water was quite ready, and the cook was waiting only that the child should die, that his corpse might be used as the principal dish. The whole scene was rendered more animated by the brilliant rays of a dazzling sun.

In the same region the Monbouttous, a pastoral and agricultural people, inhabiting an extremely fertile country, and belonging also to a superior kind of black race, are inveterate cannibals. They eat little food but the bodies of their captives. They are in a state of perpetual warfare with the inferior tribes around them, in order to procure for themselves human flesh. They cut up the dead on the field of battle, and drive home before them as a flock of sheep the prisoners whom they intend to reserve for future meals.

Cannibalism is a general feature in primitive manners. It does exist, or has existed, all over the world, and among all races of men. The facts which we have just quoted will dispel any doubt as to its practice in Melanesia and in Africa; and nothing can be easier than to collect equally strong proofs of its existence in Polynesia.

IV.

Cannibalism in Polynesia and in Malay.

The study of cannibalism in Polynesia is particularly interesting. We should here alone have matter enough to prove that the law of social progress is not an illusion, as the modern amateurs of pessimism think they believe. When the voyages of European navigators first made us acquainted with the Polynesian islands, cannibalism was then everywhere more or less the frequent custom, we could then study the different periods of its evolution. In some of the archipelagoes it was carried on in all its primitive brutality;

in others it was rare, and a thing of chance; in others it no longer existed, but still evident traces of it might be found both in the language of the people and in the forms of their religion.

In New Zealand, a country close adjacent to Melanesia, inhabited by men who were ignorant of agriculture, and having no other domestic animal than the dog, anthropophagy was practised without the slightest feeling of shame. The New Zealand tribes, constantly at war one with the other, would go hundreds of miles into the interior to fight, merely that they might feed themselves upon human flesh, or to capture slaves, generally intended to appear as the *plat de résistance* at the great feasts which they prepared for their parents or their friends, either immediately before setting out upon a campaign, or else upon some occasion of unusual rejoicing. The New Zealanders were very partial to the flesh of women and of children. On the field of battle they would pull to pieces one of the enemy taken prisoner or wounded, without even waiting until he was dead, or giving themselves the trouble to kill him. They had progressed so far that they did not eat the bodies of men of their own tribe, and they would not touch those who had died of sickness; but they most unconscientiously made use of the whole of the body they did take, and they were careful also to perforate the cranium so as to extract the cerebral substance. Contrary to the customs in the other archipelagoes, the New Zealand women used often to take part in these cannibal festivals.

This custom, which appears to be horrible to Europeans of our own time, was natural enough to the New Zealanders. To eat one's enemies, who, they said, would have eaten them instead if they had been victorious, was in their eyes legitimate enough. Judicial laws and religious ideas formed part of this theory of cannibalism. If a chief was killed, the people had the right to exact that his wife should share the same fate. She was given over to the conquerors, who put her to death. Then the bodies, when they were roasted, were eaten with great unction under the direction of the priests or the arikis, who first tasted some of the dainty bits cut out of the victims. They attached great importance to eating the left eye of their conquered enemy; for the soul of the departed, the waidowa, was placed in this eye, and therefore in eating that they doubled their own existence. In New Zealand religion had sanctified cannibalism, as it everywhere sanctifies the dominant inclinations of any race of people; but the fundamental reason for cannibalism in New Zealand was the desire to eat meat. To these poor islanders, often half starved with hunger, as they had no mammalia to kill for the purposes of food, flesh, in whatever form it came, seemed to be delightful. A soft mannered and most affable chief said to the traveller Earle: "Human flesh is as soft as paper."

Disgusting as these manners may appear to us, we should be wrong in concluding that the New Zealanders were incapable of human, or even of soft-hearted sentiments. If one of their relations or friends had been killed in battle, they felt their grief very strongly; at least they showed it in such a way that would shock the feelings of most Europeans similarly afflicted. They would cut their foreheads and their cheeks with shells or with sharp stones; and they also showed, in the same way, the joy they felt at seeing a friend who had long been absent. They used to wear, hung round their neck, little stone figures, with eyes made of mother-of-pearl, in memory of the dead whom they regretted.

As a matter of fact, human morality is very variable, and in every country, not excepting those the most civilised, it has sanctioned, and does still sanction, acts much more blamable than cannibalism.

The New Zealanders, both men and women, were a cannibal people; they practised it openly, and with no sort of feeling of shame. But the inhabitants of the Marquesas islands began to have some scruples as to eating human flesh. The women (and this may be noted as an example of their notions of shame, and is altogether to their honour) of these islands felt a great repugnance to anthropophagy, which, indeed, was forbidden to them by the customs of the country; for in ordinary times only the chiefs, the high priests, and the old men had the privilege of cannibalism. It was only in times of war that this right became extended to the populace, to the *kikinos*. Even at the end of the last century the men in the Marquesas islands began to have their doubts as to the morality of this custom. An old chief boasted to Porter, somewhat ostentatiously, that neither he nor any member of his family had ever eaten human flesh, or flesh of a pig, either stolen or one who had died of sickness. But even up to a late date cannibalism has been in vogue in the Marquesas—perhaps it still continues. The desire to procure some roasted human food was the cause of very many minor wars and skirmishes. In latter times the people have begun to reproach each other mutually, one tribe against another, with the practice of cannibalism; but these timid protestations of public opinion were not strong enough to prevail against an old and long established custom. The pulling to pieces of the bodies, and the dividing out of shares, was done most methodically according to hierarchical order. The victim was usually strangled, as was the practice with other animals, so as not to lose the blood. Here, as in New Zealand, the eyes were considered to be the most choice morsels, and they were offered to the warriors. The heart was eaten raw. The rest of the body, stuffed with the leaves of the *ti*, was cooked in the Oceanian oven. Then when the carcase was sufficiently roasted it was cut up with a sharp-pointed stick. The feet, the hands, and the ribs were reserved to the chiefs. The thighs and the most delicate parts were considered the property of the high priest.

In the Friendly islands, and in the Sandwich islands, the people used to eat their enemies without the least scruple; a chief in the Sandwich islands told Cook, laughing as he spoke, that human flesh was the most savoury of all food. At Bow island they used also to eat their enemies, and even those of their own side who fell in battle; in general everyone who died of a violent death, and at last the assassins. This is the only spot in Polynesia where the existence of judicial anthropophagy has been proved. People here were perhaps more greedy than elsewhere after the taste of human meat, and especially of the female flesh, which the islanders declared to be more tender. But Bow island was one of those small coral reefs so numerous in the Pacific ocean. The animal kingdom was here very poorly represented, and it was more difficult than in the larger islands to find food rich with azote matter.

In the Tahitian archipelago, where feculent fruits and fish were plentiful, where there were dogs and pigs, cannibalism was little more than traditional. By chance, or from a feeling of vengeance, they might roast and be glad to eat a piece of a conquered foe, but in general cannibalism was condemned by public morality. It had formerly been the custom, and evident traces of it were to be found. For instance, in the human sacrifices the priest offered to the chief the eye of the victim, and if it was refused he then presented it to the gods with the rest of the body. These gods, the Tahitians thought, were very fond of human flesh. After an offering of this sort, the people might ask of them anything, and their request was granted.

Certain names recall to our minds the ancient custom, for instance Aimata, the name that Pomare bore before his accession, and which signified "eating the eye." Some of the vernacular phrases also give us proofs of the old habit: in Tahiti a time of famine was spoken of as "the man-eating season."

In the Javanese archipelago there were some very curious forms of anthropophagy: there was anthropophagy from feelings of filial piety, and judicial anthropophagy.

The Battas in Sumatra, a numerous and well-governed people, whose habits were agricultural, who possessed a regular system of laws and of government, an alphabet and a literature of their own, used religiously and ceremoniously to eat their old relations, taking care, however, to choose for their feasting a season of the year when lemons were plentiful, and also when salt was cheap. On the appointed day the old man destined to be eaten would get up on to a tree, at the foot of which his friends and relations were standing about in groups. They sang a funeral hymn, beating the trunk of the tree as they did so in cadence. The purport of their song was: "The season has come, the fruit is ripe, let it now fall down." The old man descended from his perch, his nearest relations would kill him carefully, and those standing by would devour him.

The Battas practised also anthropophagy of the most exalted kind: judicial anthropophagy. With them the adulterer, the night thief, those who had treacherously attacked a town, a village, or a particular person, were condemned to be eaten by the people. They were tied to three posts, their legs and their arms stretched out in the shape of a St. Andrew's cross, and then when a signal was given the populace rushed upon the body and cut it into shivers with hatchets or with knives, or perhaps more simply with their nails and their teeth. The strips so torn off were devoured instantly, all raw and bloody; they were merely dipped into a cocoa-nut bowl containing a sauce prepared beforehand made of lemon-juice and salt. In the case of adultery the outraged husband had the right of choosing first what piece he liked best. The guests invited to the feast performed this work with so much ardour that they often tore and hurt each other. Though judicial anthropophagy may be the most scientific form of cannibalism it cannot nevertheless be practised without awakening in man all his most savage instincts, of which it is the last remnant.

V.

Cannibalism in America.

Cannibalism exists, and has existed, in America from the Arctic regions down to the island of Terra del Fuego. In this latter island its existence has been proclaimed by Magellan, Candish, and Fitzroy. The Moxos and many of the Guarani tribes were inveterate cannibals, as we have already seen in speaking of their custom of fattening the prisoners and treating them well before eating them. The Mexicans, who were relatively civilised, had similar customs, which became sanctified by their religion; for in their sacrament they used to eat and fight for pieces of a statue made of maize kneaded together with the blood of a child.

The first French missionaries among the Red Skins found cannibalism to be the custom. Father Brébeuf saw the Hurons eat one of their neophytes, and Charlevoix relates the story of twenty-two Hurons devoured by the Iroquois. Again, in 1833, Captain Back's expedition gave us another proof of cannibalism among the Red Skins in North America.

The Nootka-Columbians, who form a sort of connecting link between the Red Skins and the Esquimaux, will also freely resort to anthropophagy. Some of them offered to Cook the roasted skull and the hands of a man already half eaten. One of their chiefs was so partial to human flesh that at every new moon he caused a slave to be killed, to be eaten at a banquet which he gave to the other chiefs of a lower rank. The affair was conducted with much ceremony. The guests began by singing a war song, and by dancing round a flaming fire which was kept alive by oil being constantly poured on to it. Then the chief had a handkerchief tied over his eyes, and for a while he played a game of blind-man's-buff with a few slaves who had been collected; and the unhappy wretch who was caught by the chief was instantly slaughtered and pulled to pieces, and his smoking flesh was distributed to those invited to the feast.

In the time of famine, the Esquimaux are not more scrupulous than the Nootka-Columbians. If one of their hordes is very much in want of food, they will rush furiously upon another horde, massacre them and pull to pieces the dead bodies, and then devour the raw flesh in its half-frozen state. We must say, however, that anthropophagy from mere feelings of gluttony, as practised by the Nootka-Columbians, appears to be unknown to the Esquimaux. These latter do not become cannibal except during times of famine. It would seem also that, formerly, they were more than once used as game by their implacable enemies the Red Skins; for they call them *Irkily*, a name given by the Greenlanders to anthropophagous men—their wild imagination making them think that they were men with dog's heads. This mythological conception would alone prove that cannibalism was no longer their constant practice.

VI.

Cannibalism among the Mongolian and the White Races.

As we have said, and already half proved, cannibalism is not a necessary part of the existence of any human race. It is only more common in proportion as the social condition is inferior.

Among the superior races, the Mongolian and the white people, it is no longer a general and habitual custom; but it has existed in times gone by, and reappears every now and then accidentally.

An ancient Hindoo traveller relates that the inhabitants of Bhotan used formerly to eat the livers of those of their enemies whom they had killed, seasoning them with butter and sugar, that they used to convert their skulls into cups and bind them round with silver, and their bones into jewels and musical instruments. Quite latterly, during the Taiping Chinese War, an English merchant in Shang Hai met his servant bringing home the liver of one of the rebels, to eat it, not from any idea of gluttony, or of want, but from a higher moral motive—to give himself courage.

The white race is no more than the other races indemnified against cannibalism, which appears to be a sort of original sin in humanity. The Mongolian people in Eastern Europe were the first to set the example to the ancient Greeks. Herodotus tells us that the Massagetæ used *from compassion* to knock down and afterwards eat their old parents. With them, the old men who were allowed to die a natural death were considered as impious persons, and their bodies were thrown to the wild beasts. The same manners prevailed among the Issedons in the eastern part of Scythia. In the early days of Grecian history cannibalism was reprobated, but the legends of Atreus and of Lycaon show us plainly that the ancient custom had not long fallen into disuse.

In the first centuries of our era instances of cannibalism were observed in Europe, and St. Jerome says that he saw in Gaul anthropophagous Scotchmen who were particularly fond of the breasts of young girls and of the thighs of young boys.

The Semitic people, less civilised than many Indo-Europeans belonging to the white race, used to, and would again very soon, fall back into the practice of anthropophagy. Josephus has related with much hypocritical and foolish rhetoric, the story of a Jewish mother who cooked and ate her child, while the victorious Titus was mercilessly besieging in Jerusalem the last defenders of the Jewish independence. The Arab historian, Abd-Allatif, makes up a whole anthology of cannibal history in speaking of a famine which desolated Egypt in the 597th year of the Hegira (1200). In all the towns of Egypt, at Alexandria, Es Sous, Damietta, Kous, etc., anthropophagy was practised upon a large scale. People devoted themselves to eating men and children, the latter especially, for a child roasted was reputed to be an excellent dish. Punishment by fire, which was inflicted upon the eaters, was so little prohibitory, that when the tortured men had been roasted, they were occasionally devoured. Cannibalism, which seemed at first to be horrible, became at last engrafted in the manners; people became fond of it. A woman pregnant with child made human flesh her habitual food; a grocer had stowed away a large provision of human flesh, which he had salted. We need not be astonished at this, for quite recently in Aralda, at the time of the last famine, we saw the Arabs occasionally resorting to anthropophagy, following the example of their ancestors.

Moreover, similar facts are not very uncommon in the modern history of the European nations. Schiller tells us that the Saxons had become cannibal towards the end of the Thirty Years War. In France, in 1030, during a famine which lasted for three years, men ate their fellow-creatures as did the contemporaries of Abd-Allatif. A man was condemned to be burnt who had put up for sale human flesh in the market-place at Tournay. In his quaint chronicles, Pierre de l'Estoile, giving us other interesting details, speaks of the cannibalism of the Parisians during the blockade of Paris by Henri IV.—the good king Henri—in 1590: A rich lady, when her two children died of hunger, made her servant salt the bodies—this being done she ate them; also the lansquenets used to hunt down men in the streets of Paris, and hold their cannibal festivals at the Hôtel Saint Denis, at the Hôtel Palaiseau, and elsewhere.

A little later, the day after the assassination of the Maréchal d'Ancre, the people dug up his corpse, and one of them cooked the heart upon burning coal, and ate it after seasoning it with vinegar.

We may perceive that we should be wrong to be over proud of our present civilisation, which is yet so imperfect. The beast is not so very far behind us, he is even in us in a latent form. But this anthropophagical review which we have now concluded has at least one consoling side: it shows us, after its own fashion, that the evolution of the human mind is progressive. Like other creatures, man begins by being an animal, and he is not, moreover, the least ferocious. In this poor, half-starved, coarse creature hunger dominates over every other want; all kinds of meat to him are good, even that of his fellow-creatures who are nearest to him, his wife, his children, and the members of his own family. After a while he eats only his enemies, that is, his rivals, the people of the neighbouring tribes. He is cannibal from vengeance or from gluttony; but he does not gloat over his passion except when he is eating prisoners or slaves. Then at a later date his cannibalism becomes a religious or judicial ceremony, and therefore of much less frequent occurrence. From that condition it gradually comes to be condemned and scouted by public morality, and man does not go back to it except in cases of extreme famine, or in a state of madness, when his intelligence and his morality have foundered, and the beast once more shows itself to be the master.

But it more often happens that the Europeans of our own times, when they are in the most dire distress of hunger, in cases of shipwreck, for instance, will rather die than go back to the cannibalism of their ancestors. Poor and imperfect as man is he is a perfectible creature, a conclusion that may be to us both consoling and gratifying.

CHAPTER XIII.

FUNERAL RITES.

I.

The Idea of Death.

WE cannot with certainty deny that the idea of death has been given to some animals. M. Houzeau relates that in Arkansas a woman had been killed by the Indians in battle, and that her dog persistently remained by her side and allowed himself to die of hunger. M. Houzeau also reminds us of the fact quoted by Cuvier of a dog caressing a lioness as they lived together in the Jardin des Plantes in Paris, and that when the lioness died the dog also allowed himself to die of inanition. We know that the parrots which we call "the inseparables" die of grief in case of widowhood. These facts will show beyond doubt that certain animals are endowed with a great moral sensibility, but they do not prove that though animals give way under their affliction they have therefore a clear idea of the death that is awaiting them. That is not the foreseeing and determined suicide of which we see many examples in the human kind, and very often too among the savage races, contrary to the generally received opinion. But even were it well established that no animal was gifted with a definite notion of death, we must guard ourselves against recognising in this fact a very important difference between man and the animal; for, as we shall see, the idea of a natural death is foreign to the ethnical groups of men taken in a mass. The difference being rather in the existence or in the non-existence of funereal rites; and even here, between these two extremes, the movement is very slowly and very gradually accomplished.

A kind of lama (the *auchenia guanaco*) seems to have an idea of death and also a feeble notion of making his own grave, for all animals belonging to this kind who are free in their motions go to the same spot to die, there to pile up their bones. We find a stronger instance among the ants, for after their battles they bring home with them the dead bodies of their warriors. According to Battel the anthropomorphous (*gorilla gina*) always take care to cover over with branches and dead wood the corpses of the animals of their own kind. On the other hand, many hordes of human creatures abandon without a thought the dead bodies of those of their own tribes. We must nevertheless declare that the forsaking of the dead seems to be uncommon in the human race. The funereal rite is a humane feature generally adopted by all races, and those learned gentlemen who are beset with the notions of digging a gulf between man and the rest of the animal kingdom might, with some show of reason, make of man a "funereal kingdom."

As we shall see in our sketch of the principal human races, the rites and the funebrial ceremonies offer a very considerable variety; but the rites are nearly always closely connected with the idea that men have conceived of the existence beyond this actual visible life. In one place the dead are abandoned, in another they are eaten. Many people throw their dead bodies into a hole in the ground, and then cover them over with earth and with stones. In some countries people hide them in the naturally made grottoes, or in caverns of different forms, which they shut up and cover over as well as they can, to preserve the dead bodies from the wild beasts. Other people or other tribes have ideas diametrically opposite: instead of protecting their dead they take care that they are eaten by the wild beasts, by birds of prey, or by dogs brought up specially for the purpose. One people will dry and mummify their dead; another will leave them on a sort of platform, or in a boat, put up among the branches of a tree.

If we except bestial abandonment, all these customs have been prompted by the same sentiment, by a pious care for the future of the deceased. Man does not ordinarily believe that all human personality vanishes away in the tomb until he has learnt the harsh lessons which science only can teach him. The primitive or ignorant man looks upon death merely as an accident, a shock which gives a new phase to existence. It is to him but a vapour, a shadow of the departed spirit, something that emancipates itself from the body in a state of putrefaction; and this something is the conscient me of the individual, which then begins afresh a new life, more or less in imitation of the old one. Sometimes he imagines that the absent one will one day return back into the body, which is but temporarily abandoned, and he endeavours to keep for it as far as possible its ancient resting-place. Assuredly, the daily phenomena, such as sleep, dreams, the syncope, etc. have given great assistance towards misleading on this point the infantine judgment of primitive man. To this coarse-minded creature it is inconceivable that he should pass from the exuberant drunkenness of life to the absolute nothingness of death. For in the early ages of the human evolution man is far from having tested accurately the conscient and the unconscient sentiment. Man then readily lends to inanimate objects which impress him in so many different ways, ideas, sentiments, and passions analogous to those which he himself feels. As we have just said, and as the poets have so often warbled to us, sleep seems to be really the brother of death; they are scarcely distinguished one from the other. Sleep appears to be so very like a temporary death, during which time the mind frees itself and wanders vaguely here and there; and death, too, on its side, does not differ much in appearance from a long sleep, which must also lend itself to the idea of a double existence.

It is necessary to bear in mind these general views in order to understand the customary funereal rites, very various, and sometimes oddly strange, to which the different groups of men have bound themselves. We shall now mention some of these customs, but we shall have to rest satisfied with recording only the most typical.

II.

Funereal Rites in Melanesia.

Even in Melanesia, among the Australians and the Tasmanians, the lowest of the Melanesian type, so similar in everything one to the other, we find differences in the funereal rites. With them the corpse was sometimes buried, or sometimes placed where there had been a slight excavation. In the case of inhumation the body was generally set in a crouched position, the knees bent up against the chest, and the arms crossed. The cloak of the dead man was fastened around him. This funereal attitude was or is still practised among many people, among the Andamanites, the Peruvians, the ancient Scotchmen, etc. and it has given rise to much conjecturing; the most simple reason ought, however, to be the most probable. In the imagination of the greater part of primitive men, death is but a long sleep. Granting this supposition, nothing seems more natural then than to give to the body the same attitude of repose which a man would naturally fall into by the fireside in the evening after a day's hunting, or after a fighting expedition.

Sometimes also the Melanesians of whom we are speaking would place their dead either in the trunks of hollow trees, or in a coffin made of bark. In any case they would lay close beside the deceased his warlike and his hunting implements.

The burying-places were generally isolated and destined only for one individual; but the Australians wished sometimes that their bodies should be placed close to each other. This was a custom intended principally for the young people. For the older men more ceremony was practised; and instead of burying them they burnt them. They afterwards collected together the ashes of the bones to make amulets, which should protect them against sickness and assure their success in hunting or in war. All these customs show that the Australian and the Tasmanian believed in another life beyond the grave. It seems also that there, as in many other countries, the something which they imagined to remain after death, the manes, in point of fact, was regarded much more as an object of terror than as an object of affection. After the death of a man his friends avoided speaking of him, and all the members of the tribe who bore the same name were then considered bound to change it. A similar custom exists elsewhere in other races, notably in Polynesia.

The Tasmanians and the Melanesians used also, when they lost certain of their own relations, to wound themselves; they would break a bone in their finger. Did they thereby wish to give a sign of their grief, or did they intend to appease the angry manes of the deceased? We may say with certainty that they did not believe in natural death. In their minds death had been brought about by some malicious device invented by an enemy; and they thought therefore every death ought to be avenged by the near relations of the deceased. It was held to be a strict act of duty to kill the presumed assassins, who ordinarily belonged to one of the neighbouring tribes; and the amount of bloodshed was in proportion to the rank of the deceased, or to the affection and esteem in which he was held. An Australian, who wished to show to Father R. Salvado the tenderness that he felt for him, promised him that if he died, he (the Australian) would kill at least half-a-dozen of his own countrymen. This, among others, is one of the unhappy results coming, not from the doctrine of the immortality of the soul, which slowly disclosed itself to the human conscience, but from the belief in a temporary resurrection after death—an idea that is very common in all races of men.

It does not seem that incineration was practised with the Papuans and the New Caledonians. But sometimes the Papuans will dry and mummify their dead, and then carefully preserve the corpses in their huts. Some New Caledonian tribes will allow the corpses to get putrid; they will then take up the bones and put them either in a cranny of a rock, or in a small cavern dug out in the middle of a wood. They will generally bury them in a cemetery belonging to the tribe, which place is regarded as a sacred spot. Some of the New Caledonian tribes bury only their chiefs, and content themselves with placing the bodies of the common people on to the branches of trees, or in tying them with their backs against the trunks of the trees. In every case they put beside the remains of the deceased all his utensils which may be useful or agreeable to him—his lance, his pagnays, his jewellery, etc.

The death of a chief being considered, in New Caledonia, as a public calamity, and also the population having always an unfortunate tendency to exceed the extent of their provisions, it is considered obligatory after a loss of this kind to abstain from conjugal intercourse for a fortnight or even for a month. As we see, it is not here a question of court mourning. At the end of the term fixed a commemorative festival is held to indicate its expiration.

From the foregoing facts we may conclude that the Melanesians, and even the lowest types belonging to the race, believe in a resurrection of some sort after death; and for a time, more or less long, after the death of those near to them they are moved with sentiments either of affection or of cruelty.

111.

Funereal Rites in Africa.

An affectionate regard for the dead exists also among the Africans, and even amongst their most humble races. It is difficult for us to guess if the Bushmen have any ideas as to a future state, and what those ideas are. However this may be, they are sensible of the death of their friends, for they energetically show their grief by breaking the bone of their little finger. Some of the men impose upon themselves this unhappy practice; but even among the Bushmen the woman appears to have a more affectionate nature than the man, for with her this mutilation of the finger is much more common. This partial amputation of the little finger seems, in the minds of the women of this race, to be a sort of sacrifice to which they attribute very various effects; they will sometimes inflict this punishment upon their children to prevent them from dying.

The Bushmen rarely bury their dead; but the Hottentots, who are more civilised, lay their dead bodies into a shallow ditch. Like many of the Melanesians, they place the corpse in a curled-up position, winding round it a *kross*, or a cloak, and being a thrifty-minded people, they are careful to select for this purpose the worst cloak they can find.

As regards funereal rites, the Kafirs, who are hereditary enemies to the Hottentots, do not differ much from their neighbours. We may have already observed that the Kafir is not ordinarily a soft-hearted man; their different tribes give little thought to their dead. The nomad Kafir will usually not do more than throw the corpse of his relative into an open ditch, common to all the tribe, and situated at a certain distance from the kraal. The corpse is then left to the care of the hyenas and the jackals. One may not unfrequently see a son unceremoniously dragging along towards the common ditch the dead body of his father, or that of his mother. The chiefs only are buried in the public place, and with some show of respect. This public place was in the inclosure where the men of the tribe usually met together. Their corpses were covered over with a cairn of stones.

We find a great uniformity in the funereal rites among the negroes of Central Africa. The law of the people is to bury their dead; they put them in a ditch in the form of a pit, and the body is laid in a curled-up position. Clapperton found this custom practised at Yourriba, at Koulfa, at Borgou, and at other places. Schweinfurth found similar customs among the Bongos on the Upper Nile, but with a studied refinement on their part. The Bongos carefully dig out of the wall forming part of the funereal pit a hole large enough to receive the bent up corpse. Their intention, prompted by pious solicitude, is to avoid the pressure of the earth upon the body, for they afterwards fill up the pit by throwing in fresh earth. This is a pious care springing evidently from the idea that the dead man is still sensible to pain. Fears of the same kind have troubled the minds of many other races. "Let the earth be light," is an established saying, even in Europe, in the funebrial orations. In a Vedic hymn addressed to Death we find the same idea poetically expressed: "Oh Earth! cover him as a mother covers her child with the lappet of her dress."

As we have already said, in the imagination of the primitive or poorly-developed man, death is generally only another form of life. At Koulfa, in Equatorial Africa, the people carefully bore a hole in the top of the tomb, dug very often just outside the door of the deceased person. Near to this hole they place the scarf, belts, and various objects, praying the deceased to give them to this or that person who had died previously. The Niam-Niams, who often also bury their dead in a sitting position, carefully first deck them out with feathers and the skins of animals, as though for a festival. They paint them red, for most savage people consider red to be the finest of all colours.

In the western side of tropical Africa, among the Timmanis, there are in the towns mortuary houses where the remains of the kings and of the chiefs are deposited. These houses are never opened; but in the walls narrow holes are bored, through which, at certain stated times, food and palm wine are introduced.

And the Timmanis, before they begin to eat, are careful to subtract for the dead a small portion of their meal, which they will throw down on the ground. This is also a custom among the Fantis, the Ashantis, and others. We find similar customs almost all over the earth. It is the idea of a resurrection in its earliest form. But that is very far from the idea of an eternal immortality which the majority of Europeans believe, or pretend they believe. The something, which these savage people naïvely suppose to remain after death, has all the wants, all the objects, all the good qualities, belonging to the deceased when he was alive.

Nothing can be more innocent than the offerings of food, of arms, of ornaments, etc., but we cannot say as much for the funereal sacrifices proceeding from this same hypothesis of a future life, of which our priests, our moralists, and our professors of philosophy do not cease to vaunt the salutary effects.

The funereal sacrifices are perhaps, in certain cases, nothing more than an extension of the custom of funereal mutilations in vogue among so many people, more or less savage, whose grief, affected or sincere, is shown by their inflicting upon themselves wounds and mutilations. The Malanesians, the Hottentots, and others will often break a bone in their little finger upon the death of a near relative. Elsewhere people content themselves with lacerating their skin, or making incisions in their flesh, more or less deep. At the time of Bruce's travels the Abyssinians showed the grief that they felt at the death of a relative or of a lover, by slightly cutting the skin upon their temples with the nail of the little finger, which they purposely allowed to grow for this reason.

From the notion of causing suffering to oneself we soon come to that of causing suffering, or of bringing about the sacrifice of others. In this way we may account for a practice prevailing in Ashanti. There, at the death of the king, his sons, his brothers, and his nephews, under the influence of a feigned madness, rush out of the royal palace, and discharge their guns indiscriminately at any person they may chance to meet. But the ceremony is usually conducted more methodically, with greater solemnity, and also with more appearance of reason. When it is supposed that the deceased has merely passed from this sublunary world to an invisible but similar state, what can be more natural than to give to him for his companions the creatures that he most cared about while he was alive upon this earth! If the dead man has filled any position here in this life, may it not be supposed that he will arrive in the life beyond, loved, caressed, surrounded, and waited upon, as was his daily custom here below! Therefore, upon his tomb will be sacrificed his familiar animals, his horse, or his dog—as was the practice in Borgou—or else his wife, his nearest relations, and his slaves.

This barbarous but perfectly logical custom is in vogue in many districts in Equatorial Africa. In Ashanti the death of the king is followed by whole hecatombs of slaves. It would indeed be most indecorous if an Ashanti king went into the future state without an escort proportionate to his illustrious rank. At Katunga in Guinea, when the king dies, the *caboceir*, or Djammah chieftain, three other *caboceirs*, four wives of the late monarch, and a quantity of favourite slaves are all obliged to poison themselves. The poison is given to them in a parrot's egg, and if by chance it does not take effect the patients are considered bound to hang themselves in their own houses. At Jenna, in Dahomey, when the governor dies, one or two of his wives kill themselves upon the same day, so that the deceased may have agreeable companions in the post-mortem government of which he has gone to take possession. And also at Katunga, when the king dies, his eldest son, his first wife, and the principal personages in his kingdom are strictly bound to poison themselves upon his tomb, so that they may be buried with him. This custom has also a political bearing, for it prevents the consequences, often very baneful, of an hereditary monarchy. Thanks to it, at Katunga, the king is always elected, and his son can never succeed him.

When the funebrial services are over—that is, on the supposition that they exist—the people not unfrequently think about some sort of funereal memento. In this respect there is a very long gradation of ideas. The Bushmen appear generally to abandon their dead.

The Hottentots bury them in a shallow ditch, and then cover them over, as may happen, with some earth and a few stones. The Kafirs throw the common people into the public open ditch, a sort of sewer, and bury only their chiefs, raising over the grave a pile of stones in a conical form. In Equatorial Africa they also bury the people of rank or position in a cylindrical pit about six feet in depth. The ditch is often indicated by a mound of earth, or by the erection of a stone about one foot eight inches or two feet in height, something similar to the *menhirs* (druidical stones) or *celtiques*. The Niam-Niams and the Bongos are also careful to lay the body from east to west. The former people lay the face toward the east if it be a man and towards the west if it be a woman. It would be most disrespectful to lay a woman so that she should face the rising sun! But the Bongos, their neighbours, have ideas diametrically opposite. With them the privilege of facing the east is allowed only to the women.

The sun, the dazzling sun, under which man must exist and have his being, has played a great part in every system of mythology, and people belonging to all races have given much thought as to whether or not they ought to lay their dead bodies facing the east.

We have just seen that the celtiform menhir was invented in Central Africa. It has also been invented elsewhere, in Hindostan, in the Fiji islands, and in other places; we may also say the same of another funereal construction, still more celtiform, the *dolmen*.

The Hovas at Tananarivo, in Madagascar, lay their dead under real funereal dolmens, consisting of five flat stones, four vertical and one horizontal; then they cover over the whole with flint stones, so making a tumulus. The flagstone at the top is often of enormous size. M. Dupré saw one of which the side measured thirteen metres, and contained altogether ninety cubic metres.

Man is really a sheeplike animal. The similarity between the funereal monuments of the Hovas and those of the prehistoric Europeans of the age of polished stone would be almost sufficient proof, if so many other analogies, certainly spontaneous, to be found here and there all over the earth, did not otherwise decide the question so as to leave no room for further doubt. In numerous circumstances of the same kind, similar ideas have been found in very many men of every race. This is an encouraging fact for the sociologists, and allows them to hope that at some future date the vast subject which is now engaging their thoughts may furnish them with the conditions necessary for a pure and reasoning science.

IV.

Funereal Rites in Polynesia.

The funereal rites, as well as many other customs in Polynesia, present a great uniformity of ideas, not excepting, however, mere local differences.

Ordinarily the Polynesian dead man was not buried, but carefully dried in the open air, then set in a curled-up position, rolled in bandages made of paper stuffs, and so preserved in a particular *morai*. The operation of drying the dead man was long and full of risks. The corpse was first laid out in the open air and upon a framework upheld by four posts, somewhat similar in construction to the central flag in the double Polynesian canoes. Sometimes, for instance at Noukahiva, the framework was replaced by the trunk of a bread-fruit tree scooped out in the form of a boat, and again covered over after the drying with another trunk scooped out like the first and fitting on to it hermetically. The majority of the islanders imagined that the life beyond was a very distant island, and that the voyage there was very long. They were careful, therefore, to place beside the dead body his arms and his implements; they gave him his club, a cocoa-nut shell for drawing water, food to eat, water to drink, fish, and some of the bread-fruit; all these were intended to sustain the shade of the deceased, which was supposed to wander for a long time round about the body.

In order to dry the body they often extracted the intestines through the anus, then every night they seated the man upright and rubbed him with cocoa-nut oil. When the operation was successful they had only to enrol the mummy in his bandages.

In the Gambier islands the corpse was prepared in the same way, but it was not set in a curled-up position. Once dried it was laid horizontally, the arms fastened to the sides, in a funereal grotto. In Easter island the inhabitants buried their dead under the large flat stones which bore the celebrated colossal statues.

The New Zealanders also buried their dead, but not until three days after their decease, and when they had well rubbed the body with oil and placed it in a crouching position. They then put a pile of stones over the tomb, and on the stones they laid a few provisions.

A singular fact observed by Cook in New Zealand shows us how carefully we ought to interpret the ethnical similitudes, even when they are of a special kind. The New Zealanders had erected a cross upon a tombstone and decorated it with feathers, very similar to the cross ordinarily used by Catholics.

We have seen that the inhabitants of the Gambier islands place their dead in the natural grottoes. At Tonga, as in other places in the world, over the bodies of men of distinction is raised an artificial grotto, or sort of dolmen, formed by large slabs of stone.

In many countries the grief or the regret felt by the survivors after the loss of a relation or a friend has given rise to the idea, not only of erecting a funereal memento but also of engraving or painting certain emblems. The Polynesians were not an artistic people, but they had practised that kind of funereal art to some small extent. In a mortuary *morai* at Tahiti Cook saw some strips of wood on which had been cut forms of men and of animals, notably the figure of a cock, and to give the cock a more realistic appearance it had been painted red and yellow. In another place the people had cut a small figure out of stone. The large statues in Easter island were probably formed with the same idea, and whether they were or not cut by the present race of people, and with their rough tools made out of volcanic glass, of which M. Pinart has recently presented some specimens to the Anthropological Society, we may nevertheless observe that sculpture in stone is an art almost unknown to the Polynesians.

In certain archipelagoes, the custom of wearing mourning, even for a long period, was commonly practised. At Tahiti the women used to wear a head-dress of feathers of some special colour, and they would also cover their face with a veil. The women who had washed, anointed, and prepared the corpse were subjected to a most rigorous taboo, lasting for five months when it was the body of a chief. During all this time they were not allowed to touch food with their fingers; everything they ate had to be put into their mouths. And everywhere, as a sign of mourning, on the funeral grounds people used to plant casuarinas trees, bearing no leaves—a gloomy-looking plant something like the shavegrass. Singing and a show of lamentation were held imperative. In the Marquesas islands, if the deceased were a man, the lamentation of the women was accompanied by the most curious sort of mimicry. The widow and a few young girls used to jump in cadence round about the corpse, performing lascivious gestures; and then bending over the dead man so as to examine him, they would cry: "He did not move, he is quite still. Alas! he is no longer in this world." But funereal grief in Polynesia was shown not only in moaning and in ceremonies; lacerations, mutilations, and sacrifices were also held necessary. Moral grief was not considered sufficient; there must also have been some shedding of blood.

These sanguinary customs were general all over Polynesia, and they were very uniform. Nearly everywhere the people used to tear their faces with a shark's tooth or with a sharp stone. But at Tongatabou funereal grief, real or affected, was shown in a more cruel manner than at any other place. The marks of sorrow were shown in exact measure according to the social position of the deceased. After the death of a chief the people used to shave their hair, to lacerate their face and their body; they used to torture themselves by burning their skin, by driving sharp points into their thighs, their sides, their cheeks; they broke the bones of their little finger, and also those of the ring finger, as was the custom in Australia and in other places.

However violent may have been the grief of the Polynesians, it was not always sincere, for at Noukahiva Porter saw a widow, whose husband had been devoured by a shark, prostitute herself to some American sailors, though at the time her chest, her neck, and her arms were all covered with gaping funereal wounds. Hypocrisy is not a vice peculiar only to civilised races; but we may recollect that the Polynesians are endowed with most infantine fickleness of disposition, and that the feeling of shame is quite unknown to them.

In New Zealand one outburst of grief was not considered sufficient. Sometimes they would dig up the dead at certain periods of the year, and go through a fresh bout of weeping, which was also accompanied with fresh lacerations, cut deeply, and inflicted quite voluntarily.

In addition to this manifestation of grief by these torturings and lacerations, a desire was often felt not to allow the deceased to go alone into the world beyond; and to gratify this holy wish other human beings were sacrificed.

Doubtless the New Zealander's notion of morality did not always compel a woman to outlive her husband; but if she spontaneously hung herself on a tree, her conduct was thought to be worthy of great praise. In certain tribes this moral obligation was held to be a strict duty; and at the death of a chief it was customary to strangle all his widows over his tomb. Customs quite as barbarous as these were common in the Friendly islands and in other places, and often concurrently with practices of a totally opposite nature. The New Zealanders, who regarded their dead with such feelings of reverence, would sometimes eat their own relations who were killed in battle. Children have been known to eat their mother, and fathers their children. We may also add that these same islanders, when one of their chiefs died, after they had piously and ceremoniously moaned over him, would rob and steal everything that he possessed.

Slaves were often sacrificed upon the tomb of the deceased. A New Zealand mother, whose child had been drowned, insisted that a female slave should be put to death, so that she might accompany and take care of her little one on his voyage to the country beyond the grave—to the *Reinga*.

In the Marquesas islands they used sometimes to sacrifice two servants, two *kikinas*, one of whom was charged to carry the belt of the departed and the other the pig's head served at the funeral repast. This precaution was important, for in Noukahiva the guardian of the tomb beyond would have mercilessly driven back with a shower of stones any new arrivals if they did not present themselves becomingly according to the prescribed rites. Sometimes the funeral victims, men or women, were carried away by men of a neighbouring tribe who had lain in ambush for them.

Human sacrifices were also ordained by law in the Sandwich islands when the deceased had been an important personage, and funereal suicide was also customary. On the death of Tamehameha several persons who had been warmly attached to him killed themselves, so as to accompany him into the other world, and that was without prejudice to the obligatory victims and to voluntary mutilations. In addition, in after years, on the anniversary of this Hawaian Napoleon, people used to commemorate the sad event by pulling out an incisor tooth. Nomalianna, the wife of Tamehameha, had caused to be tattooed upon her right arm words which in her language signified "Our good King Tamehameha died on the 6th of May, 1819." Going to a still greater extremity, some few of the islanders caused the same operation to be performed upon their tongues. At the funeral banquet of this great prince the number of pigs obliged to be killed was so great that, after the event, pork was a meat scarcely to be met with in the island. Though it may perhaps have been excessive, this admiration of the Hawaians for their conquerors will hardly cause much surprise in Europe except to a few morose-minded persons.

We will here conclude our short enumeration of the funereal customs in Polynesia, but before pursuing the subject any further it may be well to remark that these funebrial rites, more or less bloody, rarely take place until after the decease of the men whom it is intended to honour, and that this is also the case among nearly all savage races. On the death of the husbands the wives are often sacrificed, but we do not know of even a single reciprocal instance. As regards feminine obsequies, the majority of travellers maintain an eloquent silence. We may conclude, therefore, that the woman is nearly everywhere buried without much ceremony. This is another particularity to be added to the already large number of facts showing the disdain in which woman is generally held among all the primitive races of men.

V.

Funereal Rites in America.

As regards funeral rites the vast American continent may be divided into three large districts, northern, central, and southern. These districts will no doubt be divided roughly, they will often notch one into the other, and in each, tribes will be found who have their own special customs. But we may say generally that burial is the custom in South America, and that there the corpse is often set in a seated or in a curled-up position. In the central regions, in Mexico, the dead who when alive were men of distinction, were subjected to cremation—a practice unknown for the most part in Melanesia, in Africa, in Polynesia, and in South America. The Northern American usually neither burns nor buries his dead; he places them, according to the Polynesian custom, upon a sort of platform, and then afterwards gathers together the bones more or less carefully. The instances which we propose to give will show the general characteristics peculiar to each country.

The Patagonians, the Araucanians, the Pampas, the Puelches, and the Charruas, all bury their dead generally in a curled-up position; and they are careful to put by the side of the deceased his clothes, his ornaments, his arms, his arrows, sometimes painted over with red, and some provisions. They will often burn the rest of the objects which may have belonged to him, and kill upon his tomb the domestic animals of which he had made use. The Chiquitos, the Araucanians, and the Patagonians, have still great difficulty in believing in natural death. The death of their chief is often attributed to spiteful ill-will; hence the cause of vengeance, murder, and interminable wars between the tribes.

The Charruas and the tribes of the Grand Chaco do not confine themselves in their funereal ceremonies to the sacrifice of domestic animals; but, as in Polynesia, the relations of the dead man wound themselves upon the arms, upon the sides, and upon the chest; the women break two bones of their finger, and they ought in addition to impose upon themselves severe fasting. We may remark that the funereal amputation of a finger is also a Polynesian custom.

With the Guaranis the dead man is buried in a sitting position, but the body has first been put into a large funereal vase. The deceased is sometimes buried in his own house, and every morning for a long time afterwards his family extol his virtues, pouring forth lamentations as they do so.

Funereal customs very similar in reality were practised by the ancient Peruvians, especially towards the south, in the kingdom of Cusco. The dead man was buried clothed and seated, surrounded by his own familiar objects; some provisions were also placed by his side. He was buried either in a cavern adjoining the house or else in the public cemetery. Putrefaction was avoided as far as was possible, either by drying the body, like the Polynesians, or else in using resin, as was the custom with the ancient Egyptians. They often drew out the entrails; this was also a Polynesian custom. Human sacrifices, voluntary or other, were frequent in Peru on the death of the grand personages. Balboa reports that at the death of the Inca Yupanqui many of the courtiers were sacrificed. At the death of Huayna-Capac more than a thousand persons voluntarily put themselves to death.

Cremation was practised by the ancient Mexicans; but the custom was far from being general—it was a privilege reserved only to men of distinction. The body, dressed in this or in that manner, according to the divinity which the man had worshipped, was first strewed over with pieces of paper covered with hieroglyphs serving as protecting talismans. Then, after incineration, the cinders, collected in a vase, were preserved in the house or else buried, either in the open field or in a consecrated building. The remains of the kings and of the high personages were ordinarily placed in the towers of the temples. The dead who were not burned were placed in deep ditches constructed with masonry; they were placed on low seats called *icpallis*, and the people took care to set beside them the implements belonging to their profession. By the side of a soldier they would put a shield and a sword; by the side of a woman a shuttle and a spindle. There, as in so many other countries, the personages of rank could not go alone into the country beyond the grave, and slaves more or less numerous were sacrificed upon their tombs. In Zacatecas people were persuaded that during a certain number of years the shades of the dead came back to visit their family; and therefore upon a given day in every year they prepared for them a banquet at which all the relations were present; but they remained silent, immovable, with downcast eyes, so as not to disturb the repose of the invisible guests.

In Columbia the funeral rites are now beginning to vary a little. Some people, the Tracos for instance, still purposely bury their dead, laying down their arms beside them, and also taking care to envelop them in a thick bed of banana leaves, so as to prevent the contact of the earth; but the other tribes on the bank of the Orinoco have very different customs. Their desire is to have the skeleton well prepared as soon as possible, they therefore tie a strong rope round the body and throw it into the river. In one or two days the fish do what is expected of them: they eat off all the flesh. Then the bones are separated one from the other, they are artistically arranged in a basket suspended from the roof of the house. These are funebrial rites of an aërial kind, much practised in the northern parts of America. Among the Caribs the funeral basket has already become customary. But they do not collect the bones until after the decomposition of the corpse, which at first remains, for a greater or less length of time, stretched in a hammock under the charge of the wives of the deceased; with the exception of one wife, who is often sacrificed if the dead body be that of a chief.

As we have already remarked, there is no accurately drawn line of demarcation between the different American districts where this or that funeral rite is practised. The custom of cremation does not appear to have been adopted upon a large scale anywhere but in Mexico; but nowadays the Roucouyennes, Indians of Guiana, often burn their dead after having first painted and decorated them.

Cremation was also in use here and there in North America, especially towards the south. In many of the mounds in the Southern States we find funereal urns containing beds of wooden charcoal. Quite recently, the Shoshoni used to burn their dead, and used to put with them all the objects that belonged to them. The Indians on the bay of San Francisco used to do the same. Incineration of the dead is also a custom with the Tahkali; but they proceed after a very ceremonious fashion, in presence of their doctor of medicine, who by his gesticulations and contortions is supposed to make the soul or the shade of the departed pass into the body of one of the assistants, who then becomes the inheritor of the name and of the rank of the deceased.

In Sitka island, cremation and exposition of the body without burial are combined. The corpses are burnt; then the ashes collected in boxes are deposited in small funeral buildings.

The placing of the corpse upon a scaffolding, and then exposing it to the open air, is a custom much practised among the Red Skins. The Assineboins and many other tribes expose their dead in the same way, either on the branches of trees or on a funereal scaffolding high enough to keep them out of the reach of four-footed wild beasts. After a stated time, the bones are collected and piled up in a depository specially kept for the purpose; and in case of emigration the tribe will, as far as possible, carry with them the remains of their dead, or else hide them in a cavern, or bury them in the ground. In the northern part of America, in New Albion, the corpses are deposited, with the bones and broken splinters, in the boats, which are afterwards suspended from the trees at about ten or twelve feet from the ground; the boats are covered over with a large plank of wood. The corpses of children are put into baskets, which are also hung from the trees, and into these baskets are often put little square boxes containing some alimentary substance. As among nearly every primitive race, these people think that in the shade of the dead still remain all the wants of the living man. The Red Skin thinks that the country beyond the grave is altogether similar to his earthly dwelling-place. His imagination leads him to believe it to be a promised land full of buffaloes and fine roebucks, where there are trees and flowers and perpetual spring; or else it is a deserted icebound region, in which man is in constant suffering from cold, hunger, and want of every kind. For the dreams of man can only be a reflection, either more beautiful or more gloomy, of his actual life; and, as we shall see, all over the earth and among all races of men, the future life has always been imagined to be an entire imitation of our life here upon earth.

In the extreme north of America, where large trees will not grow, the people often bury their dead on the tops of the hills, erecting a little mound of sand on the tomb. This is the custom at Unalaska. Still more to the north, among the real Esquimaux, they place their dead either under stones or else in the snow. Sometimes they collect the skulls to hang them round about their dwelling-houses, together with the heads of the bears and seals. It must be admitted that they do not care more for the one sort of skull than for the other, for the Esquimaux is not a sensitive man, nor is he superstitious. In Sitka island, among the Koloshes, two slaves are usually put to death when their master dies, to wait upon him in the other world.

Our review of the funereal rites in America is now terminated. We may observe, but without insisting too strongly, that there are certain similarities between the American customs and those in Polynesia; and this fact, taken in connection with others, may be urged in support of the theory as to the American origin of the Polynesian islanders. We may also remark that the custom of cremation is very widely spread in Central America. We shall also very frequently find this custom upon the Asiatic-European continent, of which we are now going to speak; we shall also have to question ourselves as to its signification.

VI.

Funereal Rites in Asia and in Malay.

The funeral customs in the Javanese archipelago are of the most various kind. Formerly, the Battas used piously and ceremoniously to eat their old relations, as was also the practice, according to Marco Polo, of certain people in India, and also of the Derbices of Europe, according to Herodotus. At present, inhumation is the customary mode of burial in the Javanese archipelago. The cemeteries are situated on a hill and shaded over with funeral trees (*plumeria acutifolia*). Formerly the dead were left at the foot of a tree in the forest, or thrown into the water, or burnt, together with one or more women, who had been previously slaughtered, sometimes by blows from the kriss. This was plainly an imitation of the Hindoo suttee. Or else the people laid the body upon planks, as is the Polynesian and the American custom. The corpse, laid in a bier, is lifted up on to posts. The natives of Poulo-Nias then arrange around the whole construction creeping plants, which soon form a shroud of green leaves round about the coffin. The Kayans of Borneo do very much the same; but they first keep the corpse for several days in their houses, offering to it food, and placing lights all round about it, while the women are weeping and mourning. With the dead man they bury everything that he possessed, and also very often the corpse of a slave killed specially for the occasion (O. Beccari), for the shade of the departed must be suitably accompanied into the other world. It is the same spiritualistic idea that urges the Borneo Dyaks to be so eager in their desire for procuring men's heads. The Dyaks firmly believe that each decapitation represents the acquisition of a slave in the life to come. They wear mourning for one of their deceased relatives until they have succeeded in procuring a man's head; that is to say, sending a slave to the departed. When a father has lost his child, he kills the first man he meets as he goes out of his house; this is to him an act of duty. No young man can marry until he has brought home a head. To lay traps for people, so as to decapitate them, was, and is still, among the Dyaks a national custom which they imagine to be very praiseworthy. Wallace says they are a good sort of people. We do not wish to doubt it; but we may observe that, with the Dyaks as with other people, the feeling of duty may lead them into atrocious crimes, when that same feeling is not enlightened by humane intelligence.

Among the Mongolians and the Mongoloid people of continental Asia we find a great uniformity of custom in the funereal rites. It would appear that all the ethnical branches of this large race began by leaving their dead, either letting them lie in the fields, or else throwing them into the sea or into the river. Then they bethought themselves of burning the people of rank; and at last, as is now the custom in China, cremation has been followed by burial. Here and there we find also, at least in a symbolical form, funereal sacrifices either of persons or of objects.

Incineration is a long and costly process, which can nowhere be within the means of the common people. Therefore the abandonment of their dead is common enough among the poorer classes of the Mongolians and the Mongoloid races. The Siamese common people throw their dead into the water without performing any ceremony.

The Thibetans in the same way allow their dead to be devoured by the crows and by the vultures. So do the majority of the Mongolians, consulting their lamas beforehand as to the direction in which the corpse ought to be laid. The dead children are enveloped in sacks of leather with a suitable provision of butter and other food, and then left on the wayside. The notion is that the young shade, prematurely taken from its body, may thus find the chance to reincarnate itself in the breast of one of the women passing along the road.

It would seem that among the yellow races the abandonment of their dead was followed by cremation. The two customs were sometimes joined together; for instance, in Siam, before burying a corpse, they would cut off the fleshy parts and leave them to the jackals and the vultures. Many Tartars who are ignorant of cremation cut up the corpses of their dead and hand them over to the dogs. But in Asiatic Mongolia cremation is frequently practised in the case of men of rank or position. In Siam the bodies of the chiefs are burned at a great expense upon a fastuously-arranged funeral pile, erected at great cost both of time and of money, and the corpse is also carefully embalmed beforehand. The people of low condition are also occasionally embalmed, but always at a respectful distance from their distinguished superiors. As the poor people cannot always embalm their dead, they cut up into little bits all the soft parts, which they afterwards, following the practice of the Tartars, throw to the dogs and to the vultures. The rich Mongolians hold it to be a point of honour to burn the corpses of their parents in a furnace built for the purpose; and during the ceremony the paternosters are read by the lamas. The Thibetans also incinerate their corpses in a handsome bier, music being played as an accompaniment to the ceremony. The service is conducted by priests, who naturally enough expect to be paid for their work. The Chinese now seem to have given up the practice of cremation; but it was their custom at the time of Marco Polo's travels, at least in certain parts of the empire.

Inhumation is also much practised in Mongolian Asia. The Siamese bury only the children who have died before they have cut their teeth, and women who are big with child; and these latter, after a few months, are dug up to be burnt. The Burmese sometimes cremate and sometimes exhume their bodies. The nomad Mongolians bury the greater part of their corpses which are not left deserted. The Mongolian kings and princes are sometimes buried at a great expense in a large mortuary cavern, and also with a large sacrifice of human life attending their burial. A large building adorned with Buddhist statues is placed in front of the cavern of these great earthly personages; the royal clothes are spread out; there is also laid there a great display of precious stones, and large sums of money both in gold and in silver. Round about the dead body of the great man, placed in the attitude of meditation common among the Buddhists, are the children poisoned specially for the occasion, one holding a pipe, another the fan, and others various objects of the deceased.

We know that the Chinese attach much importance to the ceremonies accompanying a burial, and that the inhabitants of the Celestial Empire have none of the puerile feelings of terror of death which are ordinarily felt by us Europeans. It is a most pleasing thing for an affectionate Chinaman to be able to give a handsome burial to his old parents. And on their side the father and mother are quite delighted with the present; for the majority of the Chinese look upon death with the utmost coolness, and are quite exempt from our sombre creed as to the future life. The Chinese think that the funebrial ceremonies cannot be too magnificent, and families will often seriously cripple themselves in their means to bury a dead man.

The funereal sacrifices still practised among certain Tartar families of high rank have long formed part of the traditions and customs of the Mongolian race. At the time of Marco Polo, when a Tartar nobleman was being carried to his last resting-place, the assistants used to put to death everyone they met on the road as the hearse was going on its way. They said simply: "Go and join your lord in the other world." They used also to kill the best horse of the deceased so that he might in the next life have that animal to ride.

The imagination of human creatures in their primitive state has nearly everywhere led them to believe that life continued after death in a condition very similar to that of their then actual existence. Nothing then is more natural than the idea of causing the material and ethereal shadow of the departed to be followed by other shadows equally ethereal, by those of whom he had been fond and who had served him during his lifetime, by his domestic animals, his arms, and other things which he had used during his visible existence. This unhappy belief has assuredly cost the life of millions of human beings; and we still find scattered remains of the same custom in every human society.

The Chinese, the least religious of all nations, have had, like other races, their phase of funereal superstitions. We have seen that in the time of Marco Polo bloody sacrifices were still held in certain provinces of China, but even then they had in other districts preserved only the symbols. Horses ready saddled, armoury, golden flags, etc., were economically replaced by imitations cut out in parchment, which were burnt together with the corpse.

At the same time and in the same country, they used occasionally to keep the corpse shut up in the bier for six months, and always every day offering to it some food to eat. And even now the Chinese hold their funereal repast, at which the dead man is offered food as though he were alive. This is only a symbolical ceremony, preserved on account of the extreme respect which the Chinese profess for their relations. After the death of their father or their mother they are considered bound to wear mourning for three years. But as during this time of mourning the public functionaries are bound to quit their employment, the prescribed duration of the term has been reduced to twenty-seven months.

The custom of funereal food is very general all over the earth, and it has evidently sprung from the notion that life has not been interrupted by the slight accident of death. In Bhotan the deceased was kept for three days before being placed on the funeral pile, and during this time the priests offered him food daily. The inhabitants of Russian Finland have nowadays similar customs. In Siberia the Ostiaks make little figures out of carved wood to represent the dead bodies of men of distinction, and during their commemorative funeral repasts they conscientiously put some aside for the funebrial doll. The Ostiak widows also make little figures of the same kind representing the bodies of their deceased husbands; they take these images to bed with them, and the relations of the deceased offer to them food.

Elsewhere, in Siam for instance, and in Tartary, people collect the ashes of the burnt corpses and make of them a paste to mould into small Buddhist images or into disks, which they afterwards put on to the top of a pyramid. The corpse thus transformed becomes the lares and penates, and they are carefully kept, evidently as the supposed dwelling-place of the shade of the departed.

VII.

Funereal Rites among the White Races.

The majority of the funereal rites which we have described have existed, or still continue, among the Indo-European races.

The custom of allowing the corpses to be devoured by wild beasts, or, in other words, of voluntary abandonment, was also followed by many people in ancient times. In Hyrcania, Cicero says, dogs were kept specially for the purpose of eating the dead. The Bactrians also had their scavenger dogs, who devoured not only the corpses but also the bodies of people enfeebled by age or sickness. The Hindoos on the banks of the Ganges threw their dead into the sacred river, and allowed the fish to perform the office of the undertakers. The Callatians in ancient India used to eat their deceased parents, and some of them wept loudly when Darius asked them at what price they would consent to burn their dead. The Parsees of Bombay even now leave their corpses to the vultures, but they take much care to spy out which eye will be first extracted, for this particularity enables them to infer whether the shade or the soul of the defunct will be happy or unhappy in the other world.

Nearly all the Semites of our own times bury their dead. This was also the custom with the ancient Persians, whose present descendants push their love of inhumation to such excess that they do not hesitate to pay large ransoms to the Turkoman plunderers for the body, or even for a portion of the body, of their relations, so as to bury the precious remains not yet polluted by the unfaithful.

As we have already seen, the Aryan Vedahs bury their dead, recommending them to the care of the earth. In India at the present day the funereal rites of distinguished men are performed by cremation, and that is the custom as far as Nepaul. However, it is not very long since that many devout people went to Benares to drown themselves, so as to be more certain of assuring their salvation.

We know also that the ancient Germans used often to burn their dead, and prehistoric archæology teaches us that our ancestors in the stone age used sometimes to bury their dead in caverns or in mounds in the earth, and sometimes they used to burn them.

Among the white races, as among the others, funereal sacrifices were very largely practised. In spite of the silence of the Menu Code, which does not prescribe human sacrifices at the funerals, the custom to burn widows over the pile of their husbands became very general among the Brahmins, and it has persisted down to our own times. It was more or less the custom with all the people of antiquity, the Persians, the Greeks, and the Romans, and others, to sacrifice human beings at their funeral rites. The Germans, too, with their dead used to burn the horses of the deceased and his armour, and did not allow their great personages to go into the next life without a suitable escort of slaughtered prisoners.

As regards these sanguinary customs there is a great analogy of ideas all over the earth. We find nearly everywhere the same infatuation, and the same cruelty, among all races, from the ancient Germans to the inhabitants of Dahomey, who, not satisfied with massacring hundreds of women, eunuchs, singers, soldiers, and others, when their king dies, despatch periodically to the invisible kingdom of Dahomey fresh servants charged to carry messages to the king who has left them. They do this quite naturally and simply, to prove to the deceased the filial affection of his successor.

The ancient Greeks and Romans also used to believe firmly, as do now many savage races, that the shades of the dead really partook of the food offered to them by the survivors. Lucian makes a widower relate that his wife came to him to ask for a sandal that she had forgotten to burn with his body and with his other ornaments. Among the Greeks or Romans, as among many other people, the spirits of the dead were often looked upon as dangerous and wicked beings. It was especially the spirits of those who were deprived of sepulture, or those who died of a violent death, that were animated with these perverse instincts. The doctrines of Epicurus fortunately relieved the most sensible minded from these chimerical tortures. There are a certain number of Latin epitaphs which tell us plainly enough that, as regards the personality of the individual, death is the end of all things, that it is everlasting peace. But these too rational doctrines were believed in only by a very small minority; the masses still troubled themselves with their notion of Charon and hell, thus preparing the way for Christianity, which brought to a paroxysm the fear of post-mortem torment.

VIII.

From this long enumeration a few general facts may be deduced. Man is so intoxicated with his feverish desire for life, that he cannot bring home to himself any idea of death. Quite at first he does not believe in natural death, but always imagines some malicious act to be the cause of the decease. Also, he generally looks upon death merely as a metamorphosis, and pictures to himself another life based upon that life which is familiar to him. Hence spring all the funereal rites which, in spite of their differences in point of detail, may be classed under a small number of heads.

At the outset of the social life man thinks no more of his dead than do the majority of animals; he abandons them without any scruple to the beasts or birds of prey. And sometimes this form, or rather this absence of funereal rites, is seen in societies that are fairly civilised; but then superstition comes in, ceremonial forms are practised, the pulling to pieces of the corpse is done only by certain animals, or it is held obligatory that the body should be eaten by the fish in certain rivers. In other countries the bodies are exposed on planks upon the trees. But in all these cases, the abandonment of the human corpse is no longer bestial.

Other people bury their dead, either in grottoes or in tombs, often built after the model of the habitations of the living, or else they bury them in the ground in a fertile mound, from which all animal life, directly or indirectly, feeds itself. Afterwards comes incineration, which, of all funereal customs, seems to be the most luxurious and the most coveted.

Whatever be the custom adopted, man always furnishes the

dead body with his arms and necessary food, under the hypothesis of a continuation of life after death. And like the human body, these objects have also their duplicate, their soul, their shade, which is destined to subsist and to serve their purpose in the kingdom of the dead. No doubt incineration was practised in so many different countries under the idea of emancipating more quickly and more completely these invisible effigies from the dead bodies and also from the inanimate objects.

In all these customs we may trace a sort of logical sequence of ideas; the people were loath to see the dead man depart without providing him with means of nourishment and also with weapons of defence, they therefore thought it wise and necessary not to allow him to go alone on the perilous journey beyond the grave. They sacrificed on his tomb, or threw on to his funeral pile, his favourite domestic animals, his slaves, and very often, if the deceased was a man, his wife or some of his wives. Hence were innumerable murders, seas of blood were flowing everywhere all over the earth, during these millenary epochs which our spiritual teachers and our moralists always forget when they speak in the most grand and glorious terms of the all-hallowed notion of the immortality of the soul. We shall now see that many people believe only, not in the immortality of the soul, but in a later mortality, in a temporary continuation of life after death.

CHAPTER XIV.

RELIGION IN GENERAL.

For many people, and some even of the best educated, the word "religion" carries with it a sound very different to that of any other word in the language. It is a magic word which instantly awakens a vast region of affective impressionability. Nothing is more natural, for there are connected with it a whole world of memories, a very large collection of ideas which have been acquired or which have been born in us hereditarily. There are few of us who, in our infancy or in our early youth, have not undergone more or less—and often more than less—the training of a religious education. How many of the most confused impressions of our childhood are not due to the whole scenery of catholicism, a coarsely-imagined scenery perhaps, but therefore the better adapted to leave its mark upon the mind of the child, so like, in this respect, to the mind of the savage? During these long early years, in which our moral nature is being formed, for well or for ill, our memory as yet unspotted, our frank and credulous nature has been impregnated, so to say, with sacred stories which have appeared to us as wonderful as fairy tales. And during this period of development, when every day some new notion as to our future personality was taking root in our brain, we were taught to imagine the universe peopled with mythic personages, until we did at last believe in their actual existence. At the same time we were told terrifying tales of the future life, we were made to believe in everlasting flames in which there were horned demons always grinning at us. After awhile, as we grew older, these sensual features in catholicism, against which our fuller intelligence would probably have rebelled, were less strongly urged; but still taking advantage of our affective nature, our teachers tried to connect indissolubly our highest sentiments, and our most noble aspirations, with these mystical doctrines; they showed us that religion is the necessary appanage of man, the mainstay of his morality, and that therein lies his glory and his strength. Finally, when it became necessary to appeal to our reason, the metaphysicians in office, continuing the work of our spiritual advisers, employed all their scholastic lore to prove to us the dualism of our being, the spirituality of our conscient life, the existence of an immaterial God, reduced to be nothing more than an unfathomable treasury of a certain ideal kind. We may also add that the spirit of our ancestors, reappearing in us to a greater or less extent, has been ground into powder in the same way, whence a sort of religious instinct remains impressed in the cells of our brain.

From all this it results that there exists in the brain of many intelligent persons—sometimes the highest natures, from a moral point of view—a special and unconquerable domain containing their religionary ideas. To these people the word religion has the effect of enchantment, they are disposed to respect everything which bears the impress of this mark. We think we have shown elsewhere that a separate religionary faculty does not exist in the human brain, though no doubt we may easily discover under this somewhat pompous expression groups of affective and intellectual acts which do not essentially differ from other conscious or cerebral actions.

We have now to consider the religious beliefs among the different human races. No study is more instructive or better adapted to dispel the halo attaching to the word "religion." A scholarly interpretation of the history of the principal Indo-European religions may still deceive us; but their formation and development is rendered very much clearer if read with the help of the formation and development of the primitive religions; and if this examination be made conscientiously, without preconceived ideas, and without blinking our eyes to actual facts, it will carry with it a death-blow to all supernatural ideas.

In this interesting study we shall have to be very brief, to confine ourselves to the most typical facts, which indeed are less numerous than might at first appear; for man's religious conceptions, more than his worldly thoughts, have nearly everywhere shown a great deal of uniformity. To abridge, as far as possible, and collect together facts of an analogous kind, we will group the so-called religious ideas under three heads:

I. Future life, and man's different conceptions regarding it.

II. The Gods.

III. The forms of worship and the priesthood.

When this threefold inquiry is completed we shall then have finished the present portion of our work.

CHAPTER XV.

OF THE FUTURE LIFE.

THE feeble intelligence of the primitive man is generally unable to conceive the idea of absolute death, the annihilation of his own personality. During this stage of mental infancy man does not often believe in a natural death. As we have already remarked, he thinks that the decease of his relatives or of his friends must be attributed to sorcery or to some malicious act. Every death is to him an assassination, and he often thinks it his duty to punish the suspected assassin. But, whatever may be his ideas as regards the causes of death, he absolutely refuses to consider it as the end of individual existence; he generally regards it as a lengthened sleep, during which the shade, the spirit, etc., quits the body, as it seems to do in a dream, to continue somewhere else an invisible existence, not eternally, but for a greater or less length of time. All man's ideas coming to him from experience, and his imagination being nothing more than memory capriciously broken by his intelligence, his future life, when he believes in it, is invariably planned in imitation of his life here upon earth; so that given the kind of life he leads here, and the conditions of the ethnical group to which he belongs, one may easily infer what he will believe as to the world beyond. This general view is applicable to all human races, commencing from the Melanesians, with whom we will begin our study.

I.

Future Life according to the Melanesians.

We have just indicated how the belief in a future life is first created; but this belief, though very common, is not necessary, nor is it always born in man. It is foreign to certain ethnical groups, and especially to many indiyiduals

In Tasmania men's minds were much divided on this point. According to the missionary Clark, many Tasmanians, especially those in the western part of the island, had not the least idea of a future life. They used to say, "they died like the kangaroos." Others thought that they were destined after death to go and live in the stars, as in an island, where they would meet their ancestors and be turned into white men.

According to Davis, this belief in a white reincarnation came directly from anthropophagy; it was suggested to the islanders by the fact that the Tasmanian flesh when skinned and roasted takes a whitish hue.

The Australian, who has never been able to bring home to himself the idea of a natural death, also believes that he will continue to live in the life beyond the grave, and that he will then become a white man, and in this post-mortem existence he will enjoy what he now considers to be his supreme happiness: being able to smoke as much tobacco as he pleases. Such is, at least, the belief of the Australian tribes near Cape York.

The Papuans believe in a future life, but of different kinds, varying according to the islands and even according to the tribes. Some of the Papuans in New Guinea imagine they will reappear as certain of the animals in their own island. The cassowary and the emu are the most remarkable animals that they know of; they have lodged in them the shades of their ancestors, and consequently the people abstain from eating them.

Among the Fijians the mythological imagination is very strongly developed. They often endow man with two spirits; but they do not stop there, for they hold that every object, animate and inanimate, possesses a shade, a spirit, a soul, an invisible emanation, which will go with the sojourn of the departed into *Bolotou*. An axe that one breaks, a house that one pulls down, a cocoa-nut that one cracks, have all got their double existence, and their soul will find its place in *Bolotou*. For man, this double existence may operate during his life, and sometimes in his sleep it happens that the spirit of the Fijian quits the body and goes to torment other persons who are also sleeping. Dreams have everywhere had considerable influence upon the formation of religious ideas, and especially upon the belief of particular persons.

In Fiji the soul is regarded quite as a material object, subject to the same laws as the living body, and having to struggle hard to gain the paradisiacal *Bolotoo*. After death the soul of the Fijian goes first of all to the eastern extremity of *Vanna Levou*; and during this voyage it is most important that it should hold in its hand the soul of the tooth of a spermaceti whale, for this tooth ought to grow into a tree, and the soul of the poor human creature climb up to the top of this tree. When it is perched up there it is obliged to await the arrival of the souls of his wives, who have been religiously strangled to serve as escort to their master. Unless all these and many other precautions are taken, the soul of the deceased Fijian remains mournfully seated upon the fatal bough until the arrival of the god *Ravuyalo*, who kills him once and for all, and leaves him without means of escape.

The New Caledonian is not less religious after his manner than the Fijian. For him there is no such thing as hell; he believes only in a paradise into which all men of his race will go without distinction, without any difference of moral worth. This paradise is in a forest, in some neighbouring island, or perhaps under the sea. It is a place full of delights, of eatable fruits, where there is perpetual feasting and dancing. Man there becomes a superior being, especially if he is a chief. He can then avenge himself upon his enemies; he can heap wealth upon his friends; he can fertilise the fields or make them sterile; he can in battle give assistance to one side and weaken the other, or turn the victory which way he pleases. In short, he possesses the faculty for gratifying every desire which was denied to him in this world below.

In a coarse way the religious ideal of the New Caledonian has a strong hold upon him. The concentration of the thoughts upon this elementary mythology will often put a man into a trance, into a fit of religious delirium, during which the inspired creature has visions, sees the shades of the dead, partakes of all their feasts, etc. If he is a Christian it is the Catholic hell which appears before him; he is a prey to a sort of demonomania which spreads itself. This happens also sometimes in Europe, merely from the force of example.

II.

Future Life according to the Negroes in Africa.

As we have seen, there is a certain foundation of religion among the New Caledonians, and the Catholic missionaries may hope to gain a few neophytes. The African negroes, in this respect, are less imaginative, specially in the South of Africa. According to Levaillant, Thompson, and Campbell, the Hottentots have no idea of a future life, nor of a god, nor even of requiting gods. Campbell says, "they think that they die absolutely, just as beasts."

The Kafirs, far superior to the Hottentots, do believe in a certain survival after death. According to them, man dies leaving after him a sort of smoke, very like the shadow which his living body will always cast before it, a sort of spirit having no corpse of its own. To make a guardian angel after their own idea they will often choose the spirit of a chief or of a friend; they will invoke his assistance in critical moments, will thank him for services done by offering him a portion of an ox they have killed, or some game, or some corn. They believe this shade to be possessed of every quality, and to have ready at hand all the wants of the man to which it formerly belonged. These people, like other primitive races, scarcely believe in natural death. For them there are only three ways by which death can happen—hunger, violence, or magic. And even the death of an old man is often the cause of murders and massacres, for they always consider themselves bound to avenge it.

In the middle regions of Africa the belief in a future life is either very small, or is absolutely wanting. The negroes in Gaboon have a horrible fear of death; they cannot bring themselves to believe that it comes naturally. How can a man, they think, die now, who was perfectly well a fortnight ago, unless some sorcerer has had a hand in the matter? Schweinfurth reports the same prejudice to exist in the basin of the Upper Nile. The negroes in this region think, as do the Kafirs, that a man cannot die except from hunger, sorcery, or violence. Woe, therefore, to the old people who, after the decease of a member of the tribe, are found holding in their hands herbs or suspected roots! Were they the father or the mother of the dead, their death is certain.

Many of the tribes in Equatorial Africa have no idea of any survival after death. After the death of a friend or a relation, the Eastern Africans sing plaintively: "Everything is finished, and for evermore!" In Gaboon, when there was a feast held, the women used to perform very lascivious dances and would sing: "While we are alive and well let us be happy; let us sing and dance and laugh. For after life death comes, the body rots, the worm eats it, and everything is finished." "Everything is finished!" they all cry in a melancholy tone, when one of their own family dies. Some of them believe that as he dies man leaves a shadow behind him, but only for a short time. The shade or the mind of the deceased remains, they think, close to the grave where the corpse has been buried. This shadow is generally evil minded, and they often fly away from it in changing their place of abode. The manes last as long as they keep their recollection of the departed. There is no need, for instance, for them to trouble themselves about the spirit of their great-grandfather; he is annihilated.

Schweinfurth tells us that the Bongos on the Upper Nile have not the smallest notion of any future life, "no more," he says, "than of the existence of the ocean."

The Bambarrans, the Mandingos, more civilised races, and among whom Islamism has penetrated more or less, believe in a resurrection after death. The Bambarrans pray for the departed spirits of their ancestors. The Mandingos talk of a future life, they are led to aspire to it when surrounded by troubles and hardships; but they declare that they have no sort of idea as to what it may be. In Congo men's ideas are more strongly formed, for there a son will kill his mother, so that, transformed into a powerful spirit, she will give him aid and assistance. This is one of the many misfortunes to which belief in a future life has given rise. And these parricides have quite as much show of reason as the human sacrifices intended to assure an escort to the departed. This practice has been common nearly everywhere, provided that the dead person was worth the trouble; for it was not customary to show so much honour to people holding no rank or position, nor to women, or slaves.

This idea of the necessary survival of important personages is the cause of the celebrated human hecatombs common among the Ashantis. Like the majority of people, the natives of Ashantee believe that the future life is merely a continuation of that life which they see before them. They think that after death their kings and their great dignitaries take their places beside the gods, still keeping up their show of worldly splendour. Therefore they consider it to be their duty to sacrifice a befitting number of individuals of both sexes to attend upon their masters and to contribute to their pleasures. It is the same creed that prompts them to cut into slices the hearts of their principal prisoners, to season these slices with sacred herbs, and then to make those men of their own tribe eat them who have not as yet killed a foe. They think that there is no other means to prevent the spirit of the dead from breaking the force and enervating the courage of their young warriors. If the enemy whom they have captured be a man well known, then his heart is specially reserved for their king and their great dignitaries. All these puerile beliefs, leading to the most atrocious acts, show plainly that in those parts of Africa where the black are predominant, the conception of the soul and of future life are of the coarsest possible kind, and immeasurably far distant from that pure conception of the soul which has been formed by our modern metaphysicians.

III.

Future Life according to the Egyptian Mythology.

We find, nevertheless, that creeds as puerile as those we have been considering are at the foundation of Egyptian metaphysics, the depths of which it is now so much the fashion to admire.

The ancient Egyptian had several spirits and several souls; one relatively coarse, a sort of refined body, having the same colour, the same features, the same form as the individual. M. Maspero calls this corporeal soul a *double*. It was an ethereal facsimile of the body, identical with the departed spirits, the shades of the dead, in which the Africans still continue to believe. This *double* had all the wants of the living man; it lay beside the mummy in the same tomb, or in a particular corner of the tomb. To take the place of the mummy, who by degrees became distorted, they gave to the *double* a certain number of statues, cast in the image of the dead man, and they placed them beside the tomb. They were careful also not to confine this unhappy *double* too closely; his room communicated with the outside world by means of a small square opening, for the *double* wanted to breathe the fresh air. He had also many other wants. The prayers of the parents were given to him; priests were paid to offer up sacrifices to him; he possessed animals and land which supplied him with provisions. People offered to him bread, oxen, geese, milk, wine, beer, clothing, and perfume—sometimes in reality, but often they only made a pretence of doing so. For this *double*, who at first was thought to consume, in fact, the shades, the souls, and the *doubles* of the provisions offered, ended by being satisfied with hearing them named. There are examples of these pious and economical subtleties elsewhere than in ancient Egypt. In the sixteenth century the sisters in a convent in Florence offered, in this imaginary way, and with an equal show of splendour, a precious casket to the Virgin Mary. They rivalled each other in this offering with promises of diamonds, emeralds, turquoises, etc. The Virgin had good reason to be satisfied! The Egyptians used to do the same, and an inscription placed on a funereal stone enjoins those who read it to repeat: "Offering to Ammon, lord of Karnak, praying him to send thousands of loaves, thousands of geese, thousands of dresses, thousands of everything that is good and pure, to the *double* of the prince of Entew."

But the conceptions of the Egyptians became afterwards more refined. In addition to the *double* they imagined a soul of a more ethereal nature, serving as a sort of covering for a particle of divine fire or divine intelligence, and which might be divided from it. This soul was born with life, then it tried to be born again after it had travelled about with the sun for twelve hours in the night, under ground, through the long and sombre passages, where the demons were torturing the souls of the wicked. In this respect their creed was not uniform. Some believed that the soul formed itself as it pleased, came to pay a visit to its body, to its *double*, went up to heaven, and came down again to the earth. It would appear, according to M. Maspéro, that every individual pictures to himself a future life according to his own fancy, as is the case everywhere with primitive man.

But with the Egyptians, as with every people who have created for themselves a fairly advanced civilisation, the religious beliefs become associated with ideas upon moral subjects. In the other world the soul gave an account of its life, and it ought to arrive there laden with charitable works. On every mummy was placed a copy of the book of the dead, which said: "I have given bread to the hungry; I have given water to the thirsty; I have given clothing to those who were naked. I have not spoken ill of the slave to his master." It was from the purest of souls that the souls of the kings ought to have been chosen.

This mythology is most interesting; for we are thus enabled to trace the sequence of ideas from the time when men believed in the rude theory of departed spirits, to the theory of the existing Kafirs who believe in the smoke of their ancestors, and even to the theory of a spiritual soul, similar to the soul professed by the Catholics.

IV.

Future Life according to the Polynesians.

As regards future life, the Polynesians had everywhere very uniform creeds, and it will be easy for us to resume them shortly.

In their opinion man had at least one spirit similar to the *anima* of the ancient Latins residing also in the breathing organs. At night the Polynesian used to invoke his gods and say: "Oh my God, let me and my spirit live and repose in peace this night." When this spirit, in spite of the pains taken by the relations of the dying man to close as carefully as they could his nose and his mouth, had once gone out of the body it remained for some short time, generally for three days, close by the corpse and heard everything that was being said. In many islands it was believed that this spirit dwelt mainly in the left eye, and in New Zealand men always ate the left eye of a conquered enemy. At Tahiti, in the human sacrifices the left eye of the victim was always offered to the chief presiding over the ceremony, which, at least when Cook was in the island, it was his custom to refuse. But in New Zealand they were still convinced that in eating the left eye they doubled their own soul by incorporating with it that of the conquered man. It was thought by some people in the same archipelago that a spirit used to dwell in both eyes. The spirit in the left eye, the most choice spirit, would change itself into a star; the other went into the New Zealand paradise, of which we shall very soon have occasion to speak.

These spirits did not always dwell in the body after death. In many islands such an occurrence was the privilege only of the chiefs, or of the priests, or of men of note. The common people died once and for all. Such was the general creed at Tonga. The New Zealanders thought that they destroyed, or at any rate absorbed, all the spiritual breath in a man when they ate him. At Nookahiva the spirit of a dead man could not reach the sojourn of his ancestors and of the gods unless the sacred funereal rites were performed over his body. If he was buried with no ceremony, or simply thrown into the sea, the spirit always remained in the body. To avoid such a misfortune people who had no children of their own adopted those of others, handsomely rewarding the real parents. These spirits were not generally supposed to dwell merely in man; all the utensils, all the inanimate objects, all the animals, were also equally provided. When a Polynesian soul quitted this world below, it was accompanied by the souls of all the objects, all the utensils, etc., which had attended him at his funeral rites. It was ordinarily the custom to kill these objects by breaking them.

Everywhere the soul, when it existed, went after death into a sort of paradise, modelled, like every paradise, upon real life by those who had imagined it. Sometimes this sojourn of the shades of the dead was placed at the bottom of the sea, sometimes in heaven, sometimes it was merely a distant and mysterious island. If we except minor differences of detail, these puerile thoughts were the same everywhere in the Polynesian islands. Paradise was always the sojourn of the gods, of the *eatouas*. According to the Tahitian creed, the souls of men went up to the *eatouas*, and were occasionally devoured by them. But the privileged souls, those of the chiefs, and especially those of the priests, became *eatouas* in their turn. In Polynesia, where Herbert Spencer, after many fruitless attempts, endeavours at any cost to place the basis of all religions, ephemerism was generally allowed, but only for the inferior gods, as we shall presently see. Future life here, as everywhere, was an idealised picture of man's actual existence. The Tahitian paradise, the *Rohoutou noa noa* (the perfumed *Rohoutou*), was placed up in the air above the high mountain of Raiatea. The priests, the chiefs, and especially the members of the celebrated society of the Areois, went there without any trouble. The friends of the chiefs, and even some individuals, might hope that they could just be able to get there, on the condition that they made handsome offerings to the priests, who had the power to transmigrate the souls from the sojourn of darkness (*Po*) into the happy *Rohoutou*. But that was such an expensive luxury that the masses of the people did not flatter themselves of ever being able to enter the empyreal kingdom. As we may expect, the paradise imagined by those sensual Tahitians was to them a place full also of sensual delights. The sun shone brightly, the air was pure and embalmed, flowers there were always fresh, fruit was always ripe, food was savoury and was plentiful. Old age, sickness, and melancholy were unknown. Man's existence was made always delightful to him by songs, dancing, and endless feasting. We may imagine also that they conceived their greatest delight to be in amorous pleasures. They supposed the women to be always young and always beautiful. It was, in a word, the voluptuous life of the Areois, transported into a heaven of their own imagining. Husbands saw their wives again, and the wives again bore children, as upon earth. Enemies would also meet each other in *Rohoutou*, and they could re-commence their fighting. In this charming country they ate of the bread-fruit; they ate also pork, and had not the trouble of cooking it. This comfortable paradise, so well adapted to the manners and customs of the Polynesian islanders, was not given up without a struggle; for when the Tahitians were christianised at the point of the sword by the English missionaries, the sect of the Mamaia replanned their ancient *Rohoutou*, according to the ideas they found in the Bible consistent with the tenor of their lives. They pictured to themselves principally the polygamy of Solomon.

The future life of the Noukahivans was passed also in an island exquisitely provided, and situated in the clouds. The spirit of a man killed in war went up into the island, provided that his body had been taken away by his friends, and that a canoe and provisions had been placed at his disposition. If the corpse remained with the enemy, the spirit could not reach the island of paradise unless his friends had been able to kill a large body of the enemy to direct his canoe through the water. The Noukahivan heaven was peopled with their gods, their aristocratic families, their warriors who fell on the field of battle, women who died in childbirth, and those who had committed suicide. People there crammed themselves with popoi, with pork, and with fish. Beautiful women were also abundant.

The souls of the Sandwich islanders, too, went to join the *akuas* especially the souls of the chiefs, priests, or of the heroes of war. We may remark that the Hawaiians, like all the Polynesian islanders, imagined the soul to dwell in the breath.

In Tonga, the paradise, *Bolotou*, was a large island a long way off, and of very difficult access. It was also a charming dwelling-place, full of useful plants, which as they were plucked always gained fresh birth. *Bolotou* was situated to the north-west of the Tonga archipelago. It was specially reserved to the chief, and to persons of distinction, who became the servants of the gods, the intermediaries between them and the men who were alive. The second life of the common people, of the *Touas*, was always regarded as very doubtful.

Three days after death the souls of the New Zealand chiefs went to *Reinga*, a mountain situated near North Cape, and from thence they went to their future dwelling-place, which sometimes was in heaven, sometimes under the sea. The *Coukis*, or the common people, died absolutely once and for all.

Paradise was specially reserved for the great warriors, for the conquerors. Men spent their days in perpetual warfare, interrupted only by great banquets, at which they over-gorged themselves with fish and sweet potato. An old warrior chief hearing a Wesleyan missionary describe the future life of the Christians, protested most energetically, declaring that he did not want that sort of heaven, still less that sort of hell, where there was nothing but fire to eat, and that he meant to go into the New Zealand *Po* to enjoy himself there with his old friends upon sweet potatoes.

The spiritual breath which outlived the Polynesian men of rank did not always go into the region of the departed spirits; but in any case it often came down to mix with those who were actually living. At Tahiti, this spirit, *Tii*, dwelt frequently in the wooden images placed round about the cemeteries. In Noukahiva, the spirits, who had become inferior *artouas*, the shades of men who had been celebrated during their lifetime for their muscular strength, and the shades of priests, used to take a pleasure in tormenting human beings by lying down at night across the roads so as to trip up any traveller who might pass. They would at once strangle him, for they still kept in the future life the same hatred and the same passions which had animated them in their life upon earth. The New Zealanders, who feared the spirits of the dead, hoped to prevent them from returning among them by sacrificing slaves at their funeral rites, so as to appease them and assuage their cruelty.

It seems that in Polynesia, as in other places, men began by not believing in natural death; for many maladies were attributed to envy or to the malice of the spirits. These often returned to their native island, taking the form of animals. In New Zealand, these animal *artouas* would often find their way into the bodies of living beings and gnaw away their bowels. Many mortal diseases were explained in this way. In the Sandwich islands men exorcised these evil spirits. The priests had also the power of putting wicked spirits into the bodies of those whom they wished to punish or to destroy. In this way were explained all delirious and convulsive maladies. A sorcerer, too, was often consulted, and he would sometimes impute the cause of the evil to one of the members of the family, who would instantly go in a terrified state of mind to *Morai*, and there, with a cord round his neck, he would implore the intervention of the gods.

Nothing can be more simple minded and less sublime than these primitive beliefs. The soul is conceived to be a material breath, a shadow, which man, animals, and things, all possess equally. The human soul can even pass into the body of beasts. For instance, in the Hawaian islands men were sometimes given to the sharks to be devoured. The souls of the victims became incorporated with that of the animal, and he thus became more leniently disposed towards the relations of those whom he had devoured.

These infantine conceptions were in no way connected with any notions of moral duty. They arose spontaneously in the imagination of the people; they amused the Polynesian islanders, and furnished them with matter for thought; they often tormented them and drove them into committing atrocious actions; but they did not exercise the slightest influence upon their moral or their intellectual development.

V.

Future Life according to American Mythology.

The ideas of primitive man as to his soul and as to his future life are so uniform that in examining them among the different human races, we are necessarily compelled to frequent repetition. Nearly everywhere man imagines that at his death a material spirit will separate itself from the body, and will in some invisible country lead an existence like that which he had formerly passed upon earth.

Everywhere, too, he finds it very difficult to admit and to

understand the notion of a natural death. Many of the aborigines of South America, notably those who are still roaming about in the Pampas, always attribute the death of one of their own tribe to evil devices. After a man's death the people collect together to resolve what could have been the cause of the homicide, for which they will have to show their vengeance. Man has everywhere been convinced that his earthly life was continued beyond the grave, and hence the reason for the offering of arms, of utensils, of provisions, of sacrifices of animals, and of human beings, for which each funereal ceremony used to, and still continues to furnish a pretext.

Everywhere, also, the future life offered to the deceased the enjoyments in which he had most delighted here below. There was not, ordinarily, mixed up with these superstitions any idea of reward or of punishment; but the pleasures of the future life were generally reserved in preference for the best warrior or for the most skilled huntsman. The Patagonians, the Araucanians, the Aucas, the Chiquitos, the Guarayos, and others, hope after death to lead a life of pleasure in a land where the game to be killed is very abundant. Sometimes the spirits of the dead came back among the living in an animal form. For instance, the Abipones used to think that the little ducks who flew about at night wailing plaintively were the spirits of dead men.

The soul of certain Columbian Indians wanders about in the same woods that the deceased had frequented during his lifetime; or it crosses a lake to reach an enchanted land where there is perpetual dancing and perpetual drinking. According to these same Indians, the animals have a soul, just as men have, and as drinking is their supreme pleasure, they pour intoxicating liquor down the throats of the animals they have killed. The soul of the animal drinks this divine liquor, and it imparts to other animals of its kind the pleasure that it has enjoyed. The men hope that by this device other animals may in their turn manifest a wish that they also may be killed.

This belief in a future life was not generally universal in America. Certain Californians expected after their death to go either into the clouds or into the recesses in the mountains, but others, living in the Sacramento valley, or in the San Joaquin valley, declared that a future life existed only for the white men. As regards their dead they used to say that when they were burnt they became annihilated.

The Indian Red Skins, like so many other people, used to attribute most of their maladies to the fate decreed to them by their medicine-men. They used to practise necromancy, as did our ancestors of the Middle Ages. Charlevoix says that they considered their souls as shadows, or as animated images of the body. After death these souls went into a promised land, a vast prairie, where there was perpetual spring, where buffaloes and roebucks were abundant, whose flesh was delicious and very tender, and which they might always kill without any shedding of blood. Everybody was not allowed entrance into this blessed land; the best places in it were reserved for the most adroit sportsman or the most fortunate warrior. Those who possessed no excellence in this life went after death to a northern region covered with snow and ice, and there they suffered the pangs of hunger, thirst, cold, and want. The Osages tried to fasten the scalp of one of their enemies to a pole planted in front of one of the mortuary tombs; and by that means the spirit of the scalp became the servant of the deceased in the next world.

Among the Red Skins there is a tendency to classify the spirits in the future life according to their earthly merits. The Esquimaux have also the same idea. The souls of all good Esquimaux, after death, go into a world below, where the sun is ever shining, where seals, fish, and sea-birds swim about in limpid waters, and complacently allow themselves to be captured. Many of these good folk are already happy in the thought that they are actually being boiled in hot caldrons. But this Elysium of delight was only reserved for those who, while they were alive had killed a great many seals, or had gone through great dangers, or had been drowned in the sea. Women who died in childbirth also enjoyed the same privilege as of right. The souls of the bad Esquimaux were less favoured; for they went into a world above, where they are in perpetual suffering from cold and hunger. These creeds do not seem common to all the Esquimaux; for one of them, questioned on the matter by Ross, had no belief at all in a future life.

The belief in sorceries and sortilegious devices seems general among the Esquimaux. Their sorcerers, the *angekoks*, have spirits always at their orders. They command the elements, they drive away or attract the seals, they bring sickness and they can also cure it.

Spiritualism among the Esquimaux is very widely spread, as among most savage people, for it is not confined only to man. All the animals have their spirits, and the spirits of men can enter into the bodies of animals. Every object, too, has its spirit. The spirit of the object calls itself a "possessor," and it governs the object of which it is the image.

All this spiritualistic mythology is doubtless puerile, and the mythology of the ancient Peruvians and the ancient Mexicans, though it was more complicated, was scarcely more intelligent. It was always a picture of real life imagined to take place beyond the grave.

The Incas, after death, rejoined the sun, their father. The Peruvian vassals, in the next life, continued to serve their masters as they had done here below. The Peruvians, a very civilised people, had also imagined different dwelling-places for the good and for the bad. The good lived in the other world surrounded by voluptuous ease, they rested themselves from the hard work of this world below; but the bad spirits had to go through perpetual hard labour. The care taken by the Peruvians in the drying of their dead has given rise to the thought that the ancient Peruvians believed in the resurrection of the body, but this argument is very far from being conclusive proof.

The Mexicans were somewhat more imaginative than the ancient Peruvians in their dream as to a future life. Their soul might, after death, go into three distinct dwelling-places. The chosen, that is the warriors who had died in battle, or the victims sacrificed to the gods, joined the sun immediately, and accompanied it on its glorious way across the heavens, dancing and singing all the while. Thus after a few years of this radiant existence they went to live in gardens rich with sweet-scented flowers, or they were transformed into very beautiful birds, and lived always in the clouds.

As we see, in America, from Patagonia up to ancient Peru and ancient Mexico, where the people were relatively very far civilised, human imagination has conceived the future life to be merely a prolongation of this early life. We do not find conceptions of a higher kind which approach more closely to scientific truth until we come to the great Asiatic religions, to Brahminism, and more especially to Buddhism.

VI.

Future Life according to the Asiatic Mythologies.

The vast Asiatic-European continent, with its numerous collection of islands dependent upon it, is the great workshop of humanity, and geographically it is larger than any other. There have been formed the most numerous agglomerations of men, there the most complex languages have been devised, there the most intelligent races have increased and multiplied—the races who have brought art, science, and philosophy to the highest point of perfection. Among these superior races metaphysical religion has, like everything else, attained to a degree of complication and elevation unknown elsewhere. Such are the great results coming from the labour and the thought of the most eminent individuals of the Aryan and Mongolian races. But these two great races did not arrive all at once at the zenith of their development, for amongst themselves the masses are far from being able to keep pace with the too-rapid strides of their leaders. Many instances of inferior people are still to be found in Asia. When we speak of Asia it becomes absolutely necessary to make divisions and subdivisions. No doubt we are now concerned principally with the future life of the great Aryan and Semitic people; but we cannot pass over in silence the ideas belonging to the other Asiatic races. We shall therefore be obliged to mention the Vedic mythology, from which the Brahmin and the Buddhist religions have sprung, and we shall endeavour, therefore, to resume as shortly as possible all these religious and metaphysical speculations.

Some of them are indeed very rude. In the Ladrone islands the people thought that the spirits of the dead simply went into the bodies of the fish, and, therefore, to make better use of these precious spirits, they burnt the soft portions of the dead body and swallowed the cinders, which they let float on the top of their cocoa-nut wine. The fish were therefore deprived of receiving the human souls, for which indeed their bodies would seem to be physically so ill adapted. At Sumatra, as in Polynesia, the popular belief is that the future life is the privilege of the rich and the powerful; the poor people die once and for all. This idea would be sufficient to show that the Brahmin and Mahomedan religions are far from having penetrated deeply into the masses of the Malay population; for everywhere, great religions possessing complex metaphysical notions, are shared only by the minority.

A most primitive polytheism still prevails among the Mongoloids of Northern Asia, among the Kamtschadales and the Siberians. An endless number of divinities dwell in the mountains, in the forests, in the torrents, etc.; and the schamans or the sorcerers act as mediums between the gods and the human creatures. The Kamtschadales believe in a resurrection after death. According to them the invisible world is made like the visible world; the difference is that man works less, he works to better advantage, and he is never hungry. The Siberians in the neighbourhood of Tobolsk are very uneasy in their mind as to their future state, for wicked and diabolical spirits lie in wait for their shade as it leaves their body; they are therefore very careful to call to the bedside of the dying man the kam (the sorcerer), who beats his magic drum, and negotiates for acceptable conditions with the evil spirits. The Kamtschadales believe so firmly in the happy idleness of a future life that they will often commit suicide, or will make their children strangle them, so that they may get their enjoyment the sooner.

Many large religions exist in the Japanese archipelago. The principal are Sintoïsm, Buddhism, and the religion of Confucius.

The first only is native to the soil, the others have been imported. We shall have to speak again of the Sintoic polytheism; at present we need only concern ourselves with the idea of the future life according to this primitive religion. Among the Japanese, as among the greater part of intelligent races, the belief in a future life has received the mythical sanction of morality. The soul of the Japanese Sintoist outlives the body, and divine judges decide as to its fate after death. The soul of the virtuous goes into a sort of paradise, where it is deified, when it becomes *kami*; the soul of the wicked, on the other hand, is hurled into the kingdom of roots. Here we see at least an ethical utilisation of religious beliefs.

It does not appear that the Vedic Aryans have imagined anything at all similar. Their hymns say nothing either of reward or of punishment after death. From a Vedic phrase that we have already quoted, one might think that the Vedic Aryans believed that the corpse itself was not totally deprived of conscience or of sensibility. From another phrase it would seem that they believed in a dividing of the different parts of the body after death: the animation shown in the face went to the sun, the breath went to the winds, the members of the body went to the earth; an immortal portion, dedicated to Agni, went to the world where all the good men go. This paradise (*Paradêça*) was situated above the clouds. Man was there perfectly happy; every wish was instantly gratified. The Vedic soul was not immaterial. It was an ethereal but corporeal substance, as has been conceived by every primitive race of people.

And what other conception can we form of the soul, unless one is a professor of official philosophy in France—in other words, condemned to unnatural absurdity!

The materialism of the soul is also admitted by the subtle Brahmin metaphysics, which are evidently grafted upon the Vedic doctrine. According to this great religion, of which the first idea, setting aside all metaphysics, is not irreconcilable with materialistic science, there are two individual souls, emanating from the supreme soul, the soul of the world, as sparks emanate from a brazier. The soul is an elementary form capable always of greater extension. It is shut up in the body, as in a scabbard. During sleep the soul retires, and returns when man awakes. At death it leaves its envelope, and begins to wander. The sinning souls fall into the infernal regions, and endure a thousand torments. The virtuous souls go to the moon, there to receive the reward for their good actions, after which they come down again to the earth to give life to fresh bodies. The wise men only go up higher than the moon; they go up into the dwelling-place and to the court of Brahma, and there, if their liberation is complete, they enter into the divine essence; they become mixed up in the soul of the world, as a drop of water is mixed up in the ocean.

The doctrine of pantheism, of emanation, of incarnations or transfigurations, the absorption in the great whole of the *Nirvána*, used to exist therefore in the Brahmin religion, and Buddhism has only brought it out in stronger relief. According to the legend, Çakya-Mouni, the founder of Buddhism, recalled to his mind the recollection of his former incarnations, and he then recognised that the total number of bones to which he had successively given life made a heap materially larger than that of all the planets. He perceived, too, that the blood which he had shed in the numberless decapitations as punishment for his crimes, during his incarnations, was equal in quantity to the total amount of water in the universe. All beings undergo similar transmigrations; it is only by the force of virtue that they can succeed at last in freeing themselves from this unbearable and interminable cycle of personal existence, and bury themselves in supreme absorption, in the *Nirvána*. But everyone cannot reach this so ardently wished for annihilation; the inferior natures, those who give way in the struggle towards what is good, are punished by incessant incarnations, of a lower and lower nature, in proportion to their increasing perversity. They can even incorporate themselves into inanimate objects. Buddha himself, in his self-humiliation, leaves behind him emanations, superior men, Buddhisatwas, who complete the work first undertaken by him. The dalai-lamas of Thibet are Buddhisatwas, and there are also many others. Every important brotherhood has its Buddhisatwa. According to the belief of this brotherhood, these living emanations of Buddha are immortal, in the sense that they reincarnate themselves immediately after their death in the body of a child who succeeds them. In speaking of the clergy and the Buddhist form of worship, we shall have to return to curious and fruitful dogma, which the Catholics have been so wrong in not adopting.

But at what a distance do the Brahmin and the Buddhist metaphysics throw into the shade the puerile mythologies of the Christian and Semitic religions of which we have now to speak! If we exchange words for things, if we put aside all the paraphernalia of clerical subtleties and popular superstitions, if instead of the vague idea of a divine essence we conceive the scientific notion of a material substance, always variable and movable, of a substantial universe ever changing its form, we shall arrive directly at the truth, we shall then come to the great materialistic conception grown out of the depth of science itself, and destined to annihilate every form of religion.

By the side of Brahminism, of Buddhism, with which it would not be difficult to connect the religion of *Zoroaster*, the monotheistical religions of Moses, of Mahomet, and of Christ are but poor creatures of the human fancy. Our opinions as to future life are so well known that it will be sufficient to recall them rapidly.

The small Hebrew people, the people of God, had much trouble in believing in the notion of the survival of the soul, in the divine *breath*. But the Jews at a very early date believed in a subterraneous dwelling-place, a *Scheol*, a dark country, inclosed by gates, with valleys in the midst. The witch of Endor evoked Samuel before Saul; she made him rise from the country under the earth. It would appear that the inhabitants of *Scheol* lived there in a state of profound torpor. The Jews, unmetaphysical by nature, did not expect punishment or reward except in their earthly life; they did not understand the notion of a survival except in the gross and palpable form of a resurrection of the body. It was not until very late, and after long contact with the unfaithful, that the dualist doctrine penetrated into their mind. We find it clearly expressed nowhere but in Ecclesiastes: "Then shall the dust return to the earth as it was, and the spirit shall return unto God who gave it."

Nothing can be more material than the soul according to the primitive mythology of the Greeks, as it is naïvely described in the eleventh book of the Odyssey. The shades with whom the prudent Ulysses converses have preserved all the needs and all the passions felt by the different personages during their lifetime. In Hades the shade of Ajax is as hot with anger against Ulysses as the warrior had been when actually alive. Orion chases on the lawns of hell the wild beasts which his fearful truncheon had already felled among the mountains in his own country. The waters of the lake in which Tantalus is standing flow away from his thirsty lips: the fruits of the pear trees, of the pomegranate, the oranges, the figs, and the olives are all blown away as he endeavours to snatch them. At last, all the shades with whom Ulysses is conversing become animate; they rush headlong to drink the blood of the victims which the explorer of the sombre Hades has collected in a hole dug with the point of his sword; Penelope's husband makes them stand aside merely by threatening them with his dagger, and they do not consent, or do not succeed in speaking until they have lapped up some living blood. We may say, at all events, that the dwelling-place beyond the grave, where the Greek souls took their flight when the flame of the funereal pile had consumed their flesh, was not more intelligently conceived than the paradise imagined by the Polynesians.

In Christianity, an unoriginal doctrine, in which are confusedly mixed up the ancient religions of Central Asia, Judaism, the mythical conceptions of Egypt, the Græco-Roman polytheism, and the popular superstitions, the soul was for a long time considered as perfectly material. It was a shadow, a body more refined than the living body, but having the same form. "The soul," says Tertullian, "is material, composed of a substance different to the body, and particular. It has all the qualities of matter, but it is immortal. It has a figure like the body. It is born at the same time as the flesh, and receives an individuality of character which it never loses." The coarsely-imagined torments of the Christian hell, the insipid enjoyments of paradise, which may be put as a counterpart, imply in the most absolute manner the belief in the materialism of the soul; and it was necessary to urge the dreamy wanderings of Plato, and the madness of his followers the Neo-Platonists, to introduce into the Christian theory of metaphysics the unintelligible conception of the immaterialism of the soul.

More simple, more logical, more impregnated with the common-place good sense of Judaism, Mahomedanism also conceives the soul to be a very concrete substance. According to the Mussulmans two exterminating angels come to examine and even to chastise severely the corpse in the tomb, and the corpse must remain seated and submit to the punishment. We know also well enough that the Koran only promises to the faithful, in the next life pleasures and pains of a very sensual kind. Rivers of milk and wine and honey flow abundantly in paradise. Beautiful virgins, whose skin is of the colour of an ostrich's eggs, fondle the chosen people. The sinners on the other hand are thrown into the fire, they have given them to drink boiling water which burns their entrails. However, like all the great religions of central Asia, like Christianity which has sprung from them, Islamism has imagined the future life to be a moral instrument either of reward or of punishment: far superior in that way to the coarse primitive Judaism.

In terminating this short notice of the different chimeras conceived by mankind as to his future life, we will mention our own savage European ancestors. In the Scandinavian Wallhalla men cut each other in pieces in the morning to rise up again and drink hydromel poured out into skulls by the Valkyries. The Gauls, more civilised, had imagined, or had received from Central Asia, a theory of metempsychosis; but in these infantine religions we find no trace of any new metaphysical thought. The Wallhalla closely resembles the paradise imagined by the New Zealanders, and the Gallic metaphysics are very poor beside those of Brahminism and Buddhism.

This review which we have undertaken would be slight even if expanded into a large volume; repetition, too, would be abundant if we wished to mention all the particularities conceived by every race of people and every tribe on the subject of the future life. The numerous facts collected here and there which we have thought it our duty to mention, will be amply sufficient to give a general idea of all this mythology, to connect it with the causes from which it has sprung, and also to show to us how unimportant they all are.

VII.

The Evolution in Ideas of Future Life.

With the assistance of ethnography and of history it is easy to trace the genesis and the evolution of ideas in the human race as to a future life. The soddened intellect of the primitive man will not allow him to understand that there should be such a thing as natural death. How can man, unless the victim of some wicked device, pass from all the boiling heat of life into the cold immovableness of death? But is death really the extinction of life? In spite of the decomposition of the corpse, the personality of the dead man has not wholly disappeared. The recollection of the dead man still remains in the memory of those living; and further, they see him, they speak to him, sometimes through hallucination, sometimes in a dream. Death is, therefore, only apparent. It is a simple dissociation of two principles. When life seems to fade away, it is but only a light body, a shade that separates itself from the invisible body, and wanders about on the rocks, through the forests, over the mountains, still feeling the wants, the desires, the passions which animated it formerly. Such is the first stage of belief in the doctrine of survivance.

Later, man conceives the idea of reuniting these wandering shades in an invisible dwelling-place—some place beyond—fashioned upon his real life. Henceforward the belief in a future existence acquires a real importance. The dwelling-place of the dead becomes a beautiful image of terrestrial life, a supreme refuge where man enjoys without effort all the good things which he has vainly struggled for here below.

When the human mind has once fully conceived this consoling idea it clings to it with most indomitable energy. Man feels it to be, in the midst of all the trials of life, a most comforting thought; or even much better, it is a sort of intellectual opium which consoles him as it deadens his faculties.

When once the moral sense is born, when man has ideas of justice ill requited upon this earth, new and powerful motives come to strengthen his belief in the life beyond. The idea of future life then gives its assent to morality; after death every one is judged according to his works. For the wicked man an abyss of pain is opened before him; a voluptuous paradise welcomes and consoles the good. There is no system of religion, however little complex, which does not show itself disposed to lean towards this ethical side of belief as to the future life. In dreaming of the supreme delights awaiting them on the other side of the grave, the unhappy and the wretched grow patient and take courage. "Leave us the best part of this lower world," the happy and the powerful say to them, "you will be well rewarded in the next."

Everything will turn as man wishes so long as the human understanding is so little developed as to pay itself in this imaginary coin; but in proportion as his intelligence increases, science will establish itself, and in this or in a super-terrestrial universe, the piercing eye of knowledge will no longer find a place to allow of the sojourning of souls. We can no longer refuse to see in the conscient life a function of nervous centres at once unstable and perishable. We have come to consider life as the sound of a harp, of which death will cruelly break the cords. Then, in order that it may subsist, metaphysics are obliged to grow more and more subtle. The soul ceases to be an ethereal image of a real body, a shadow; it becomes the verbal entity of metaphysicians, a nothing, so impalpable and inconceivable that every strong and free mind refuses to believe in it. The human being then knows that his poor personality is but a passing existence, since it springs only from the ephemeral grouping of indestructible atoms, which the shock of death will one day disperse. From this moment man is really a man; the field of his activity becomes brightened and smaller in extent; he brings his dreams of happiness and his aspirations of retributive justice down from heaven on to the earth; he knows in what direction he ought to aspire, and he manfully resigns himself to that which is inevitable.

CHAPTER XVI.

THE GODS.

I.

Mythology in General.

MYTHOLOGY, which, at a certain period of the social evolution, wears such an imposing aspect and plays such an all-important part in the life of nations, is very humble at that time which we may call its ovular period. It is then nothing more than the reflection of very simple emotions common to man and to superior animals, and it sometimes makes a clumsy attempt to explain the natural phenomena. In both cases man only represents naïvely the creations of his own imagination, and these creations are of the rudest kind. It is not an easy thing to picture to oneself the mental condition of the primitive man. To assist us in doing so we ought first to go back as far as possible to the years of childhood and observe the manners of children; we ought to analyse dreams, delirium, or other phantom objects. At the outset of his mental evolution man, unskilled as yet in the art of observation, feels, beyond all comparison, much more than he thinks; he is inexperienced in testing the subjective with the objective phenomena, he is perpetually confounding what is real with what is imaginary. He cannot for an instant doubt the reality of the beings who appear before him in his dreams. These beings may be invisible to others, but he has seen them. There exist, therefore, *spirits* who habitually fly away from the eyes of man. Again, the primitive man ill distinguishes the animate from the inanimate objects; he is inevitably inclined to endow with conscience and force of will, to anthropomorphise, or to zoomorphise, to vivify all the natural agents which serve his purpose, or which are prejudicial to it. He will gladly lend to them emotions and ideas similar to those which they awaken in himself. As regards the animals, he does not consider them as being essentially inferior to himself, but there are many of which he thinks very highly, which he fears, which he venerates, which he even looks upon as more powerful than himself; for as yet he is but a weak and ill-armed creature, and therefore often feels his powerless condition when he sees in front of him their claws, their teeth, or their venomous poison.

It is not until very late, after many efforts and much experience, that man, regulating and controlling somewhat his conscient life, feels that he is gradually becoming less credulous, and that his mythology is also gradually diminishing. His gods gradually appear to him to become more powerful and less numerous, more spiritualised, less real. As he has gained ground in his battle against the animal world he begins to disdain his half-conquered rivals; his gods are then to him anthropomorphous, and they always tend to diminish in number. The bright glare of mythology grows pale and becomes confused; philosophy gradually intervenes, and ends at last by looking upon all forms of religion as dreams of childhood, and as the mere cradle of humanity.

If we could read the brain of the superior animals we should undoubtedly find there a rudimentary mythology. Many mammalia, dogs for instance, have, like man, dreams and hallucinations; they can connect certain facts with their real or their imaginary causes. We do not want more evidence than this to arrive at the puerile conceptions of primitive mythology. There is no essential difference in a mental point of view between the African negro who worships the crocodile, and who will probably eat him afterwards, and the dog fawning at the foot of his master and licking the hand that has beaten him.

It is a matter of course that religion so understood should exist more or less not in all men, but among the majority of the ethnical groups. Nevertheless certain people, certain tribes are very poorly endowed in this respect: we have enumerated them elsewhere.

In spite of the vulgar prejudice, it is certain that belief in imaginary beings worthy of the name of "gods" is far from universal. There are two principal causes which keep certain tribes or races outside of or above these errors: either a brain so ill-developed that it is incapable of any speculation, or else our own clear practical sense, or an innate common sense too strongly born. Sometimes among the Kafir Makololos, or among the Basutos, a vague belief in the departed spirits of their ancestors will comprise all their mythical creed; sometimes, also, at the other end of the metaphysical ladder, the Buddhist thinkers make of their religion a vast mythological system resting upon an atheistical metempsycosis.

It is nevertheless beyond doubt that man more or less peoples the cosmical life, in the middle of which he lives surrounded, with fictitious creatures of his own imagination. We should fill volumes if we were to go through all the dreams of mythical speculation; but in mentioning only the most typical facts, in comparing them together, the work becomes very much shorter. For the differences are rather matters of detail, as to the colour, the form, and the number of the myths, of which the genesis and the evolution are everywhere more or less very similar.

All this collection of mythical creations may be classified and subdivided in many ways. The gradation most commonly adopted goes from fetichism to polytheism, and from that to monotheism; but we must also add pantheism, which will include all the great Asiatic religions. This classification is convenient, but open to criticism, for all these mythologies result from one mental process, which Mr. Tylor has called *animism*. This is nothing more than placing a me similar to the human me in the breasts of certain beings of the outside world. For the primitive man finds difficulty in allowing that there should be movement and action without will and without knowledge. He imagines that everything is animated by nature; the field of this imaginary life, at first indefinite, then gradually contracts before him in proportion as he observes and reasons to better advantage. This general conclusion will clearly show itself in the short mythological review which is now before us.

II.

Myths in Melanesia.

The Tasmanians, who in the scale of human races hold one of the lowest positions, were also at the bottom of the ladder as regards any mythological imagination. The Rev. Mr. Bonwick says that "they had no idea of the divinity." That means probably that they did not believe in anything analogous to the God of the Anglican church, for Dr. Milligan reports that they had peopled with spirits the crevasses, the rocks, and the mountains. These spirits, created by an unhappy race, who were ever painfully struggling for their existence, were generally evil minded, and the people rendered to them no sort of worship. The Australians, also, who were so similar to the Tasmanians, had no other religion than a vague fear of evil spirits, whom they did not even dream of worshipping. During a storm they would curse these wicked beings, they would call them by hard names, and spit upwards towards the heavens as though they were spitting at them. The Australian gods were generally anthropomorphous—sometimes also they took an animal form. Certain tribes believed in the existence of a mythical serpent, who hid himself in the pools and in the rivers, and tried to catch hold of those who came to quench their thirst. The idea of these evil spirits is very simple. The Australian does not ordinarily doubt as to the reality of his dreams. He supposes that the beings who may be invisible to others, but which visit him during his sleep, actually do exist, and with them he peoples the forests, the rocks, the grottoes, etc. This same notion has been all over the earth one of the most fruitful sources of mythology.

In Australia we find ourselves, so to say, at the creation of mythology; but in Fiji we see a mythology already made, more rich, more complex, but yet not essentially different from the other. There is a whole world of Fijian gods. Many of them are merely the incarnation of the passions, of the instincts of their worshippers—the adulterer, the night-ravisher of rich women, the quarreller, the bully, the murderer, the man coming out of a slaughter-house, etc. These personages are classed according to a divine hierarchy, similar to the Fijian hierarchy. A master-god, *Dengei*, concerns himself more or less with all the acts of human life; immediately below him are his two sons, who make known to their august father the wishes of the Fijian people. And below them are the plebeian divinities—the gods of the fishermen, the gods of the carpenters, the gods of war, the national gods, the gods of certain districts, and the family gods. Each chief has his own familiar god, whom he consults upon important occasions. Each one of these gods has an earthly dwelling-place in the Fijian archipelago. The great *Dengei* dwells in a serpent. The other gods live, some in a plant, some in a bird, some in a shark, in an eel, in a hen, etc. Each one has his faithful followers—the worshippers of the eel god never allow themselves to eat of that fish, the peculiar property of their own god. Certain gods dwell in the uplifted stones, similar to our druidical stones, and to whom animals are occasionally offered up as sacrifice, for even the gods want their food. In this rich pantheon there is room enough for deified men, as they have in their lifetime excited the imagination of their fellow countrymen. As it is natural, these last especially are animated with all the human passions, those which they felt when they were struggling for their own existence.

The Fijian mythology is most interesting because of its extreme simplicity. All these coarsely-imagined gods, these deified crimes, are manifestly the exterior and personified images of the desires, the emotions, the fears of the islanders who worship them. Such is, and such always has been, the process of mythical creation; but it has not always been so simple and so self-evident. We have often to read the signification hidden under the layer of ornamentation, of accessory matter, of transformations of a more or less subtle kind.

This is not yet the case with the New Caledonians, so nearly allied by race to the Fijians, and so susceptible to religious emotions that they are subject to visions, to a sort of ecstasy. Like the Fijians they have also a great number of invisible gods, who govern the elements; and like them they deify the departed spirits of their ancestors, especially those of their chiefs. There exists also a certain hierarchy among these New Caledonian divinities, and some of their hordes have given a supreme chief to the tribe of the gods. This Melanesian Jupiter is a spirit of the earth, having the supreme command over the elements.

Under other names, and with differences of detail, we shall find almost everywhere this same primitive mythology.

III.

African Religions.

Putting aside the religious ideas that have been introduced into the country, we may say that the African negro has not got beyond the lowest stage of animism—that which has been called feticism —to which in certain regions we associate the belief in departed spirits and in the shades of the dead. The majority of the negroes believe in the existence of an invisible spirit, of a conscient me, similar to their own, and lodged in different portable objects, very capriciously chosen, and which the Europeans call phantom or fetich. In the same manner they vivify different objects which they see daily round about them, such as trees, rivers, animals, etc. In considering, from a general point of view, all these attempts in the African mythology, we may follow one long gradation of ideas which goes from almost a total absence of religion to the old religion of the ancient Egyptians. We will mention the principal characteristics.

In the southern parts of Africa, among the Hottentots and the Kafirs, the religious feeling is very small indeed. If we may believe Levaillant, the Hottentots are completely devoid of any such feeling. Some of them, it is true, believe that the dead leave behind them spirits which are generally very ill disposed. A Bushman, after he had killed a sorceress, broke her head, buried it, and then lit a large fire over the spot to prevent her shade from afterwards coming out and tormenting him. This belief in the survival, for a greater or less length of time, of the departed spirits of the deceased appears to be the only mythical idea among the Kaffrarian tribes, and it is not at all fully established that even this is to be found everywhere. These shades will wander about in a calm and silent way; they may be good, or they may be evil, and they will sometimes interest themselves in the fate of their descendants. The people curse them and call them by bad names when they are hurtful; they deceive them whenever they can. The Basutos, when they are going to steal their neighbours' cattle, whistle gently, as though they were conducting their own flocks, so as to deceive the *morimos* of the tribe whom they are going to rob. None of these tribes have any idol or form of worship. Their religion is reduced to the lowest possible condition. But among some of them, for instance the Bechuanas, there are traces of zoolatry, for they call themselves after the names of the animals: there are the crocodile tribes, buffalo tribes, monkey tribes, elephant tribes, lion tribes, and others of the same kind. And the Bechuanas refrain from eating the flesh, or from clothing themselves with the skin, of that animal who is the patron saint of their tribe.

The negroes of Eastern Africa, near neighbours to the Kafirs, believe in the existence of evil spirits, but spirits who are mortal, and who may be killed. When Burton spoke to them of God, they asked him where was this god, that they might go and kill him. "It is he," they said, "who devastates our houses, who kills our wives and our cattle."

In Equatorial Africa, the mythic malady is very much more intense. This is the classical country of fetichism. Here the people worship the serpents, birds, rocks, peaks of mountains, feathers, teeth, etc. One sees hideous idols, and the chief of every family has his idol in Gaboon. These inferior gods live exactly as men do; they walk, and drink, and eat; the people paint them and adorn them. For certain wandering spirits houses are built that they may repose themselves. Here we see the most primitive notion of a church. These nomad gods are sometimes very ill disposed. There are some who squat all day in their caverns, and go out at night to seize and devour travellers. They sometimes enter the body of a man or of a woman, and do a thousand bad things, beating and knocking down everyone they may chance to meet. Sometimes a man may struggle against them and even kill them; but then he must be very careful to burn their body and to leave nothing remaining, for they will come to life again if the smallest bone is spared. The negroes of this region surround the objects they have imagined with the most childlike fancies. They will give life to, or make a divinity out of everything. Du Chaillu's clock was to them an all-powerful spirit carefully watching over the traveller. They did not at all dispute the existence of a Biblical God, of whom Du Chaillu spoke to them; but they did not care to trouble themselves about the matter. "He was the god of the white men who had sent them many good things; but he had nothing to do with the black man, who had fetiches and idols of their own." When the tribes possess sorcerers, properly so called—the primitive priests—they make these important personages hallow and *bless* their phantoms and their talismans; so do also the anthropophagous tribes of the Fantis.

The Ashantis, who are more civilised, have not got beyond this first degree of mythology. They have their fetiches and their numberless idols; they deify their kings, their chiefs, and the dignitaries of the kingdom, whom they will not allow to go into the next world without a large accompaniment of victims sacrificed for them; they worship animals, serpents, and vultures. Each family has its domestic fetiches, its own household gods. They never drink without offering a libation to the fetish, throwing some of their liquor on to the ground. They have even fetich houses, men-fetiches living in the sacred house, that is, in their temple, they have priests: they are altogether a most pious-minded people.

We find similar creeds of belief, neither more nor less exalted, in all the middle part of Africa, where the Mahomedan religion does not yet prevail: in Senegambia, in Guinea, in Soudan, and as far as Abyssinia. Facts are abundant, and they generally repeat themselves. We will choose a few, and so make a short fetich anthology.

In Guinea on the Gold Coast, the people worship vultures, crocodiles, etc. In Yarriba (basin of the Niger) they have fetich trees, and a quantity of phantom spirits, fruits, calabashes, feathers, egg-shells, bones of animals, etc. Any object, no matter how capriciously chosen, may serve as a dwelling-place for a spirit, or rather may become a spirit. The people venerate the fetich trees; they do not fasten animals to them, but they hang up rags, and tatters, and bandages. The rivers are often deified or feticised. A guide given to R. Lander by the king of Kiama begged him not to mention the name of any river in the presence of the river Moussa, who was a woman married to Niger, and jealous of her husband, who was disputed to her by other rivers. She would incessantly reproach her husband river on account of the familiarities he took with other rivers, her rivals; and at the confluence of the waters there was always a violent and very brawling conjugal dispute between them. The king of Boussa, before he let R. and J. Lander embark upon the Niger, consulted the river, and obtained a promise "to conduct the travellers safe and sound all the way down to its opening into the sea." The conductor of a canoe who went down the Niger with these same travellers shrieked loudly at each bend in the river, and whenever he heard an echo respond to his cry he poured into the water half a glass of rum, and also threw in a piece of yam and a bit of fish. He said this was to feed the fetich, who otherwise might prove dangerous to them.

The fetiches are not pure spirits; they have all the wants of men. R. and J. Lander were advised to roast a bull that they had killed, under the nose of a fetich who dwelt in a small temple covered over with thatch, so that the god might inhale the smell of the roast meat and eat a little of it if he was so minded. Through all this region sacred edifices are coarsely constructed; the people make sacrifices to the fetiches, they even practise the art of divination.

The temple is of the most rudimentary kind. It is a hut, a dwelling-place devoted to the fetich, and containing very often several coarse wooden carvings, representing men, alligators, boars, tortoises, etc. The people worship the fetiches in prostrating themselves; they sometimes offer them cowries, small white shells, representing the current money of the country. The worshippers pray the fetiches not to desert them in the time of need, to assist them in their enterprises, and even in their acts of vengeance. Laing one day heard a negro in a fetich hut make an imprecation in all respects similar to the formula of the Catholic excommunication. The devout man was praying for the death of one who had violated the tomb of his father, and he had beforehand offered up a fowl and a little palm wine as sacrifice. The negro was heard to say: "If he eats may his pork suffocate him; if he walks may the brambles tear him to pieces; if he bathes may the alligators swallow him; if he goes in a canoe may he tumble into the water." The offering of the sacrifice of a fowl is very common. This is the practice of the Bambarrans when they want to draw an augury from the gods. They cut the neck of the animal half through, and then throw him into the fetich hut. They consult the god by a Yes or by a No. If the hen as she dies throws her head backwards a Yes is signified; a No is meant if her head falls in a forward direction. Sometimes they beat the fetich when he has not granted the prayers asked of him.

They do not always confine themselves to sacrificing hens, cows, and sheep to the fetich. In Yarriba, the fetich-man, or primitive priest, will sometimes declare that a human sacrifice is necessary.

By degrees fetichism becomes complicated and organises itself into a system. It is at first animism in its basest form: the attribution of superior powers to any object or to any animal; afterwards the people build a house for the god or for his emblem; at last they appoint a fetich-man as keeper of the consecrated house. Priesthood is thus constituted. There are appointed sorcerers, divine men, causing the fetich to speak, men who know better than anyone else his desires and his intentions; these are the parasite mediators between the gods and the devout people.

On the Upper Nile, among the Niam-Niams and the Bongos, Schweinfurth found similar creeds of belief. The negroes in this country believe in the existence of spirits, always evil minded, and inimical to mankind. These spirits are hidden in the depths of the woods, and their language is interpreted by the rustling of the leaves. Thanks to some magic roots, the people can guarantee themselves against these dangerous phantoms; they even make use of the roots to hunt the evil spirits. They consult their conjurors before going to battle, to discover a guilty person, and for many other reasons.

The same primitive mythology is seen in the valley of the Upper Nile, as far as the banks of the lake Albert Nyanza, but with local differences. In the vast monarchy of the king of M'tesa, in the Ouganda, near the Nyanza lake, talismans and magic horns were common, also fetishes, sorcerers, and sorceresses. The people believed in the spirits of the lakes and the forests communicating with man through the medium of the clergy—a body of men handsomely paid out of the mortmain funds. In Ounyoro the people believe very strongly in magic, in auguries drawn from the peristaltic movements of the intestines of embowelled hens. The Obbos, more to the north, have very little faith in anything except in magic whistling, compelling the clouds to answer them by sending rain. A total want of religious feeling, an absolute atheism seems to exist everywhere among the Latoukas, according to the interesting conversation that Sir S. Baker held with the king of this country. We have already quoted this dialogue elsewhere.

We may say that fetichism prevails generally all over this vast region of Africa, which we have now been considering. But upon its northern frontier Islamism has made itself felt. The Arabs have been the Islam missionaries, and the Pulah negroes have been most ardent neophytes. It is somewhat curious to study the contrast between the dry and simple monotheism of the Mahomedans and the multiform fetichism of the negroes. The examination will give us another proof of the fact that the conversion of an inferior race to the religion of a superior people is only apparent. Like every great intellectual and moral manifestation, the religious feeling in any race is the expression of the mental condition peculiar to this race—resulting from its degree of development, from its normal changes, from the habitual tenor of its life, through all of which it has ever carried on its battle for existence.

The Fellatahs repeat, in Arabic, their formulas and their prayers, but the majority of them do not understand a word of what they are saying. They say their prayers five times a-day, and are firmly convinced that the goods, the wives, and the children of the unfaithful belong to them; that it is perfectly lawful to rob or to kill an unbeliever. But still, of all the people of whom we have been speaking, the Fellatahs are the most thoroughly converted. Nearly everywhere else the people worship Allah, and they worship their fetish also; Allah is for them only an additional fetish. The Bambarrans call the god Nallah. And in addition we find aboriginal gods to whom bears are sacrificed, and to whom cooked millet is offered.

At Kimma, on the banks of the Nile, the people call themselves Mahomedans, but they, nevertheless, place fetiches at the doors of their houses for protection.

It is impossible for them to live without their phantom spirit. The Bambarrans adore phantoms in every imaginable form, either roots, egg-shells, horns, stones, teeth, pieces of dried leather, and especially a fragment of umbilical cord; they will tie on to this a few verses of the Koran written by the Almoravides. These last phantom spirits are the dearest and are the most highly valued. And the Nubians of Sennaar, who are nearer to the great stronghold of Islamism, worship the moon, trees, and stones. The Abyssinians, in spite of their Christianity, worship raised stones similar to our druidical stones, and they cover them with amulets, offerings of butter, votive threads, and the peritoneums of animals. They worship the serpents, they pray to them, they consult them in their important affairs. They respect the Blue Nile and abstain from bathing or even washing their clothes in it.

In Madagascar we find pure fetichism without any admixture. The people there believe in wicked spirits, they have idols to whom they offer up sacrifices of animals, to whom they pray, but only when they have any service to ask of them. Some of these idols have an official existence, they have houses, priests, and appanages of their own. Madagascar is very African from a mythological point of view.

From this short review it would seem to be clearly proved that fetichism, or animism of the poorest kind, is everywhere at the bottom of the African mythology. The black races in this vast continent have not, of their own accord, yet got beyond this first phase of religious evolution. The worship of the ancient Egyptians cannot be urged as an exception, modified as it was by the introduction of certain Asiatic myths supporting a whole theory of metaphysics and polytheism divided into geometrical sections.

In this singular country, the cradle of so many arts and sciences, animal worship has gone to greater lengths of extravagance than in any other. Like many contemporary negro tribes, each locality in ancient Egypt had its sacred animals. The inhabitants of Mendes worshipped their goats and ate their sheep; those of Thebes honoured their sheep and ate their goats. Near to the lake of Mœris crocodiles were venerated; at Elephantine they exterminated them. The murderer, however involuntarily, of a sacred animal, was tortured and pulled to pieces by the people. They fed in the most delicate way a certain number of these divine animals in parks specially kept for the purpose, they adorned them with jewellery, and they scented them with perfumes. Large revenues were set aside for their maintenance. Personages of high rank took care of them and endeavoured to make their life pleasant to them. In the case of fire the father of the family would first endeavour to save his cat, he would then try to quench the fire. The dynasty of the oxen of Apis is well known.

Anthropomorphism became afterwards mixed up with these primitive forms of worship, but it was always more or less zoolatrous. Horus carried the head of his sacred falcon, Athor possessed a cow's head, and Typhon the body of a hippopotamus. Astrolatry was joined to all this; then the adoration of the chief generators: a custom so widely spread in the East and in all classical antiquity. First it was Isis, *magna mater*, the mother of Horus; then Osiris, the fruitful chief, mortal god, commanded by his wife Isis, who was immortal, and queen of all the earth. Language, music, writing, architecture, had been taught to the ancestors by the god Thoth. At last, under Ammon Ra, their supreme god, the Egyptians, who believed in every form of worship, ended by inclining to monotheism; but as they were a most conservative people they kept both their ancient and their new gods, the sacred animals worshipped by their ancient ancestors, as well as the simplified and subtle gods who sprung from the sacerdotal system of metaphysics. They afterwards attempted to introduce some more systematic kind of worship into their incongruous pantheon, by dividing the gods into triads, in strict hierarchical order, connecting them one with another, and making them ascend in rank so that the most important in each division should be placed nearest to their special chief Isis, Osiris, or Horus.

In all this motley religion, among all these various divinities, zoolatrous, astrolatrous, anthropomorphical, and metaphysical, each Egyptian had no trouble in finding a god suitable for his own peculiar fancy. Everything had been religiously preserved. The Egyptian mythology, therefore, somewhat resembles a vast necropolis of embalmed creeds. But fetichism may be seen at the bottom of it all, a fetichism conceived on a larger scale than has elsewhere been known, and in this respect the Egyptian forms of worship do not differ from the others all over the African continent.

IV.

Religions in South America.

In our endeavour to form a schedule of the various systems of mythology imagined by mankind, we are often obliged to repeat ourselves; for in every corner of the globe religious speculation among the primitive races has been very unfruitful and very monotonous. Man has everywhere worshipped animals; he has everywhere peopled the forests with spirits, either of an anthropomorphous or zoomorphous kind, and these spirits have generally been evil-minded; he has often deified the stars, the rivers, and the high mountains. Primitive man has everywhere willed that his desires, his passions, and his emotions should be seen outwardly; he has endowed nature with his own personal sentiments.

Taking this as a general rule, we may be brief in sketching the religious condition of the human race in America. Darwin has seen in Patagonia a sacred tree which the people honoured by shooting at it. The Patagonians, the Araucanians, the Puelches, the Charruas, and others, believe in the existence of evil spirits who are hostile to man; and also that there are other spirits, better natured, who take pleasure in assisting poor humanity. But these people do not abase themselves to pray to either one set of spirits or the other. The Moxos had their gods of harvest, of fishing, and of hunting; they also deified thunder. In this the Yuracares and the greater part of the aborigines of Brazil imitated them. We can easily imagine that they should have made a god of what was to them such a noisy and such a startling phenomenon. The Yuracares had also a god of war, a ravisher god, who laid in wait for them as they were wandering through the woods.

But one of the forms of worship the most widely spread in South America was the jaguar worship (*felis onca*). This god was unhappily too real, for he had with his claws driven a religious fear for his person into the hearts of the Indians. The terror inspired by this divinity was so great that the Moxos first began to worship him to appease his anger. They built altars to him, they made him offerings; they fasted rigorously to obtain the priesthood—a favour given in preference to those men who had fallen into his power and had been fortunate enough to escape.

Many other animals in America were deified, notably the toad. The Indians on the banks of the Orinoco attribute to him the power of sending rain; and they beat him when he does not grant their request.

The Guarayus were anthropomorphous. They worshipped Tamoi, the grandfather, the old god in heaven. He was their first ancestor, and he had taught them agriculture. They built temples to him in an octagonal form, and they went there to ask him to send them rain, good harvests, and other things.

In many tribes the people worshipped the stars; and this

astrolatry became the more common as one got nearer to Peru, where the religion was firmly established.

The Chiquitos used to call the sun their mother, and at every eclipse of the sun they would shoot their arrows so as to wound it; they would let loose their dogs, who they thought went instantly to devour the moon. In Columbia the Indians used to worship the sun. The inhabitants of Bogota worshipped both the sun and the moon; but as their civilisation was already well advanced, their animism was of a perfected kind. They had temples, altars, priests, religious ceremonies, and they used also to offer up human sacrifices. We find all over the earth that gods of every shape and form have ever been greedy for human blood.

V.

Religions in Central and in Northern America.

We have seen that the mythical conceptions of the people all over South America, from Patagonia to Central America, take their rise in the multiform fetichism common to all primitive races, then simplify themselves but without changing their essential character, until at Bogota they form a system of astrolatry in a comparatively learned form. In North America we find the same gradations, starting from the Arctic regions, and going down southwards as far as Mexico. Many hypotheses, devoid of any serious foundation, have been made, attempting to connect the religions of ancient Mexico with those of the old continent. Writers have laboured to show that the biblical Eden must have been the cradle of the human kind. But as we shall presently see, the Yucatan, the Peruvian, and the Mexican mythologies do not in reality differ from fetich animism. They were simply forms of zoolatry, of naturalism and belief in spirits, all of a more or less coarse kind.

In North America, as in South America, we find the foundation of all these creeds, in a form all the more rudimentary as the particular race of people is less civilised, and as they are more remote from the great empires of equatorial America.

The Greenlander and the Esquimaux had similar religious beliefs:

faith in invisible spirits was always their predominant characteristic. Their most powerful spirit was Torngarsuk; he governed a world of inferior spirits, and used sometimes to impart some of their power to the sorcerers, or the angkoks—the intermediary beings between mankind and himself. They had also wonderful fetiches and amulets, which gave to their proprietor the faculty of taking the form of the animal with the skin of which they themselves were made. The people could even create for themselves magic animals. For instance: cut out of a bear's skin the form of a bear, and then enjoin this fetich to go and kill an enemy.

More to the south, among the Red Skins, the worship of animals was very common. They venerated the bear, the bison, the hare. The Mundans used to adore serpents. The Selishes and the Sahaptins deified the wolves on their prairies. Whenever the Red Skins reach the banks of a large lake or a large river they make an offering to the spirit of the waters. Nearly everywhere we find spirits clothed in the human form. A chief of the Red Skins, frightened by a violent storm, offered some tobacco to the thunder, beseeching him not to make any more noise. The more-civilised tribes believed in a spirit more powerful than the others—a grand spirit—of human form, like the majority of his subordinates. Besides these invisible gods they had fetiches, manitous, of whom they used to ask for help in critical moments. In the south, astrology prevailed, and it became the stronger in going southward towards Mexico. The Cumanches of Texas worshipped the sun, the moon, and the earth. The Natches venerated especially the sun; they kept up a perpetual fire in his honour; they built temples to him, in which priests used to perform the services. And we shall see also that astrology was predominant in the large empires of ancient Central America.

VI.

Ancient Religions in Central America.

The religions of the ancient states of Central America—of Yucatan, of Peru, and the neighbouring republics—differed in rality but little from that of the primitive tribes of the Americans. The Mexican pantheon was very vast, inasmuch as the gods of the neighbouring people were everywhere gladly received.

Man worshipped serpents, the jaguar, the puma lion, etc.; images of these animals were to be seen in the temples. Syphilis, even, was deified; this fearful malady was called the god Nanahuatl. Great fêtes were held in honour of the god Tlaloc, the genius of the waters. Each month, and in the Mexican calendar there were eighteen in the year, was under the patronage of a special divinity; the tenth month was consecrated to the god of fire, the thirteenth to the genius of the mountains, the fourteenth to the god of the chase, another god to wine and drunkenness, in whose honour they made large libations of pulque and other liquors.

In addition to all these gods and many more they worshipped the sun, the moon, the stars; but the favourite god, the grand god of the Mexicans, was the god of war, the ferocious Huitzilopotchtli. Nearly all the religious Mexican festivals exacted human sacrifices. Religious madness has never been more bloody than in this country. Those were only ordinary victims who had their chests cut open with volcanic glass, and who were thrown into the fire.

At the accession of each sovereign slaves were killed until there was a sufficient quantity of human blood flowing to make a lake large and deep enough to float a boat.

But the god of war, the terrible Huitzilopotchtli, was more thirsty for blood than any other god. On the occasion of the dedication of the grand temple of this divinity in Mexico not less than 80,000 human victims were sacrificed. It has been reckoned that at least 20,000 victims were annually sacrificed in the Mexican district Anahuac. The faithful were convinced that those who were sacrificed went directly to their gods, and they often charged the victims to bear to the gods their vows and their prayers.

In spite of the fairly-advanced stage of the Mexican civilisation, of the ingenious organisation of its fearful religion, of its numerous clergy, of the immense and numerous pyramidical teocallis built in honour of the gods, and daily stained with human blood; in spite of certain similarities between the Mexican and the Catholic forms of worship, baptism and the confessional for instance; in all these we do not see anything of a higher and better nature than the confined naturalism common to all primitive people; we do not see anything which authorises us to believe that Asiatic or European civilisation was, as has been supposed, sent providentially into Central America. Traditions of this kind, which were common in Mexico at the time of the arrival of the Spaniards, are not different from other legends of the same nature to be found in every country of the world. The Mexican religion and the Mexican civilisation appear to us to be native born. At the utmost we might perhaps connect them with the more ancient societies of which we find many traces in the valleys of the Ohio, of the Mississippi, and of other rivers. These Americans are of much older date than the ancient Mexicans. They had already raised immense hillocks; some funereal, others religious; at one time circular shaped, at another elliptical, and sometimes pyramidical. It is notable that they worshipped animals, as their enormous hillocks, upon which were figures of alligators and of serpents, would seem to show. Like the Mexicans, they had their weapons, made of volcanic glass or of copper. It is probable enough that from this ancient centre (supposing it to have existed) the Mexican civilisation spread itself into Yucatan and into Peru.

The great mass of the ancient Mexican people did not rise above the coarse naturalism we have just described, and the monotheism of Nezahualcoyotl, the king of Tezcuco, who built a temple "to the unknown god, to the cause of causes," was only an individual instance.

The same "unknown god" has been looked for in the Yucatan mythology, richly peopled as it is with naturalist divinities: gods of the air, of the seas, of the rivers, of the forests; from which sprang also abstract gods—the god of death, of life, of love, and others. The natives of Yucatan also put into their pantheon the gods of their sovereigns, of whom they were very fond, or else whom they feared very much; especially the great Zamna, their legendary civiliser. They also had their temples, their priests, their vestals.

They used to make large human sacrifices; their victims, after they had been enjoined to take commissions to the gods, were thrown by hundreds into the holy pit of Chichen.

There is surely nothing in this that need excite our admiration, and the Peruvian mythology is exactly similar.

In Peru, as in Mexico, the gods were very numerous, but they had not all the same rank in the mythological hierarchy. It was customary in Peru to place among the secondary order of divinities the gods of a conquered people. Here, as everywhere, the people were fetich; they worshipped trees, animals, mountains, rivers, and the sources of rivers. Under the name of Mama-Cocha, the sea was the principal divinity of the Chinchas; but astrolatry was the official religion of the people. The sun was to them the god of gods. He was evidently imagined to be an anthropomorphous creature, for the Incas derived their genealogy from him; and at the time of his grand festival, during the summer solstice, when everyone went in great state to watch for his rising, the Inca offered him, in a large gold vase, some *maguey*, a fermented liquor made from maize. If the god-sun was the father of the Incas, the goddess-moon, his sister, was the mother; if gold was the metal consecrated to the effigies and decorations in the temple of the sun, silver was used for the same purpose in the temple of the moon. After these sovereign stars came the retinue of the smaller stars, to whom also a human form was attributed. The planet Venus, called Chasca, or "the youth with the long and curling locks," was worshipped as the page of the sun. The rainbow had also his form of worship; and so also had thunder and lightning, the ministers of vengeance of the star-king.

All these gods were worshipped in many different temples, some of which were adorned with great splendour, notably the celebrated temple of the sun at Cuzco. A whole army of priests performed the rites of worship; they presented the offerings; they sacrificed the lamas at the grand festival in the summer solstice, and they drew omens after inspecting their entrails.

The fetich gods and astrolatry were not the only objects of worship in the Peruvian pantheon. In the kingdom of Quito temples were built to the god of health. And a great spirit, Pachacamac, who was represented in figure, had a temple in the south of Peru. The expounders of the monotheistical mania have endeavoured to make the personification of their fixed idea in the god Pachacamac, who is only a secondary divinity in Peru, and probably the remains of a very ancient form of worship much anterior to the times of the Incas. It is to this antique god, according to the legendary story of Balboa, that Yupangui, an Inca, attributed the government of the world. It was he that Yupangui proclaimed in council as being the first original cause. We ought to view with some distrust all the Catholic similarities that the Spanish writers at the time of the conquest have been so anxious to find between their religion and those of Mexico and Peru. From the abundance of information collected as to the forms of religion in these two countries, we may certainly conclude that the worship did not go beyond an anthropomorphic astrolatry. It was simply the extension of primitive animism that we find everywhere, all over the world, at the commencement of any stage of civilisation. We shall have to mention many other examples before we have terminated our little journey over the mythology of the human kind. The old animism of Central America is far from being extinct, in spite of the efforts and the cruelties of Spanish orthodoxy. Dr. Bell has seen a sacred source in New Mexico. Colonel Macleod has seen sacred fire still burning in some of the valleys of Southern Mexico. Bullock has heard, in the town of Mexico, an old man of Indian extraction regret the ancient gods, in spite of "the three good Spanish gods;" and it is only by means of imposing ceremonies that the Catholic priests have been able, apparently, to convert to Catholicism the Peruvians, the Chiquitos, the Moxos, the Guarani, and others. We cannot repeat too often that the mental condition of a race does not become seriously altered until after a long period of a healthy form of culture.

VII.

The Polynesian Gods.

Though in Polynesia every island, every district, every tribe, sometimes every chief, has a different god, the foundation of the mythology is so homogeneous through all this vast and widely spread archipelago, that it is not difficult for us to describe, generally, the principal characteristics of the religious condition of the people. This same homogeneousness is assuredly one of the principal arguments that may be adduced in favour of their common origin.

Every degree of animism may be found in the Polynesian mythology, from the coarsest form of fetichism to a small number of cosmical anthropomorphous and invisible gods.

At Tonga, anything that excited a feeling of fear, of wonder, etc., was worshipped; animals, and especially the sharks, were deified. In other places reptiles were preferred. When an islander had once chosen his animal-god he confided to him his fears, he consulted him as to all his projects, and asked him for help. Portative fetiches were very common; sometimes they were red feathers, at another time a collection of wooden statuettes, a sort of divine toy, that Porter saw a Noukahivan chief arrange in front of him, singing and clapping his hands as he did so for whole hours together. In the Pomotou islands the fetich was a piece of wood ornamented with some plaiting made of human hair. As often as possible they would have in place of this piece of wood the thigh bone of an enemy or of a dead relation. To these gods they addressed their prayers, and they made no difficulty about changing their gods when the gods did not answer promptly the prayers that were addressed to them. The Noukahivans also wore, suspended to their necks, small gods cut out of human bones.

In New Zealand the people often changed the substance of which their gods were made; the men not unfrequently wore on the chest a little grinning idol cut out of a green stone.

If zoolatry is common in the world, anthropolatry, at least the deification of a living man, is less common. But it was practised in Polynesia. Cook saw at Dolabola an important old man who was the god of the country. These men-gods were not very rare in the different archipelagoes. It is but a coarse form of divine anthropomorphism. Why should they not deify men when they imagine their ancestral, naturalist, and cosmical gods, in the form of invisible spirits, having a human form?

Now, the invisible gods of the Polynesians generally had a human form. That was the reason why the Hawaians did not hesitate to worship Cook and to decree to him divine honours. Even the death of the celebrated navigator did not undeceive the islanders. His bones were religiously collected and borne in great state; and the worship of them every year served as a pretext for collecting taxes for the god Rono. We know, too, with what facility the Polynesians deified, after their death, their chiefs and their men of distinction. A chief of Samoesmo said to Hunt: "If you die before me, I will take you for my god."

They attributed a human form to many inferior divinities of a second order, with which they had peopled the universe. These gods dwelt in the water, in the woods, at the bottom of precipices, and at the tops of the mountains. Every condition and every part of man's labour had its tutelary divinity. One watched the growing of the plants; another the ripening of the fruits; they brought rain, wind, cold, and heat. These familiar gods, generally called Tiis, were often represented by coarse statues, more often cut out of wood, but sometimes out of stone, which they placed either on the banks of the morais, of which they preserved the enclosure, or on the rocks, or along the shore, so as to maintain a pleasant harmony between the earth and the sea.

Like the Icelanders, like the Guanches in Teneriffe island, the Polynesians who lived in volcanic islands deified the volcano. At Tonga a god inhabited the volcano Tofoua. He was supposed to be uncomfortable at the bottom of his crater; and every time he turned round he caused a fresh earthquake. The powerful goddess Pele, who lived in the great volcano at Hawai, is celebrated. Less than half a century ago she had her priestesses in the Sandwich islands; she often came up from her crater to possess or inspire these priestesses, who henceforward were gifted with the power of curing maladies, etc.

The inferior mythology of the Polynesians was never stationary; new gods were always being created, and the old ones were forgotten. The people readily adopted the gods of a victorious tribe, often those of the chief. There were male gods and female gods. At the time of Cook it was a goddess who governed, at Tonga, the thunder, the winds, the rain, etc.; if she was annoyed she destroyed the harvest.

There was a god, too, for each malady, nearly one for each organ of the human body. In New Zealand there was the god of headaches, of heartaches, a lizard-god, causing diseases of the chest, a god of phthisis, a god of the stomach, a god of the feet, and ever so many more.

There were the household gods, generally benevolent and peaceful minded, who as far as they were able maintained order and quiet in the different families, punishing the quarrelsome by visiting them with sickness. In certain islands the people deified the vices that Europeans think to be the most abominable: they had a special god who presided over unnatural love. Men made their offerings to Hiro, the god of thieves, whenever they wished to commit any act of theft.

Astrolatry, which predominated in the mythology of Central America, held only a secondary position in Polynesia. But in Tahiti the sun was deified; the people had placed there a very handsome anthropomorphous divinity, whose hair came down to his feet. But these were the cosmical myths which give a special character to the religious creeds of the Polynesians.

The god Rii had separated the earth from the heavens, which he had spread above as though they were curtains. The god Mahoui had drawn the earth from the bottom of the waters; he had, in regulating the course of the sun, created night and day, which was the delight of all men, for before they were living in darkness. Rou, the god of the east wind, had caused the sea to swell and had broken up the earth into numerous islands, etc. These legends, which formed the stock of Polynesian mythology, differed in each archipelago, but they all related the deeds of valour of the anthropomorphous gods; they told the way in which they had disentangled chaos, fished up the islands from the bottom of the ocean with mother-of-pearl fish-hooks, and performed other wonderful exploits. The most curious and complicated of these legends are to be found in New Zealand, and one of them bears some resemblance to the Aryan myth of Ouranos.

These superior cosmogonical gods lived ordinarily in the heavens, each one in his own place, according to hierarchical order; and Moerenhout says that in Tahiti they were all subjects to a sovereign god, "a great spirit," called Taaroa, to whom the universe was only his shell.

Admitting, to a certain extent, the belief in a future life, of which we have already spoken, the Polynesians did not attach to it any idea of recompense or of punishment. Their gods did not trouble themselves with human morality or immorality; they punished only, and always during men's earthly life, any want of proper reverence shown to them personally. The fear of offending them was ever present to the mind of the poor Polynesians. Every action of his life was connected with his worship, and was manifested by some ceremonial form. A man dared not cut a tree in Tahiti without having first gone to the morai, with the axe in his hand, to give warning to the god, and before bringing to him the first piece of the tree that had been cut. They dared not take a boat out of the timber-yard until they had said their prayers to the morai, and in the presence of a priest, accompanying the procession, who launched the boat into the sea, being very careful that it should not beforehand touch the ground. They could not receive a stranger without the consent of the gods; they would not give lodging to a friend without offering to the gods the first morsels of the meal.

The whole life of the Polynesians was impregnated with mythology, and mythology also entered very strongly into all their religious worship. The people constructed morais, or temples, where they were obliged to bring frequent offerings, for all the gods ate very largely. Sometimes they tried to deceive them by infantine tricks of cunning. They brought to their gods green fruits, promising them better the next time if they would ripen their bread-fruit trees.

The Polynesians had got beyond the primitive worship of a purely individual kind. They had a numerous and powerful clergy, in which the priesthood was hereditary; and its members had the power to taboo, to render everything inviolable, and also in many archipelagoes to designate human victims as offerings to be sacrificed to the gods.

VIII.

The Asiatic Religions.

In order that we may make a ready and tolerably easy classification of different religions in Asia, we should first recall to mind the distribution of the human races over the vast Asiatic continent and its dependencies. There is a race of black men, with curly hair, of small stature, whose facial features are more delicately and finely cut, and who belong more fully to the large Aryan race, than the negroes of Melanesia or of Africa, and who, it would appear, at first inhabited the Malay archipelago, the Malay peninsula, Ceylon, and all the southern and eastern portions of India. We even now find the remains of them in all three countries, specially among the Veddahs of Ceylon.

Two great races, the Mongolian and the Aryan, the first in the human species, came by different roads into contact with the black aborigines. The first of these, the Mongolians, who are now predominant over three-fourths of the Asiatic continent, probably formed the sub-Mongoloid races of Malay, Siam, and Cochin-China, as they mixed themselves with the primitive races. In Mongolia, in Thibet, in China, and in Japan, they have preserved more fully their own special features. In Burmah, and in the eastern portion of India, they have in them a mixture of the Aryan element. In Siberia and in Kamtschatka they may be found still in a savage state.

The other great race, the white Aryan race, occupies hardly

more than the south-western quarter of Asia; and if this race was once predominant in India, it is only in the north-west portion of this country through which it first made its way, along the valley of the Indus, that it has kept itself uncontaminated from admixture with foreign blood.

We have, therefore, now to consider these three great ethnical creations from a mythological point of view.

(A) *The Mythology of the primitive Malay Races and those of India.*

The contemporary remains of these aborigines, and certain mongrel races, the offspring of their admixture with foreign emigrants, have not yet got beyond the rudimentary phase of mythological evolution: the worship of animals, of trees, of stones, a belief in the departed spirits of their ancestors, or, at the utmost, a belief in the genii personifying this or that portion of the life which daily surrounds them.

The poor Veddahs of Ceylon still continue to offer honey, roots, the flesh of monkeys to the spirits of the dead in order to conciliate them. The inhabitants of the Mariana islands preserve in their huts the bones of their forefathers. According to Alvar de Mindana they used to incinerate the flesh and swallow the ashes floating on the top of their cocoa-nut wine. The Tikopians used to worship the sea-eel. In Sumbawa island the Orang-Daugos attribute a magical power to the sun, to the moon, to the trees, to the stones, which they identify with their own genii. Some of the Dyak tribes appoint coarse wooden idols to guard over the paths leading to their habitations, placing beside the idols a basketful of betel nuts to repay them for their trouble. Among other tribes of the same race, certainly mongrel tribes, it is forbidden to cut certain trees which are inhabited by the spirits. Some of the Siamese, in the same way, offer cakes and rice to the trees before cutting them down; the Kariens in Burmah will pray to the spirit of the tree before they begin to cut the tree down.

In the Philippine islands the aborigines, when they saw an

alligator, would throw into the water everything they had in their canoe, to appease the ferocity of the animal, and they would at the same time pray to him not to hurt them. The natives of Sumatra call tigers their "ancestors;" they always speak of them respectfully and make their excuses to them; and all the while they are laying traps to catch them.

The Dyaks also have their genii: Tapa is the creator and protector of man; Jirony presides over his birth and watches over him when he is dying.

The Khonds of India have many local gods, and they are often represented by uplifted stones. These divinities belong to a hierarchical order, under the control of a divine aristocracy, constituted in the first place by the divinised spirits of the ancestors, who also have a few sovereign gods above them: the god of rain, the god of the chase, the god of generation, the god of war, the goddess of the first fruits, and others. And still higher than all these are the god-sun, and his wife the goddess of the earth. It was to Tari-Pennon, the goddess of the earth, that the Khonds not very long since used to offer up human victims, women for the most part, the Meriahs, who were pulled in pieces alive, and whose bones and membranes so dragged to pieces were afterwards thrown into the fields.

The Kariens of Burmah build in their fields a small house, and place there for the goddess of the harvest their presents, and also two pieces of string, so that she may strangle, as he comes in, any evil-minded spirit. They then give to the goddess the instructions that they conceive to be the most fitting. "Grandmother, watch over my field," etc. "Bind together all strangers with this piece of cord." When they thrash their rice they address the same divinity. "Stir yourself, grandmother; stir yourself, that my crop of rice be as big as a hill, or as a mountain." Again, in Burmah the demon of fever dwells in the jungles, and the attacks of fever are the misdeeds performed by the other evil spirits.

All these facts, and it would be very easy to prolong the enumeration, show clearly enough that all these people of whom we are now speaking have not yet passed the most primitive form of animism. We shall find, again, similar beliefs among some poorly-developed individuals, and even among some portions of the great Mongolian race.

(B) *Mythology of the Mongolian and the Mongoloid Races.*

The Kamtschadales believed in a multitude of genii of the forests, of the mountains, of the torrents, above whom was placed a more powerful god, Koutka, or Koutkou. This, however, did not prevent them from worshipping whales, bears, and wolves. Like the Ainos of Japan, the Yakuts of Siberia worship the bear, and before the Russian tribunals the Ostiaks will swear by the head of this animal. They will drive posts into the ground upon the mountains, and they will adorn these posts with rags and then worship them. The Samoyedes pay their homage to certain stones; and also, like a great number of Tartar tribes, the Tunguzians, the Ostiaks, the Voguls, and others, they have deified the sun, though they have numerous small fetiches of their own, and though they believe in the spirits of the forest, of the rivers, of the sun, and of the moon.

In the time of Marco Polo, the Tartars had their household gods, who were the guardians of their families, their animals, and their property. They represented them by idols made of felt and calico, simulating the god, his wife, and his children. They never ate before first rubbing the mouths of these protecting divinities with the fat of their meat. Timkowski has seen among the Mongolians the same idols and the same forms of worship of which Marco Polo speaks. Gmelin has seen Tartars in Tobolsk turn themselves round towards the sun every morning and say to him: "Do not kill me!" Other people in this same region have divinised and worshipped fire (Tylor).

In Thibet, in the midst of the Buddhist country, the mountaineers have their gods consecrated to brigandage.

The Tartars of Altai have gods represented under the form of bearded old men, dressed as Russian dragoon officers. We see that in the mind of the poorly-developed man any strong emotion may, so to say, create a god.

A distinguishing feature in the primitive religions of the Mongolians is the important part played by the sorcerers, often called chamans. They serve as mediators between men and the spirits whom they evoke, usually by means of a magic drum, as they feel or feign a sort of ecstatic rapture. It has been argued that these practices have some special signification, and the Mongolian animism has been designated under the name of chamanism. But there is in it no characteristic feature except the habitual practice of sacred convulsions or falling into fits of ecstasy. All over the world, and in all races of men, the primitive creeds first gave rise to sorcerers, who, in after time, with the progress of civilisation, became transformed into the priesthood.

The boorish forms of mythology which we have just described are again seen in the large Chinese and Japanese empires; they were more ingeniously organised in the latter country, where by degrees they grew into Sintoism.

In China, a belief in the spirits of nature is very widely spread. Every chain of mountains has its divinity. The people set up idols to the god of spring, whom they imagine in the form of a young man. They believe in various nightmare demons, whom they try to drive away by the noise of gongs and of crackers. They have libations of wine to the demi-god Chinnoung, who first taught men the art of cookery. The religion of Tao-sse, or the "doctors of reason," proclaiming, as it does, the belief in primordial reason, according to the rules laid down by the famous Lao-tse, a contemporary of Confucius, admits also the existence of innumerable genii. The priests and priestesses practise magic, astrology, necromancy, etc.

The ancient naturalistic worship, evidently imported from Mongolia, is the official religion of the emperor and of educated men. The son of heaven, and he only in Pekin, worships in different temples, the heavens, the earth, the sun, and the moon, all of which are holy objects, and to which everyone else is forbidden to address his prayers. Personages of a lower station must confine themselves to offering their sacrifices to the spirits of the wind, of the rain, of the thunder, of the dragon, etc. If we put above all this mythological world, the "Heaven," or *Tien*, the only divinity vaguely preserved by Confucius, we shall then see with tolerable accuracy the religion of the Samoyedes, the Tungusians, and of many Tartar tribes, who consider the heaven as an omnipresent god, and who has delegated the government of the world to inferior gods: to the sun, the moon, the earth, the fire, etc.

To these naturalist creeds, scrupulously observed in China, may be added the worship of their ancestors, which seems to have taken a more serious hold upon the people. This also is a Mongolian relic. The Mongolians deified and worshipped Gengis Khan and his family. In the same way, when the Mantchu-Tartar dynasty ascended the throne it apotheosised Kouang-ti, the famous warrior, the Mongolian Mars, and made choice of him as tutelary spirit over the dynasty. It was ordered that he should appear in the official forms of worship; the public officers, and specially the Mandarin soldiers were compelled to worship him. The Chinese are also bound to offer sacrifices to the departed spirit of Confucius; to certain wise men, or celebrated warriors, to whom the emperor built temples; to the patrons of the towns chosen by the emperor from among the celebrated personages. This is the worship of great men inculcated again in our own day by Auguste Comte and his followers. The worship of departed spirits, properly so called, is taken more seriously by the Chinese sceptics. The rich families have in their houses a little sanctuary, in which are placed the tablets of their ancestors. The Chinese, or many of them, believe that one of these spirits in man comes after his death, to live in the tablets of their ancestors, to join together the prayers of the survivors; another spirit rests by the body, which, for this reason, they keep as long as possible in a coffin of gilt gum-lac, near to which the people pray.

In China, as elsewhere, the spirits of inferior divinities, and those of the dead, are often animated with the most wicked intentions. They often dwell in the bodies of men to injure them, or to make them fall ill. Special mediums, often of the feminine sex, have the power of driving away these demons by practices very similar to those of the Siberian chamans. The two ideas probably come from the same origin.

We have now perhaps said enough to show that all the Mongolian mythology is still preserved in China, and that the introduction of Buddhism has only introduced new elements into the forms of worship. In speaking of the Aryan religions we shall have to mention the great religion of Çakya-mouni. But the Buddhist doctrine, of which the primordial idea is so scientific, has become much altered in China from its contact with the old Mongolian creeds. In spite of the prescribed limits of this book, which prevent us from describing these alterations at length, we would wish, however, to mention one fact which does honour to the humanising spirit of the Chinese. In the Celestial Empire there is a numerous sect who worship the goddess Pity, in the form of a woman holding a child in her arms. This deification of one of the noblest of human sentiments constitutes a religion apart, counting numerous votaries who have their temples and even their female convents, and whose duty is to give aid to the faithful souls, and to visit the poor and the sick. These Chinese nuns take the vow not only of abnegation but also of virginity. They are called "the annihilated," "the absorbed," because, according to the doctrine of Buddha, they hope by reason of their sacrifices to deserve absorption and annihilation after death. The analogy between them and the Catholic sisters of charity is striking; but the Chinese doctrine is more noble, and the devotedness is more disinterested.

We have depicted the religious condition of the Chinese, we must now say a few words as to their irreligious condition. In the opinion of all travellers, missionaries and others, there is not a country in the world in which a total indifference on religious subjects is so widely spread as in China. No doubt a man taken from the great mass of the people is often loaded with the grossest superstition; he has idols whom he beats, or whom he worships, as they grant or refuse his prayers; but the various categories of educated people, "the governing classes," are generally either indifferent, or they are impious. They follow the official forms of worship, but without believing in them; and their official worship is only a civil code. The obsequious politeness of the Chinese makes it a duty to say, if they are asked, that they believe in their religion, whatever it may be. The whole matter is perfectly indifferent to them. Chinese law even inflicts a sort of civil death upon the bonzes (the priests) and the tao-sse (those who worship the Supreme Being). It forbids them from offering sacrifices to their ancestors, and forbids, also, under the penalty of a hundred strokes with a bamboo cane, from wearing mourning for their dead relations.

A short time before his accession to the throne, an emperor, Tao Kouang, addressed a proclamation to the people, in which, after he had made examination of all the religions in the empire, Christianity included, he concluded that they were all false, and that men would do well to despise them all alike. This fact, surely unique of its kind, shows the extent of public impiety in China. It is this absolute impiety which has made such an impression upon every traveller, and which paralyses all the efforts of the Catholic missionaries and causes them to despair of hope.

We are compelled, however, to praise a kind of impiety practised by the Chinese and stigmatised in their code with a precision that reflects honour upon this country so different from our own Europe. "Impiety," it is said in the code, "is the want of respect and of care for those to whom we owe our existence, to whom we owe our education, and by whom we are protected. It is impious to bring an action against any one of our near relations, to insult them, not to wear mourning for them, and not to hold their memory in respect."

In spite of the progress of Buddhism, the old form of worship practised in Cochin-China is still found in Japan. In this country the ancient national religion of Sin-Tou is officially recognised; we find traces of it in every hut and in every palace. It is animism of a coarse kind, to which is joined worship for their ancestors, the *kamis* promoted to the grade of benevolent genii, always recruiting itself from humanity; for the departed spirits of the virtuous men go to swell the number of the glorious *kamis*, to whom morning and evening the people address their prayers in their chapels.

The people have also divinised animals and natural phenomena. Following the example of the Ainos, who worship the bear, the Japanese have erected temples to the fox, and they consult him upon any intricate matter. Earthquakes are attributed to a large whale burrowing and dragging himself along under ground; the waterspouts are the flying dragons, etc. Sintoism comprises also the adoration of the celestial bodies. The goddess-sun has her temples, and she is represented quite in an anthropomorphite manner. Before praying to her, they put a clock into motion, so as to attract her attention. The origin of this worship comes, no doubt, as the Japanese say, from their ancestors, from the *kamis*, to whom they rightly attribute the fabrication of the relics in the Japan age of stone. It is altogether a collection of infantine conceptions, such as we find nearly everywhere in primitive societies among men of every race and of every colour.

Mythology of the White Races.

(*A*)

The great religions and the most complicated forms of mythology have been born among the white races; but if, as regards religious speculation, and also as regards every other form of intellectual development, the people in this human type have shown themselves to be more intelligent and more inventive than others, they assuredly have the same puerile creeds of faith. No doubt but that the animism in these races has grown to be more advanced; but they, like other races, have practised, and still continue to practise, the grossest form of feticism. And among them, as also among the others, it is very easy to discover the phases of animism.

The worship of animals may be seen nearly everywhere in the Aryan and Semitic nations. The serpent has been worshipped in India, in Phœnicia, in Babylon, in Greece, in Italy, among the Lithuanians, in Persia, and in other places. Hanuman, the monkey-god, had his temples and his idols in India. Ganesa had an elephant's head. The cow is still in India a sacred animal; according to the religion of Zoroaster, the dog was the creature worshipped. The Lombards adored a golden viper until Barbarossa took possession of it; the ancient Prussians used to place food-offerings before serpents.

In European India there is still existing a whole mythology of trees. In the course of his innumerable transmigrations Buddha has been held to be the genius of a tree forty-three different times. The Græco-Roman polytheism had its dryads and its hamadryads. According to Cato, the woodcutter before he began to cut trees in a sacred wood was bound to sacrifice a hog to the gods and goddesses of the wood. Every Hindoo village has its sacred tree (*ficus religiosa*). The ancient Slavs consulted their old oaks, and the veneration of the Celts for the same trees is well known. Holy sources are still very numerous in Europe, not counting that at Lourdes.

The worship of stones is also very widely spread in India. No doubt in many stones a divinity or a spirit is frequently lodged; for instance, Shakty, Siva, and the five Pandous. But we may believe that the early adorers had a less subtle form of worship. The Greeks had their sacred stones which they would anoint with oil, and to which they would pray with earnest devotion.

The adoration of rivers is less rare. The Hindoo has always considered the Ganges to be a sacred river, and we read in Homer that the river Scamander was deified.

The divinised winds, the Vedic Marouts, drive the angry clouds over the sea. In the Iliad, Achilles offers libations to Boreas and to Zephyrus if they will blow upon the funeral pile of Patroclus.

Even now the Carinthian peasant will place upon a tree in front of his house a wooden vase holding meat, so that the wind may eat and appease his anger.

The worship of the earth, of the stars, of the sun, of the heavens, had quite another signification. That is the second phase of animism, in which man begins to look higher and farther; he begins to show his contempt for the smaller local gods, though he has not yet altogether abandoned them.

The earth-mother, *prithivi-matar*, is deified in the Rig-Veda. The Romans also had their earth-mother, *terra mater*; and Tacitus found this divinity among the Germans. Hesiod relates that every night the heaven, Uranus, used to come down upon the earth, to water it, until Chronos, or Saturn, the god of time, who waited for a favourable opportunity, emasculated him with his scythe. A pleasing and ingenious fable, no doubt, and we may therefore take it for what it is worth.

The earth, as deified by the different Aryan mythologies, is not at all the astronomical globe rolling in space; it is a fertile country clothed with trees, corn, and harvest produce. At the same time, also, the people used to worship the sea. Cleomenes, as he directed his course towards Thyrea, sacrificed a bull to the sea before allowing his army to embark. The sea, the Poseidon of Homer, the Neptune of the Latins, was the god who shook the earth.

The sun, the moon, and the stars have been divinised nearly everywhere. We read in the Bible that these forms of worship were forbidden to the chosen people of God. The silver star Selene, Diana, was much respected both in Greece and in Rome.

The Vedic singers used to celebrate the great Sûrya, who knows everything and sees everything, before whom the stars would fly as though they were thieves. "The eye of the sun," they say, "is supreme benefaction." It is the Helios of the Greeks, the Persian Mithra, the Tyrian Baal, the Latin Apollo, of which the Christmas festival also unconsciously celebrates every year the revival. The great religion of Zoroaster, based on the struggle between light and darkness, personifying the good and the evil in the forms of Ormazd and Ahriman, is the most celebrated of all the solar religions. Though we may put on one side with becoming irreverence the solar myths of every sort devised by the ingenious spirit working in the numerous different mythologies, we may still find a great many of these same solar myths in all the antique Aryan religions, even in those of the Germans and the Celts.

By a greater effort of generalisation our Aryan ancestors divinised the whole heaven. At first they worshipped the bright sky Dyu (Dyaus), in the midst of which were placed the clouds. Then the Vedic singers, speculating after their own fashion, imagined the heaven—that is the heaven-father—to be the husband of the earth-mother. Then it was the starry heaven, Varouna with a hundred eyes, etc., which we find again as an all-powerful divinity, as a master of the gods, in the Greek Zeus and in the Latin Jupiter. Secondary divinities accompanied the great god—for instance, the Indian Indra, the god of thunder, and others.

The powerful Odin of the Scandinavians seems also to be identified with the heaven, which, as we have seen, is still the principal divinity of the great Mongolian race.

The heaven conceived as an animated living creature is certainly the myth which realises more fully than any other the idea of omnipresence. Man cannot escape from him. He is everywhere, like the subtle god imagined by the Christian metaphysics, which has probably borrowed this quality from him.

All these examples, which we might multiply tenfold, prove beyond doubt that in his mythology the primitive white man did not rise to higher ideas than men of lower races; and we know also that, like them, when he wished to represent his naturalist divinities, he conceived them to be anthropomorphous. Nor has he shown himself superior to the other human types in his conception of the nature of the manes of their ancestors, of genii or wandering spirits.

(B)

According to the Vedic mythology the departed spirits, the souls of the dead, go either to impound themselves with the spirits of the wind, forming part of the cortége of the god Indra, or else they join the gods in celestial space and partake of their existence, the amplified image of human existence upon earth. The departed spirits of the Hindoos exact, like the spirits of so many other people, that one of their descendants should bring peace-offerings. On the thirteenth day of the moon, rice boiled in milk and honey should be offered to them. For they are corporeal beings, having more or less all the wants of the living man. They have also all his passions. Like the aborigines in the Marquesas islands, and many other primitive people, the Romans used to think that the spirits of the wicked, the larvæ or lemures, used to torment man by their apparitions. Both the Greeks and the Romans used to think that the manes of men who died a violent death, or those who were not buried, would pursue and ruthlessly destroy the innocent. To charm away the evil intentions of the manes, who often drew their own near relations to them, the people used to offer to them sacrifices. We read upon one epitaph that a husband begs his well-beloved departed wife to spare him for many years, so that he may continue to offer to her sacrifices, to bring her crowns, and to fill her sepulchral lamp with perfumed oil. We know, too, how much the apparition of the dead is thought of in the Christian mythology. The dead are often animated with very bad intentions, if we may trust to the popular opinion now current in Brittany. There are places in Brittany where there is hardly a woman who has not seen an apparition, or where the boldest man would not cross a cemetery at night without some inward fear of danger.

The belief in demons and in genii is also very common. In the south of India men do not dare to go out after sundown for fear of the spirits, or at least they provide themselves with a firebrand to drive them away. Even in the daytime people will light their lamps to drive away the demons. The Greeks and the Romans also believed in men being possessed of devils. The sick people in Homer were tortured by a horrible demon. Pythagoras thought demons were floating about in the air, and that they were the cause of sickness. The mad people, and especially the epileptic, were the creatures who were possessed. Plutarch admits the existence of a demoniacal hierarchy. According to him the manes of the best of men may become heroes, the heroes may become demons; and these demons are sometimes promoted to the rank of gods. The New Testament is full of demoniacal stories, and all the fathers of the Christian Church describe minutely the possession and the practices of exorcism. All the Middle Ages in Europe were, so to say, a sort of demoniacal monomania. The orthodox funereal piles have during many centuries devoured sorcerers by the thousand; our ancestors have been tormented by all sorts of nightmare dreams, they have fancied their blood being sucked out of them by vampires.

Like many savages, the Aryan people have sometimes metamorphosed various maladies into evil genii. The Persians have seen scarlet fever in a human form. They have said to one another: "Do you know Al? She has the look of a blushing young girl with pink cheeks and hair like flaming curls." The Jews thought that the pest was an exterminating angel; the Romans used to worship the god of fever.

The genii of antiquity, the guardian angels, and, in general, the angels of Zoroaster and of Christianity, have also very primitive conceptions. The demon of Socrates is well known in history. Menander believed in a serving demon, in a guardian angel, belonging to each man from the time of his birth. Augustus had a guardian angel to whom he used to make offerings, and by whom he would swear. According to Maximus Tyrius, a whole legion of these genii used to serve as mediators between men and the superior divinities. The characters of these genii were as various as those of men, and also there were among them many of the shadows of men. They used to cure maladies, give counsel, and sometimes offer their protection to certain well-favoured individuals. There are some, like Æsculapius, who continued to interest themselves in the same pursuits that they followed here in this life below. There were some who were extremely spiteful. The deification of the Roman emperors, alive or dead, belongs to the same order of ideas, to which is also connected the Christian hagiography: all this, as a matter of fact, comes from the early primitive superstition of man's adoration of his ancestors.

The idol worship in the Aryan races is not one bit more exalted than the coarsest rites of many of the savage tribes. Nothing can be baser, or more fetich in its nature, than the worshipping of idols; and by idol worship we mean the actual images of divinities, whatever they may be. In this way the majority of the great Aryan races fall below the level of the African fetichism. It is plausibly argued that the adorer pays his homage to the spirits themselves, whom these images are intended only to represent; but the greater part of the devout people in every religion are not so clear and distinct in their metaphysical ideas. We may advance with full certainty that it was the god, the saint in wood in plaster, or in marble, that they worshipped. In Rome the bigots who came to pray in the temples used to treat with the officiating ministers to be placed as near as possible to the ear of the idol, so that they might be better heard. The Tyrians used to chain up the statue of the sun, to prevent it from quitting the town. Augustus punished the Neptune in effigy because he had behaved badly. The ancient Arcadians used to beat their god Pan if they came back from the chase empty-handed. On the day of the death of Germanicus all the idols in Rome were broken.

When the missionary Dietrich overthrew the idols of the Esthonians, the people were very much astonished not to see blood flowing from the figures. We know that the Catholics venerate this or that statue of a saint, or of the Virgin, in preference to the others. The following prayer of a Finnish Russian will give a very good general notion of the coarse idea that the poorly enlightened Europeans have of their divine personage and of their idols: "Tell me, God Nicolas, did my neighbour little Michael whisper to you evil things of me, or do you think he will do so? If he does, don't believe him. I have done him no wrong, nor do I wish him harm. He is a terribly conceited fellow, and he is a chatterbox into the bargain. He is only a hypocrite, for he does not really respect you. I honour you from the bottom of my heart; and, look here! I will burn a taper for you."

We may feel tolerably sure that the African negro converses with his gods much in the same way; and if we call to mind the fact that the veneration for spirits is still common in a so-called superior religion, we may conclude that in a large portion of his mythology the white man is not different from men of inferior races. However, in one direction the mythology of the white races does raise itself above that of primitive men, and this we now propose to consider.

(C) *The Great Aryan and Semitic Religions.*

The principal dogmas of the great Aryan religions are so well known that it is unnecessary to repeat them here; our endeavour is principally to show the higher qualities of these mythologies

It is beyond all doubt that in the majority of instances where sentiment and intelligence are strongly shown, the white races have surpassed the others; but it is necessary, also, to see this superiority in a true light. If we look around us we shall see that the mean level of the majority of individuals in the Aryan races is hardly higher than that of the African negro. The privileged races only, those of whom we are now speaking, have furnished a very small number of chosen people, sometimes better and sometimes more intelligent than the rest of their kind. These intruders, naturally reformers, have been nearly always scoffed at, persecuted, and often put to death; they have by degrees enlarged in every sense the horizon of the human mind; and, thanks to their conquests, the common herd of men have ever led a more and more humane life.

It is in this way that religion, like everything else, has slowly become perfected. In the Aryan parts of Asia, and in Europe its dependency, the natives have followed, and do still follow, the rudest form of animism. When the Hindoos first saw a railway train they divinised "the vapour" and offered up to it garlands and melted butter. They had also many years previously deified smallpox, which became the goddess Mata-Jita. This unintelligent credulity is not peculiar to the Hindoos. Burton heard an old Arab woman of the tribe of the Ess cry out in a fit of neuralgia: "Oh Allah! May your teeth hurt you as much as mine are now hurting me." To find examples of this savage form of religion the European has only to leave his own country.

All this is incontestable, and yet the religious speculations of the Aryan races have grown to very considerable proportions. We know from the Vedahs that the ancient white tribes of Central Asia at first adored fire with complete fetich worship, thereby not differing from many primitive people who used either to worship fire or the sun; but by slow efforts of thought the descendants of these fetich Vedahs have ennobled their worship and have recognised in the god Agni the physical fire, the vital warmth, and the thinking principle. Their progress did not stop there, for by dint of simplifying and bringing their polytheism into a synthetical form they have ended by deducing from it the quasi-scientific trinity of Brahma, Vishnu, and Siva—production, conservation, dissolution, or return to the pantheism of Brahma, from whom every creature springs, and where they finally go after a series of incarnations more or less long.

In the same way the god Agni, of whom we have spoken, after being for a long time fed upon melted butter and the alcoholic liquor from the acid asclepias, the sacred Sôma, first became a glorious child, then a metaphysical divinity, a mediator living in the fathers and living again in the sons. A long mythical story has sprung from Agni, in which has been imagined, and with a great appearance of probability, all the legend of Christ.

The Parsees also, whose dogmas have penetrated so widely into Christianity, first considered fire as an anthropomorphous god, giving to man happiness and good health on condition that they kept him well supplied with wood, with perfumes, and with fat. They have now become more intelligent, they address their prayers only to the invisible spirit of the fire. This religion of the Parsees, or rather the antique religion of Zoroaster, is different from the inferior form of worship by the great simplicity of its divine personages, reduced almost to the god of light and the god of darkness, to Ormuzd and Ahriman. And also it has further advanced by suppressing all idol worship, to which the Hindoos so strongly cling.

Every superior religion, Brahminism, Mazdeism, and above all Buddhism, have in a very large measure brought moral ideas to bear upon their mythology. Buddhism has attempted quite a social revolution. It has undertaken to break the social chain of caste, and to inaugurate into the world a desire for humanity and peace. Now, though this religion be mainly professed by the Mongolian races, it is in reality of Aryan origin. The Dalai-Lama of Lassa and all the Chaberons of Thibet and Mongolia, who expound so cleverly the doctrine of transmigration and of revelation, live now under the Buddhist religion, but they would never have invented it.

This absorption of morality by religion characterises the great mythologies of which we are speaking. The creatures of these religious metaphysical systems have not been satisfied, as were the Vedic tribes, as were and still are all primitive races of men, to imagine their myths in a simple and credulous way, simply to invest their emotions with a human form, or to explain coarsely the phenomena of the outside world. They have had some ideas of social life; they have interested their gods in the ways of men, and made their religion be governed and bound by their morality.

In the same way, also, the Greek and Roman polytheisms originated, though at a later date and with much less completeness. These religions sprung evidently from the ancient creeds of Central Asia, where we now find another kind of thought; the divinisation of certain ideas, of certain superior faculties in the human brain, for instance, the deification of Time—Saturn, of Reason—Minerva, etc.

The great Semitic religions, Judaism and Islamism, inferior in so many ways to the mythological syntheses of Aryan Asia, have also narrowly joined together their moral ideas and their dogmas, banishing as completely as they could all fetiches and all idols. But their anthropomorphous god, a cruel, violent, capricious, true Oriental despot, is very far below the Brahmin and Buddhist notion of pantheism, which, as Moleschott says, expresses mythologically the great formula of all the universe: the circulation of life.

In spite of its being so barren in metaphysical qualities, Christianity, a hybrid religion, a confused mixture of Vedism, of Mazdeism, of Brahminism, of Buddhism, of Judaism, nevertheless deserves some of the praises which we have given to the great Asiatic religions. Like them, it has deeply concerned itself with moral duties, though it has borrowed from them the greater part of its lessons. But the Christian metaphysics, poor and without logical sequence of thought, distinguishes itself from the others only by the adoption of an insane idea, borrowed from Philon and the Alexandrine dreamers: the idea of creation ex nihilo.

Christianity has also lowered itself by taking note of all the coarse manifestations of the primitive religions: fetiches, idols, the worship of one's ancestors, the adoration of genii, etc. Its rites, for the most part servilely imitated from the Buddhist rites, are wholly devoid of originality. Finally, and this is a much more serious matter, Brahminism and Buddhism are not incompatible with science; Christianity is diametrically opposed to it. Scientific thought has grown and made its way in spite of Christianity, and by means of scientific thought Christianity is one day destined to perish.

II.

The Evolution of Mythology.

The mythological gradation of ideas, now generally admitted, starting first from fetichism, passing through polytheism, and ending at last in monotheism and in pantheism, is convenient for the exposition of facts; and, also, it corresponds tolerably well with the principal phases in the evolution of religious ideas. We must not, however, attach to it any definite or absolute value, and especially we must guard ourselves from thinking that between the different shades of error there are lines of demarcation clearly marked. All these degrees confound themselves together into one illusion, which Mr. Tylor has called *animism*.*

During the long periods of its infancy and of its youth, the human mind has frequently wandered out of its own province. It endows liberally with conscient faculties similar to its own everything that falls within its reach: trees, mountains, stones, animals, rivers, etc. There is, however, in this animism a gradation which corresponds to the progress of intelligence and of experience.

Vast as is the universe man's will is very much restricted by the shortness of his own sight. The savage, whose eyes are hardly opened, knows of nothing beyond the small district in which he contrives to live by fighting for his daily existence. His animism is as closely confined as he is himself. It is only to the objects

* See A. Lefèvre, "Dialogue between A and B on Survival and Animism," in the "Revue Internationale des Sciences," March, 1879.

and beings round about him that primitive man can lend a conscious life similar to his own—it is that which we call *feticism*.

Then, by degrees, as the intellectual vision becomes more and more distinct, the animal horizon grows larger; man anthropomorphises the great phenomena of nature. At the same time the crowd of smaller gods fall more and more into discredit. The divinities grow less in number and increase in importance—this is polytheism if we wish so to call it.

But even this divine aristocracy, following the general course of ideas, will not, any more than the anterior democracy, be proof against the progress of the human understanding; and, by degrees, suppressing first one and then another of these imaginary personages, we come either to monotheism or to pantheism. Finally these great and last mythical conceptions in their turn undergo the tragic destiny of the gods who lived before them. They fall under the blows of the battering-ram of science, in spite of the desperate attempts made by metaphysics to clothe man with a divine idea.

We have now run rapidly round the whole of the mythological cycle. Man now sees truly and clearly his position in the universe. The "nature of things," according to Lucretius, or, in more modern language, the understanding enlightened by science, says to man: "Poor creature, so humble and so sublime. You must now recognise that you are but the first of earthly animals. Your extraction is very small, you have sprung from very low, but by a slow series of efforts you have raised yourself above your inferior brethren. You see and understand that which they could neither see nor understand; and this same fulness gradually acquired by your conscious life has made of you a most singular being. You know your origin, but you cannot see the goal to which you are tending. Persevere and work on. The road over which you still have to travel is much longer than that left behind. Before heaven's light, Apollo, as he was called in former days, is extinguished. Generations innumerable have yet to struggle, to suffer, and to live."

But not to wander needlessly, future humanity must know that the exterior world which binds him is blind and pitiless, that to get the upper hand over it he must first overrule its laws. It is all-important for him not to forget that for each individual the conscious life is strictly confined between his earliest bleating sound and his last and latest breath. It is only in the future destinies of humanity that we can place the care of the life beyond. To improve one's own self, to work for the general welfare, ought to be the effort of every human being. Manu said: "We must pay the debts of our ancestors;" in other words, bequeath to our descendants, in a happier condition than our own, the patrimony we have received from preceding generations. That is man's great duty in life; it carries with it its own reward. We must live usefully, and die without lamenting our fate.

CHAPTER XVII.

WORSHIP AND PRIESTHOOD.

THE mental evolution from which worship and priesthood have sprung has been so much alike all over the world, that the subject matter of this chapter may be comprised in a few pages.

The very inferior races, the Fuegians, the Tasmanians, the Australians, the Hottentots, and others, have neither temples, nor priests, nor rites. In this primitive phase of human development, man's religion does not, at the most, go beyond believing in the existence of anthropomorphous or zoomorphous spirits, who haunt the rocks, the grottoes, and the trees; the idea of communicating with these beings, supposed generally to be evil minded, does not occur to anybody.

At a somewhat later stage, when man has become more thoughtful, when he has begun to use his reasoning faculties, he naturally imagines that by presents and genuflexions, etc., he will influence the decisions of the gods, made after his own image. Then the temple is built and the priest appears. At first the temple is very humble. It is a mere hut, like any other habitation. The gods are conceived to be wandering creatures very similar to men;

they have, therefore, an asylum offered to them, or a house in which they may rest themselves. The images of his gods are put in this dwelling-place, and they are often confounded with the gods themselves before man has learnt how to abstract the spirit from the idol which it is intended to represent. Nothing can be more common than these primitive temples in Equatorial Africa, in the valley of the Niger, in Ashantee, and in other places. Nothing also is more natural than that when the people should have formed a convenient lodging for their idol they should offer him food, fruit, and animals to eat. In principle, the sacrifices and the offerings have no other object but to supply the god with nourishment. It is for that reason that the African negro kills his fowls in the hut of the fetich; it is for that reason that the Polynesians allow their pigs to become putrid in their Morais.

With the temple, or often before the temple is built, the priest will appear. He is not yet the majestic and official personage of more advanced civilisations. He is simply a member of the tribe, who, with more or less good faith, pretends to be endowed with the privilege of communicating with the spirits, to serve as a medium between them and men, to cure the maladies so often caused by the ill-will or anger of the invisible powers. This chosen man is as yet no more than a modest-minded sorcerer. We find him gifted with very inferior powers among the Kafir tribes; for there his only superiority is in being able to send rain.

But when the house for the spirit has been built, the sorcerer then very readily develops into the man-fetich, the guardian, the servant of God. This is not, however, always the case. In this early period the forms of worship are often performed by the head of the family; for each family has very frequently its own peculiar gods. Among the Vedic Aryans it was the head of the family who offered up the sacrifice to Agni. The same practice prevails also in many districts in Equatorial Africa. But when there are gods belonging to each tribe, the care of communicating with them is borne by the chiefs or by special sorcerers, by priests, who, possessing the ear of the divinities, know what they ought to do to conciliate them; and these men are paid by the public for their services. By degrees, these priests, among whom the priesthood was at first merely a personal matter, ended by forming among many people an hereditary caste. We find this to be the case in the majority of the Polynesian archipelagoes, as Tahiti, in the Marquesas islands, at Tonga, and in others. These important personages henceforward concern themselves with all the acts of social life. Their important business is to present to the gods their offerings, to sacrifice both the animal and the human victims; and sometimes, as in Polynesia, to designate the human beings. For nearly everywhere the primitive gods have been large eaters; they have been very fond of blood, and often of human blood.

Once constituted into a caste apart, or even a well-defined class of men, the priests have nearly everywhere been disposed to set themselves up as civil governors, and they would have succeeded if in the primitive societies the perpetual state of war did not necessarily engender a class of military chiefs. But sometimes, as in that part of India where the Manu Code is in vogue, the warlike aristocracy itself submits, more or less docilely, to these divine men, who were ordinarily held to be the guardians of morality. Sometimes, also, an alliance, more or less intimate, was formed between the kings, the nobles, and the priests, between the men of the sword and those of the altar; the spiritual and the temporal powers mutually protect their own interests in the best way they can.

It may happen, also, that theocracy will get the upper hand; but it can hardly do so unless the country is more or less completely free from warlike invasions. The Tibetan theocracy offers to us the most remarkable instance of the absolute triumph of the religious power. There we find a whole people of lamas maintained by the laborious part of the nation, and inhabiting some thousands of convents, from which they govern the country. This clergy, hierarchically organised, has its kontoukous, whom we may compare with the Catholic cardinals, and its Dalai-Lama, the grand lama of Lassa, a pope three times holy, and having over his colleague in the Vatican the inestimable advantage of being immortal—a privilege enjoyed by all the great dignitaries of the lamaic church. None of these holy personages ever die. From time to time they transmigrate, leaving a worn-out body to be born again in a miraculous child, who has kept the recollections of his former existence, as he is able to prove by interrogatories that are put to him. Occasionally, the lamaïc dignitaries before changing their body will designate the place of their future regeneration. Many simple lamas also enjoy this faculty of revival, and in certain convents in Bhotan the share of temporal wealth assigned to each father is made proportionate according to the number of his transmigrations.

Wherever priesthood has taken a firm hold, the ritual, gradually growing more perfect, becomes at last a science; and also the dwelling-places of the gods, which were at first so humble, become changed into sumptuous buildings. But, in short, the ceremonial forms are always reduced to certain obsequious practices destined either to appease the divinities or to bring them into a good humour. The worship is nothing more than knowing how to ask favours of the divine personages. This act will vary very much, according to the race of people and the degree of their civilisation; but the foundation of it is always the offering, the spontaneous gift of presents supposed to be agreeable to the gods. At the outset these presents were very coarse. They consisted of food, drink, and perfumes, and often of animal or human sacrifices; for anthropophagy exists or has existed in every country. The gods are everywhere very like the men who made them.

In the most primitive phase men thought only of giving to the gods food to eat; then in order to please them more they invented self-privations, they imposed upon themselves sufferings, and they allowed themselves to be sacrificed as victims. Facts of this kind are over-abundant in the annals of humanity. When he was losing a battle, the king of the Moabites promised to burn his son under the walls of the town. The sacrifice of Jephthah's daughter is well known. In honour of their god Moloch the Phœnicians used to burn the children of their noblest families.

But as the intelligence in any race of people increased, their religious fervour has grown more timid and more parsimonious. Prayer and genuflexions have gradually replaced the burdensome peace-offerings and the bloody sacrifices. Sometimes, as in China, offerings are made only in effigy; paper images are substituted for the real objects, and are burned in their place. The primitive man, urged by some need or by some emotion, prays coarsely but sincerely. "Come and pray," a missionary said to an islander in Madagascar. "Pray for what? I am not in want of anything now," was the answer. After awhile prayer becomes a formula read mechanically at stated times; rites and ceremonies are performed without warmth, without earnestness, simply from habit. This change seems to indicate that the age of faith will give way to the age of examination, that the age of ignorance will have to yield to the age of knowledge.

BOOK IV.

SOCIAL LIFE.

CHAPTER I.

MARRIAGE.

I.

Union of the Sexes among Animals.

It may not perhaps be amiss, if on the subject of marriage, as well as on the majority of psychological and sociological questions, we begin by noting shortly some of the habits common in the animal kingdom.

As reproduction is the most imperative condition for the duration of any organic kind, all animals of different sex, especially those for whom coupling is necessary, seek each other and draw near together at different seasons of the year for the purpose of multiplying their own kind. Were we now concerned with psychology we should ask, whence comes and how is born in us the amorous instinct—a providential condition for the duration of any species—this tyrannical instinct, the cause of sympathy for others, and also of intense egoism! It is not our business now, in a sociological point of view, to do more than state the fact, and to enumerate the different customs.

These customs are multiple. The most inferior, and one of the commonest, is promiscuity. Many animals copulate together as the wish prompts them. They are not careful in their choice, nor do they think of inconstancy. We know, however, that among many kinds, specially among certain kinds of birds, the male courts the female, and tries to please her. They will sometimes leave each other as soon as their desire has been satisfied, or they may remain together until the young ones are fledged.

If promiscuity is frequent among animals, polygamy also is not uncommon. In that case, the male will appropriate to himself a certain number of females, and will drive away all rivals. The barn-door cock is a type of a polygamous and jealous animal. But polygamy is far from being the rule among animals. In point of fact, it is hardly possible but among the sociable animals, among those who flock together, and also where the number of the females is very much larger than that of the males. It becomes a matter of necessity in hymenopteral societies, where there are only a few males among a very large population of females.

Animal polyandry is very rare, for among all the superior kinds, the female, because of her relative weakness, is obliged to receive the caresses of the male, and she cannot form for herself and maintain a masculine seraglio. In many kinds the female seems to have a decided predilection for the stronger creature, and when the male rivals fight together as to which shall own her, she waits quietly until the conqueror declares himself, and then gives herself up to him. We should have to shut our eyes very closely not to recognise that, in the human kind, amorous selection is made in the same way, perhaps in more disguised and more various forms.

Monogamy is not very rare among animals. It has been adopted by some of the superior races of man, among men, too, of different temperaments: our moralists are pleased to regard it as the form *par excellence* of human marriage. Monogamy is compulsory among creatures who are scattered apart very widely, who can hardly live together except in couples, either because their means of subsistence is scarce, or because they are naturally of an unsociable kind. These conditions, however, are not absolutely necessary. The *macacus silenus* of India has only one female, and he is faithful to her up to his death. In the guinea-fowl kind the male confines himself to one female, no matter how large be the number of hens. The mode of intercourse will sometimes change with the kind of life. The wild duck, for instance, ordinarily a monogamous bird, becomes polygamous when he is tamed. It may be that the monogamous social animals, as the guinea-fowl, come from ancestors who have for a long time lived together in isolated couples.

We know that there are animals among whom a real and moral monogamy exists, to a greater extent even than our notion of human monogamy. In our kind the death of a husband or of a wife does not leave the other inconsolable. With the Illinois parrot (*psittacus pertinax*), widowhood and death are ordinarily synonymous; and a similar case has been observed in an ouistiti (*hapale jacchus*) in the Jardin des Plantes in Paris. In man, as in animals, the strength of the affective sentiments is not always proportionate to the degree of intelligence.

We will conclude by mentioning certain kinds of an eminently sociable nature--bees and ants—among whom the desire for the public prosperity has overruled their own personal instincts to such an extent that with them reproduction has become, through the divisioning of labour, a work set apart for particular individuals. So much self-denial has not yet been observed in any group of men; for it would seem that the celibacy of the lamas, of the Tibetans, and others, has quite other ends in view than working for the joint interests of all the members.

But among animals there are no laws or codes decreeing or prescribing the union of the sexes. Why then are the modes of this union so various? It is only in the competition for life, in the necessities brought about by the struggle for existence, that we can find the reason. The dispersion or the congregation of individuals, the proportion of the sexes in each kind, play certainly the principal part in production coming from promiscuity, from monogamy, or from polygamy. The conjugal form, which is the most sure method for the reproduction of the species, which has best adapted itself to the ways of household life, and for guarding against rivalry; —for these and other reasons the conjugal form has been necessarily adopted, it has become customary, first from habit, and afterwards by force of instinct. There can be no doubt that an unswerving monogamy, of which we have given instances, has for certain kinds, under given circumstances, a considerable advantage; it enables them to protect their young and also to preserve of them a greater number.

The same law, the same necessities, have led different human societies into this or into that connubial system. And to say the truth, man, intelligent as he may be in this respect, has not shown himself to be much more inventive than the animal. He has been less so, for, unlike bees and ants, he has not attempted to set apart a caste uniquely for the purpose of reproduction. Man has sometimes, though not always, better determined the form of sexual unions. It is mainly in the human societies that marriage exists—that is to say, a sexual union governed by social conventions—but these conventions are far from having everywhere and always any strict system of laws.

II.

On Human Marriage.

We shall see that in many primitive human societies the sexual unions are altogether of an animal kind, without law, and without restraint. Are we to suppose, then, as some sociologists would have us believe, that this bestial condition is in man a first step, from which marriage has evolved by passing through well-known phases of existence, everywhere the same? By no means. Like animals, men have to obey the severe laws of necessity, and necessity makes her wants felt in very various ways. And also, like animals, primitive men satisfy their gross desires as far as it may be in their power to do so. In one race of men, or even in one tribe, we may find different forms of sexual union, for among most savages there are no laws upholding morality, or offering protection to the weak. We have already seen how, in primitive societies, the lives of children were left merely to the caprice of their parents. The lives of women are not protected more surely, and their liberty of action is very much less certain. But, as in the long run the prosperity of any social group will depend upon the individual acts of each of its members, this or that code of manners will at last be the means of bringing about the extinction or the survival of a tribe, struggling for existence against its rivals. Under the influence of outside circumstances, customs that are socially useful will end by implanting themselves in the ethnical groups. Habits and manners are formed, and to depart from these is found to be prejudicial to man; but these manners will vary because the conditions of the battle for life are not everywhere the same. That is why the form of sexual unions—of marriage if we wish—is also variable. Promiscuity, polygamy, polyandry, partial marriages—obliging those joined together for a portion only of the week or of the month, and permitting simultaneously a dozen or twenty unions—monogamy, exogamic marriage, and endogamic marriage, all these will be found to exist capriciously in the different human societies. For instance, a never failing monogamy, which all our moralists extol as as the type *par excellence* of conjugal life, is practised by the Veddahs in Ceylon; and here we find a people not much superior to certain animals. Many of the Veddahs, like many animals, live in couples, being very much scattered and dispersed, and, consequently, polygamy as well as polyandry is impossible to them. The noblest forms of connubial life are not, in our opinion, always the sign of a high intellectual development. The facts which we shall address will show that marriage, like everything else, is governed by the necessities of existence.

III.

Marriage in Melanesia.

There used to be in Tasmania a custom, which still exists in Australia, and in the greater part of the Melanesian islands, allowing exogamic marriage, or marriage by capture. Such is, among these people, the legal form of sexual union. Later on we shall have to describe it. It is the amorous form of sexual union; for exogamic marriage has no other object in Australia than the possession of a slave, of a beast of burden, to carry wood and water, etc. In the heart of the tribe, girls from the age of ten years and upwards, and boys from the age of thirteen or fourteen, will cohabit together freely. The people hold certain festivals intended as the signal to the young people to begin their intercourse. It is also the duty of young girls to go at night and offer themselves to the guests made welcome by the tribe. Parents, too, will often unite with their children.

In point of fact, marriage in Australia does not exist; and if we look at the matter a little closely, we might dispense altogether with any theory as to exogamic marriages. That which travellers have wrongly called marriage is no more than the capture of a slave, who, no doubt, may be used by the master for his amorous pleasures, if he so pleases. She is his domestic animal, his thing, his creature, whom he has the right to beat, to hurt, to kill, or even, in case of need, to eat. We may understand that this unenviable position has been reserved to women violently taken away from rival tribes, and that the men have not the right to inflict such punishment upon those women who bear their name, or even upon any woman belonging to their own tribe. Eyre, in his "Discoveries in Central Australia," has given us a description of one of these so-called marriages. The man begins by stunning with his *douak* the woman whom he finds a long way from his own people; then he drags her along by her hair. He waits awhile until she has partially recovered her senses, and then obliges her to follow him into his tribe. He violates her, of course, if he chooses to do so. This form of rape is considered in Australia as a praiseworthy action, and children will practise it in their games. This is the most brutal form of sexual union, and assuredly it does not deserve the name of marriage. A milder form of rape, which is sometimes practised, may be called marriage with a smaller inversion of truth; for, according to the Rev. Mr. Bonwick, the woman in Tasmania was occasionally forewarned, and the rape was therefore only fictitious. This was a rare occurrence; but when the woman had been taken an arrangement was entered into with the people of the tribe to whom she belonged. On a certain day, in presence of the two tribes collected together, the ravisher had to stand as a butt to the adverse tribe, while they threw at him a certain number of darts. He was allowed to make use of his little shield as a weapon of defence. As a general rule no blood was shed; the reconciliation was effected by love-feasting on both sides. Or sometimes the people celebrated the marriage by tying the couple to the same tree, and by breaking in each of them one of their incisor teeth. From this moment the woman carried off belongs to the man; he has the right to treat her as he pleases, and also to lend her, or to hire her to any customer. As regards the woman possessed, an unauthorised infidelity on her part is forbidden to her. She is often punished for it most brutally, for the captured woman is held to be the property of her master.

As it was not always easy to arrange this pretended exogamic marriage many of the Australians remain unmarried, and the majority of them rarely try the experiment before they are thirty years of age. We can understand that polygamy is not forbidden to them. Those who do not possess wives of their own resort to endogamic promiscuity, when they do not for a few presents hire the wives of their friends.

These customs, taken together with the infanticide of girls, keep the population upon a level with the means of subsistence to be found in the country. In many tribes the women are less numerous than the men; hence the necessity for promiscuity and a natural tendency towards exogamy.

The custom of capturing women exists also in a certain number of the Melanesian islands. At Bali, an island between Java and New Guinea, the carrying away of women is practised exactly as in Australia; violation follows immediately afterwards. The ravisher then pays to the parent the price of the woman ravished, and she becomes his slave. In Fiji, too, rape, real or feigned, was a common practice. As regards the woman, either she ran to some one to implore protection, or she accepted the ravisher for her husband, and a feast given to her parents legalised the whole transaction. In Fiji, as in Australia, women are the property of their husbands; the chiefs sometimes possess several hundred, among whom some few are considered legitimate, and their children can inherit. The others are concubines, slaves, domestic animals, whom the master keeps for his warriors so as to encourage their fidelity. Even the lawful wives of the Fijian chiefs have very singular duties to perform. At the time of their marriage they choose from among the people a very young girl, whom they bring up carefully until she has reached the nubile age; then upon a given day, after having carefully washed, perfumed, and decked with flowers the hair of this girl, the wife leads her naked to her master and retires in silence.

It does not appear that the marriage by means of capture is very frequent in New Caledonia. Children are betrothed to each other when very young; but this is a matter of no consequence whatever. Here, as elsewhere in Melanesia, polygamy is lawful to all those who can practise it; it replaces domesticity, a thing quite unknown to the New Caledonians.

Marriages between close relations on the mother's side do not take place in New Caledonia, and this probably indicates a former epoch when promiscuity was common and it became difficult to recognise the father. The New Caledonian married woman is considered to be the property of her husband; he may put her to death in case of adultery, but he often contents himself with punishing her by driving her out, after he has made her undergo some process of scalping. In New Caledonia the sexual relations are curious by reason of their very brutality. They rarely take place at night, for the men and women sleep in separate huts. It is ordinarily in the daytime, in a thicket, that the man and the woman come together, like animals, *more canino*, as the theologians say. (E. Foley, "Bull. Soc. d'Anthropologie.") According to O. Beccari, similar practices exist also among the Papuans of New Guinea.

We may add that the New Caledonian customs oblige every man, married or not, to marry immediately his brother's widow; this custom is very common, and it was in vogue even among the Jews.

From what has been said, we see that the union of the two sexes in New Caledonia does not deserve the name of marriage. But the institution is just beginning, for a friendly convention takes place between the parents, establishing the right of property in the woman; and even in certain tribes, notably at Kanala, there is a kind of legal control over conjugal affairs. In this last-named tribe every individual taken in adultery is judged by a council of old men, over whom the chief presides, and the culprit is generally put to death upon the spot. This infraction upon the husband's rights is held to be a serious violation of property.

By the side of the marriage, or the so-called marriage of the New Caledonians, we may consider the promiscuity of the Andamanites. It is altogether primitive and animal. Their women belong to all the members of the tribe; resistance is held to be a grave offence, and is very severely punished. When the woman becomes big with child, a sort of temporary union sometimes subsists between her and the man; but that ceases as soon as the child is weaned.

These Melanesian customs show us how the people originally regarded the marriage ceremony. At first there was absolute promiscuity, which still exists more or less among the different ethnical groups; then the scarcity of women and the need of beasts of burden inclined them to have recourse, as far as they were able, to exogamic raptus, which they began by practising with excessive violence, as though they were beasts of prey. Then by degrees the tribes who were interested, after examination and bargaining on both sides, ratified the treaty. Later on, the rapture of women became more and more an affair of ceremony, a sort of predestined comedy. But polygamy has always been lawful; the woman has always been the property of the master; infidelity on her part was not allowed unless he ordered it; she never had the right to be jealous of her husband, and she was exposed to every kind of bad treatment which he chose to inflict upon her.

IV.

Marriage in Africa.

It is not impossible that promiscuity may have been general amongst every horde of primitive man, but we do not now find it among the Hottentots, nor do we find the exogamic rape. Of this latter phase of conjugal existence, M. Lennan would make a phase apart, as though it either does now, or has at one time, everywhere existed. The Hottentot girl belongs to her parents, who, in a friendly way, barter her for an ox or for a cow. The Hottentot women age very rapidly, and the men, whether married or not, for polygamy is not forbidden to them, generally bespeak in advance little girls, six or seven years of age, to replace their wives when they have grown too old. The Hottentot girls are nubile at the age of twelve or thirteen, and with them old age is as precocious as their nubility. The Hottentot marriage is purely commercial. It is in no way binding, and may always be cancelled at will. The Hottentot women are more numerous than the men, polygamy may therefore be considered necessary; and in agricultural districts founded by the missionaries, the introduction of a Christian form of marriage was met with the strongest objections by the people of both sexes. We find among them nothing which, even in the slightest way, resembles the form of marriage, as it is understood by our European moralists and our legislators. The Bushmen, whom we may consider as the least civilised of the Hottentot tribes, have in their language no expression to distinguish a girl from an unmarried woman.

As regards the sexual union, the manners of the Kafirs are very like those of the Hottentots. Lichtenstein tells us that love has absolutely no voice or persuasion in the matter; and the opinion of many other explorers fully bears out this assertion. The chiefs and the rich men have always several wives, and therefore women are relatively rare in the marriage market. The custom is to buy them from their parents when they are young children. With the Makololos the price paid to the father of the future wife buys also the right which the father-in-law would otherwise have over his daughter's children. When once duly bought the Kafir woman is, in the full meaning of the word, the property of her husband. The master of the woman may use her or abuse her, he has the right to kill her if she dares to lift up her hand against him. He may beat her as he pleases; he may let her out for hire to the first white man who asks for her. The Kafir women, too, appear not to know what the feeling of jealousy is; they are very anxious that their masters should buy wives younger than they are themselves. Their authority thus becomes increased and their labour lessened. Exogamy does not seem to exist as a rule among the Kafirs, unless we rigorously examine the system of rape practised by the Dammaras upon the Namaqua Hottentots, or the repugnance of the Bakalaharis to marry a woman who has been chosen for them.

In Gabon the habits are very similar. Girls are bespoken, bought, and betrothed—if we wish so to call it—when they are three or four years old; they are mothers at thirteen or fourteen; they age very rapidly, and they often die young and childless. The buyer takes possession of the creature bought when she is eight or nine years old. Polygamy exists everywhere. The women are mere beasts of burden, who cultivate the earth; they are obliged to furnish their master with food, who remains idle, and who lacerates them at his pleasure with his fearful cutting whip. As the woman is only an article of property, her adulteries, to which she is very liable, are severely punished, both upon her and upon her accomplice, especially if she be the principal wife, who is ordinarily the one first married. The paramour of this first wife is at least sold as a slave. After any man's death his heir has the right to take possession of all his wives, and to distribute them if he pleases among his relations.

Among the Ashantis, a people relatively civilised, we find the same system of polygamy, and the same custom of early betrothals. The rights of the husband are always excessive. Every act of privacy taken with the young girl betrothed is punished with a fine, of which the affianced husband has the advantage. Later on, in the case of adultery, the master may either kill his wife, or slit her nose and marry her to a slave. He has also the right to cut her upper lip if she betrays one of his secrets, or to cut her ear if he catches her eavesdropping. If the husband disappears for three years his wife may marry again, but should the first husband return all the children of the second marriage will belong to him. The daughters of the king are the only girls allowed to choose their lover or their husband; he is literally their slave, and he is bound to kill himself if his wife dies before him. In this particular case the masculine pre-eminence has to lower his flag before the prestige of monarchy.

In the basin of the Niger, in Soudan, among the Fulahs, the Iolofs, the Mandingos, the Bambarraus, and others, where Islamism exercises more or less influence, marriage is performed after a somewhat less inhuman fashion. The woman is not so fearfully ill-treated. Her wish is even sometimes consulted. If the husband is a rich man, he gives a dowry to his wife, and to his slave-wives he gives different articles for toilet use, or he gives them stones and mortar to grind their corn. Here and there, too, certain rights are granted to women: for instance, at Wow and at Houssa no woman can marry but with the consent of her grandmother as long as the grandmother be alive.

With the Soulimas the woman may leave her husband to marry another man, on condition that she gives back the money paid for her at her former marriage. The Fanti women on the Guinea coast have a similar custom; but beyond this repayment of money they have to discharge a certain debt for each of the children that their husband has been good enough to give them. We must add that in this country the husbands turn polygamy to their profit, for they make a certain yearly income by bartering their children.

In many districts of Soudan the union between the two sexes is contracted after a manner quite as brutal as among the most inferior class of negroes. The Timmani girl is not consulted in any possible way. A man buys her for so many earthenware jars of palm wine, or for stuffs, etc. The Mandingos have the same customs; when the treaty is concluded, the husband and his friends will carry the girl away with them. It is simply marriage by capture. In Yarriba the natives choose their wives with the utmost indifference, just as they would "pick a handful of corn." In Kouranko the young girls are first bought by rich old men, and when their old husbands die they are free, and consequently take their revenge by choosing for themselves a young man whom they like. In Yarriba the son inherits his father's widows. Elsewhere the women are sold on the death of their husband, if they have born no child. In certain districts of Senegambia the people have invented a rather fantastic form of administering justice. The Mumbo-Jumbo appears in the evening, dressed in a peculiar and strange costume, before all the assembled population, and he there and then chastises all the women of bad reputation.

If the African negro woman is nearly everywhere treated as a beast of burden, it would seem also, from the opinion of most travellers, that her manners are very dissolute. According to Du Chaillu and Schweinfurth, this is the case from Gaboon as far as the basin of the Upper Nile. There, too, polygamy prevails; the woman is bought for so many iron utensils, which, among the Bongos, the father is bound to return, or at least some of them, in case of divorce. If the husband keeps the children the restitution must be complete.

The property in the woman is inviolable in law, and adultery is punished with immediate death.

In nearly every district of Africa peopled by negroes the woman is the thing of her husband; he has the right to make use of her as though she were a beast of burden, and he does in reality make her work as if she were one of his oxen. "I bought her," said a Kafir, one day, speaking of his wife, "and therefore she ought to work." Among the Mussulman negroes the marabout blesses the marriage; but everywhere else marriage is merely a civil and commercial contract, into which religion in any form does not enter. Polygamy is everywhere the rule; the man may also divorce his wife by paying some necessary compensation. In Madagascar the marriage ceremony is performed before a magistrate after the payment of a certain fee; but even then the husband may, on paying the fee a second time, repudiate his wife. She then is free, unless before the expiration of twelve days the husband thinks well to change his mind.

The woman's virginity is neither thought about or cared for except in the Mussulman countries where the Moorish race has more or less penetrated. At Kaarta the women of the country come together the morning after the marriage and carefully examine the nuptial bed, and unless the woman's innocence be shown the marriage may be considered as void. But with the Sakkalaves of

Madagascar it is quite otherwise. There the young girls unflower themselves before marriage, unless their parents have already taken the same necessary precautions.

In those parts of Africa inhabited by the Arabs, the Berbers, the Nubians, the Abyssinians, the connubial customs will vary considerably, for in them people are of different races, some of them of Asiatic origin. The Egyptian, the Arabian, and the Byzantine civilisations and religions have left deep traces over all the northern regions of the African continent. The ancient Egyptians were monogamists; they gave a dowry to their daughters, and they punished adultery by inflicting on the man a thousand cuts with the rod, and by slitting the nose of the woman. None of these customs have come down to our times, but it is perhaps owing to the traditions of Egyptian manners that the Berber women now enjoy their great liberty: they are absolute mistresses over their property, their own actions, and their children who bear their name. In certain of the Sahara tribes, Berbers to all appearance, repudiation is considered as an honour for woman. They say to each other: "You are a woman, from no one knows where . . . . men have disdained you, and yet one man wished to have you as his own."

Among the Hassiniyeh Arabs of Nubia we find a very strange custom—that of three-quarter marriages—which allows the woman to dispose of her person one day out of every four. Among other Juarick and Saharah tribes, it is thought that a girl before she is married ought to gain as much by prostitution as she has hitherto cost to bring up; and she who has had the greatest success is the most eagerly sought after in marriage. The spirit of initiative movement and of conjugal liberty in conjugal life appears to be a characteristic trait among the Berbers, for the Guanches, who, it would seem once belonged to this race, had systematised polyandry in some of the Canary islands.

Marriages between brothers and sisters—which were held legal and honourable in ancient Egypt, where the queens used to boast of being both "sisters and wives of the king"—are now also customary in Darfur. In this country, where the liberty of manners is very great, the daughters of the sultan are the absolute mistresses of the man. They admit them to their bed, though polygamy, the most unrestrained concubinage, and the servitude of women are the usual habits of the country.

We have mentioned the three-quarter marriages among certain of the Nubian tribes. At the other extremity of the north of Africa, among the Jews in Morocco, we hear of temporary marriages, marriages blessed by the rabbis for three months, or for six months; the man makes a donation, and binds himself to recognise the child should a child be born during that time. At Haïti and in Abyssinia, even now free marriage is common; at the time of Herodotus certain of the Ethiopian tribes did not know what marriage meant. We know that in Abyssinia, in spite of the Christian religion—or what passes for such—people come together and leave each other as they please. Bruce saw an Abyssinian woman surrounded by seven former husbands. At Haïti, in addition to the legal monogamic marriage, there are free unions, implying no sort of dishonour. These are the unions of those who are already *placed*. People in the most respectable families contract these free unions. Children thus born enjoy all the rights of lawful born children, and separations are more rare among the placed people than divorces among those who are married.

From what we have already seen, it would be very difficult as yet to deduce any general theory as to the origin and the evolution of marriage; but we are far from having terminated our inquiry. Let us see if America, Polynesia, Asia, and Europe will furnish us with any new materials.

V.

Marriage in America.

Among the belated Fuegians, as soon as a young man has acquired a sufficient proficiency in the chase, as soon as he has learnt how to build or to steal one of the coarsely made canoes of bark used by the people of his race, he has then the right to possess a wife. He generally kidnaps her. But, according to Captain Fitzroy, it would seem that he confines his choice to women of his own tribe; consequently the kidnapping is not exogamic.

Endogamic kidnapping is practised also among the Araucanians. These women will always protect the girl as far as they can with sticks and stones, and if the parents make any opposition, the future spouse blows his horn, thus giving the signal to his friends to make a raid and so carry off the girl of his choice. This endogamic kidnapping does not at all prevent the exogamic rape. The men endeavour to keep up a harem of female prisoners, and in the midst of the tribe, the kidnapping, real or feigned, does not dispense them from paying for the woman more or less dearly. Among the Charruas polygamy prevails, as, indeed, is common among all primitive races; they also maintain the usual custom of allowing the first wife to have the upper hand over the others.

With the Indians, as with the greater part of the aborigines of South America, the union of the two sexes was purely a civil and commercial affair, which might be made to cease according to the man's caprice. He had a despotic right over his wife; the Moxos could put their wives to death in the case of abortion.

Among the Guaranis, too, we also find polygamy, and the same system of kidnapping; woman is with them reduced to the same state of servitude, she is compelled to perform all the agricultural labour. This is the reason of the greater part of their wars and of their migrations. These rapes are made simply from the desire to possess many women; love has nothing to do with the matter. On the other hand, in unions that are made from free will on both sides, endogamy and even marriage between relations is common enough among the majority of the tribes in South America.

This latter custom was frequent with the Caribs. These men would marry any one of their female relatives, with the exception of their sisters. In this respect the customs will vary in particular districts. The Indians of Guiana are exogamic. They are not allowed to marry any one in their family who bears their name; the child belongs to the mother's side and takes the maternal name. According to Thevet, the Brazilians had very different notions; with them the father in each generation played the principal part, the mother the second. Martius tells us that the Brazilians were very irregular in their matrimonial habits, that they act merely upon the necessity of the moment. In the small isolated tribes the nearest relations will marry each other; but in the populous districts exogamy becomes the law. It is the same all over the world. "Necessity has no law," says the old proverb. And though polygamy is general in South America, we must make an exception in favour of the Ottomacos, and these are a very unintelligent people.

We may also remark that among the aborigines of South America, the chastity of the girls is held to be of no consideration. They are neither cared for nor thought of until they come to be the property of the man. Then, adultery is very severely punished, sometimes with death, if the proprietor so wills it. Thevet says that the punishment falls only on the woman; the man is spared, for fear of vengeance.

In North America marriage customs are not more uniform. Among the Indians of California there is almost total promiscuity. People come together without any sort or kind of formality; there is no word in their language signifying marriage. The women seem to belong to all the men of their tribe, and the man's jealousy is not aroused until the woman chooses a lover in another horde. On the other hand, many tribes on the banks of the Gila, in Colorado, and in New Mexico are monogamous, and with them adultery is very rare and is very strongly condemned.

Polygamy is nearly everywhere allowed to any man who is rich enough to buy several wives by purchase or otherwise. Among the Red Skins the marriages are generally exogamic. Those sociologists who, following the example of M. Lennan ("Primitive Marriage"), wish to make of this mode of conjugal union a general phase common to all humanity, have drawn some of their most convincing arguments from the inhabitants of North America.

The tribes of the Red Skins are ordinarily divided into several clans, each having their animal coat-of-arms, their totem, and marriage is usually forbidden between people having the same totem. This custom, already noted by Charlevoix, has since been mentioned by many travellers. It is with them a question of an analytical and consequential form of exogamy, producing a line of maternal descendants, for the children take the totem of their mother and belong to her clan.

With many tribes, the Shyennes, the Ioways, the Kaus, the Osages, the Blackfoots, the Crees, the Minetaris, the Crows, and others, exogamy complicates itself with consanguineous polygamy; for a man has the right to marry his wife's younger sisters, and each of them considers herself the mother of the children of the eldest sister. Are we to suppose that this pacific form of exogamy has followed exogamy by means of kidnapping? It is quite possible, at least in certain regions; in Canada, for instance, where the husband, if his marriage has been blessed by the chief, carries his wife to his hut amid the acclamations of the spectators. We may remark, in passing, that here marriage is not considered as an individual act; it is a social performance which must be sanctioned by the chief of the tribe.

But this marriage is not any the less merely a commercial enterprise. The husband buys his wife, and if he is a poor man he is obliged to work to gain her. He binds himself for a stated time to the girl's parents, for whose benefit he hunts, he digs, he scoops out the canoes. In some tribes a certain period of servitude was always customary. The husband was obliged to subtract a tenth part of the game that he killed for the father and mother of his wife, and he was not exonerated until a girl was born to him, who became the property, or perhaps, occasionally, in after years, the wife of his maternal uncle.

The exogamic law, to which the majority of Red Skins are bound to conform, proves that in this race marriage is tending to become a social institution. But this tendency is as yet very weak. Among most tribes a marriage is contracted without witnesses, before no magistrate, and in presence of no priest. It is always an individual act, and is often merely a coupling together. The Tinné Indians possess no word signifying "dear," or "well beloved;" the

Algonkin idiom did not know the verb "to love." The dialects of the American tribes differ very much one from the other, and as marriage among the Red Skins is ordinarily exogamic, the husband and wife often speak different languages. They content themselves by making signs to each other. This moral intimacy between them is so slight that they will constantly live together for many years without being able to understand each other's language. Man is everywhere a polygamous animal when such a condition is possible to him.

Among the Apaches the greater number of wives a man has the more is he respected; and with them religious ideas are considered. The Chippewyans say that polygamy is agreeable to the Great Spirit. The Red Skin, like the Hottentot, women are bespoken a very long time in advance, at the age of ten or twelve years; the conjugal union lasts just so long as the master chooses. He has always the right to send away his wife for the slightest reason, or for no reason at all. Chastity is imposed only upon those married women who are kept as slaves and possessed by a master. The Natches, one of the most civilised of these tribes, would readily lend their wives to their friends; and, according to Hearne, two Algonkin friends would often be glad to exchange wives for a night. Among the Nadowessioux a free woman does herself honour by giving herself after a feast to all the principal warriors of the tribe; and an exploit such as this assures to her a husband of high rank.

A Red Skin, because he buys his wife, is not therefore always her lawful master. In many districts the husband ought to be constantly ready to refund his feminine property. Up in the far north the man has to fight for the possession of his wife, and unless he be known as a most skilful huntsman the weak man is rarely allowed to keep his wife if a stronger man than he wishes to take her. These facts will show clearly enough that in spite of exogamy, marriage is as yet far from being a soundly-established institution amongst the Red Skins. And also, to this exogamy there are many exceptions. The Chippewyans will often marry their sisters, or their daughters; they will cohabit with their mothers. The same Indians compel a man to marry his brother's widow, whether the marriage be exogamic or not. This is the famous custom of the levirat, common among many tribes and many races of men, as well as among the Jews.

This levirat custom, incestuous according to our European notion of morality, is almost necessary in the primitive phases of civilisation, where abandonment of the woman is equivalent to causing her death. From this we may gather an instructive fact: that in their most essential senses our creeds of morality are prompted by necessity, and also in savage tribes the period of fecundity in the woman is usually short and limited; therefore the levirat law must have been favourable to the primitive human groups in their struggle for existence.

The Esquimaux, whose manners are curious in many ways, would seem to have at one time practised exogamy more or less, if, as is probable, marriage by means of kidnapping may be considered as coming from this custom. The Esquimau of Cape York, and also the Greenlander, first makes his treaty for the marriage with the parents; then he affects a feigned rape of the girl, and she, on her part, also makes a sort of simulated resistance. The same customs prevail among the Kamtschadales, a people who are very similar in their manners to the American Esquimaux.

Nothing, however, can be more loose than the manners of Esquimaux in their sexual unions. The chiefs of the Nootka-Columbians used to exchange their wives. The same process was thought by the Greenlanders to be a noble action, showing a dignity of character. The loan of a wife, among the Esquimaux, is usual sometimes for several months. It is a mark of most particular affection shown by the lender to his friend. But the friend ought to give back the wife or wives lent, punctually at the appointed time, if he wishes to be considered a gentleman. Polygamy and polyandry are practised simultaneously by the Esquimaux. Ross has seen certain tribes in which polygamic marriages were held to be allowable, but only in the case of sterility. With the Esquimaux, children are considered as a source of wealth; they begin to make themselves useful at eight years of age, and they support their parents when they have grown old. A widow will re-marry all the more easily in proportion as she has a large number of children.

From the foregoing facts we may conclude that among the savage tribes in America there is no prescribed form of union between the two sexes; and throughout all these vast regions there is nothing which deserves the name of marriage. It is only in civilised societies that marriage is legalised, and is regarded as an institution. It was so considered even among the ancient Mexicans and Peruvians, and of them we now propose to speak.

VI.

Marriage in Mexico and Peru.

In the great empires of Central America, whose social structures were of a very complicated kind, sexual unions were not considered as purely individual transactions. In Peru, where the whole population belonged in some way to the Inca, where men were parcelled out into so many appointed divisions and made to work as beasts of burden, marriage was purely an administrative act. In the kingdom of Cuzco, all the individuals of both sexes who were old enough to be married were brought together into a public or open space in the towns and villages once in every year. In the town of Cuzco the Inca himself used to marry those of his own family in the public place, by joining together the hands of the different couples. The chiefs of the districts, or the *curacas*, in their own divisions used to perform the same functions with regard to people of their own rank, or even of a lower rank than their own. It is said that the consent of the parents was necessary; but that of the contracting parties was not called in question. The law prescribed the marriageable age to be from twenty-four to twenty-six for men, and from eighteen to twenty for women. Monogamy was imposed upon the masses; polygamy was a luxury allowed only to the nobles and to the Inca. The Inca made very liberal use of his privilege; for Montezuma, the last Inca, had three thousand wives, or concubines. Contrary to the exogamic habits of the North Americans, endogamy was strongly enforced in Peru. It was strictly forbidden to anyone to marry outside his own administrative division, consequently, nearly all marriages were between relatives. The Inca alone married his sister, provided, however, that she was not born of the same mother. By degrees this incestuous habit became allowed to all the nobles of the empire.

This interference of the State in the Peruvian marriages was not without certain advantages. A habitation was prepared for the newly-married couple at the expense of the State; and a certain portion of land, with clearly defined boundaries, was assigned to them for their maintenance. Authoritative communism was thus rigorously carried out; and we may be sure that this plan served as the model for the organisations of the Jesuit missions in Paraguay, and also in other parts of America. The Jesuits, however, went further than the Incas, for in the provinces of Moxos and Chiquito they used to take care that the husband and wife were awakened an hour before the time for mass, but without obliging them to get up; a curious application of the biblical precept: "Increase and multiply."

When the curaca sanctioned the marriage, he made the newly-married couple both take an oath of fidelity to each other, and this oath, according to P. Pizarro was generally kept; for the law regarded adultery as a capital crime. At Quito, the law did not intervene as between the two sexes, and the husband and wife might separate by mutual consent; but the woman taken in adultery was usually buried alive, together with her paramour. It would appear, however, that at Quito, as well as at Cuzco, the law regarded principally the conduct of married women; for in both kingdoms prostitution was tolerated.

In Mexico also monogamy was the rule. The nobles had, no doubt, more than one wife, but one only was legitimate; her children inherited their father's title and his wealth, to the exclusion of all the others. The levirat was lawful, even when the brother of the late husband was already married.

Marriage in Mexico was conducted with much ceremony; it was arranged amicably between the parents, and the civil law of the country did not interfere. In Mexico we find no trace of exogamy;

there was no kidnapping, even of a simulated kind. On the other hand, the betrothed girl was led in great pomp to the dwelling-place of her future husband, and he then left the habitation with his family, the girl and her friends following behind. When the pair met they scented each other with perfumes and incense. They then sat down upon a mat, and a priest married them by tying a strip of the girl's dress to the end of the man's cloak. Henceforward the woman belonged to her husband's family.

We may add that the marriage was not consummated until after much divination and consulting of the augurs. Strictly speaking, it was not consummated until after four days of nuptial feasting, in which the married couple took no part. In principle, the conjugal union was for life, but divorce was admitted, and there was even a tribunal specially set apart to hear and decide upon all matters relating to dissolution of the marriage. After a minute examination of all the facts, and after the parties had each appeared three times before the court, if they both persisted in their determination they were sent away without judgment being pronounced. They were therefore considered as free and for ever separated. But the law would not explicitly decree a divorce; it was content to tolerate it.

In our very short study on the subject of marriage in America we have endeavoured to mention only the most interesting and the most typical facts. What we have said is no doubt very insufficient. For the ethnical groups of men in America have no real history; it is at most only a legendary history, incomplete and silent as to bygone times. We may, however, be sure that in America the institution of marriage has not evolved itself regularly, following any clearly-defined system. Each tribe, each ethnical group, has consulted only its own wants, its own wishes; and the larger nations of Peru and Mexico, possessing nearly the same degree of civilisation, regulated their own marriage institutions, each according to their own fancy, both following a very different type of legal procedure. These and many other facts seem to show that it is at least premature to formulate sociological laws, precise and exact in their nature as the laws of science. For our part we cannot conscientiously do more than collect various facts, class them together, and so endeavour to form some general theories, far from absolute, but which, we know, must always be subject to revision.

VII.

Marriage in Polynesia.

After what we have already said of the unbridled desires of the Polynesians in their amorous passions and of their naive shamelessness, of the society of the Areois founded upon the idea of promiscuity and infanticide, we need not expect to find among them a code of marriage laws very rigorously enforced.

Nearly everywhere in Polynesia girls and unmarried women were free to give themselves to whomsoever they pleased, and from their earliest years they began to take advantage of their liberty. In the Marquesas islands, for instance, girls used to begin to debauch themselves when they were twelve years old. Owing to this kind of life, fecundity among women was not common; and a woman pregnant soon found twenty men who were willing to marry her. In some few islands however—notably at Rotouma—the virginity of girls was very highly esteemed; its absence might justify repudiation; and those who had kept themselves uncontaminated made it known by powdering their forehead with white coral, and painting their cheeks red.

It is somewhat strange that in Polynesia, where religious rites showed themselves in nearly all the acts of daily life, marriage, or rather the union of the sexes, was not attended by any form or ceremony. It was purely an individual act concluded between the man and the girl's parents, to whom he offered pigs or stuff in exchange for his purchase. The union was effected at once, in a house belonging to either family, and was followed by a banquet at which a pig was served as the *plat de résistance*. When once married, the woman, who had hitherto been so frail, was not allowed to be unfaithful to her husband unless she received his consent or his command, which, however, was easily bought. This free marriage might be cancelled at pleasure, sometimes at the will of either party; it was not unfrequently cancelled, especially if no children were born. When separation did take place, the children (if there were any) went with the father or the mother, as agreed upon beforehand. The woman who proved unfaithful without her husband's consent was chastised by him, or sometimes put to death, for she was his property. In New Zealand the father or the brother, in giving his daughter or his sister to a man, said to him: "If you are displeased with her, sell her, kill her, eat her." An unequal marriage was condemned by public opinion, more strongly even than adultery. It was regarded as a capital crime for a man to have intimate communication with a woman of a rank superior to his own; the woman, too, if she belonged to the aristocracy, was tabooed as a plebeian—she was forbidden to marry him.

After a woman had been given to a man she still belonged to her father, and he sometimes took her back if he was not satisfied with the presents he received for her; and of course he afterwards sold her again upon better terms.

Polygamy was everywhere permitted, especially to the rich, without prejudice to the numerous concubines. At Samoa, especially, the chiefs used to make for themselves harems which they would renew constantly, guided merely by their own personal pleasures. Very often these mistresses, when the nobles had grown tired of them, were attached to the service of the caravanserai. Travellers were here gratuitously lodged, food was given to them, and the women were free. The levirat was lawful in these islands, as in so many other countries.

Some of the women in the Marquesas islands were polyandrous, for the Polynesians had very few prejudices as regards sexual unions. Ellis speaks of the polyandry of certain women who were married to the chiefs. In the Hawaiian archipelago there were constitutional husbands who reigned but did not govern; they merely legitimised the children of their wives, and the wives kept their own personal property. Marriage between brothers and sisters, rare enough in most of the islands, except with nobility, to avoid an unequal union, was, however, common in the Hawaian archipelago. Often also in these islands, the brothers on one side and the wives on the other used to live in common.

The great conjugal liberty of the Polynesians co-existed in New Zealand with the practice of endogamy. It was strictly forbidden to the New Zealander to marry, or rather, to buy a wife belonging to another tribe, and leave could not be granted unless some strong political argument might be furthered by the marriage. But the New Zealand *form* of endogamy was followed by kidnapping and a simulated quarrel; in fact, by all the ceremony of marriage by capture. We may consider this as an instructive fact, destroying the theories hastily put forth as to the natural evolution of marriage. We do not agree with M. Lennan in thinking that the form of marriage by capture is necessarily a sign or a remnant of exogamy; and we are also obliged to differ from Mr. Herbert Spencer, who maintains that endogamy is peculiar to peaceful races of men, for no tribes can be more warlike than the New Zealanders.

The prudence of the serpent is a virtue which we ought not to grow tired of recommending to our present sociologists, who seem to be commissioned to found, or rather to sketch out, a theory of social science.

VIII.

Marriage in the Malay Archipelago.

We will not do more than mention some of the matrimonial customs in the Mongoloid archipelagoes of the Pacific ocean. Polygamy there is the rule; women are generally bought from their parents. Adultery is blamed and punished as an outrage upon property. In the Caroline islands the aggrieved husband may be appeased by a present suitable to his rank, nor will he make any difficulty about lending his wife or his wives to strangers. In the majority of these islands, and especially in the Palaw islands, polygamy, as in Polynesia, is purely an individual transaction.

The Akitas in the Philippine islands have still preserved the form of marriage by capture. The lover has to look for and find his future wife in the woods; she has an hour's start given to her, but the intended husband must bring her home before sundown.

In Malay the modes of marriage are more various and are more interesting. In Sumatra we find that different matrimonial customs were in vogue. Firstly, the woman might buy the man, who became henceforward the property of the family of his wife, and this family was responsible for the faults of the husband bought; the man was bound to work for the woman, he held no property of his own, and he might at any time be dismissed. Secondly, the woman and the man might marry upon equal terms. Thirdly, the man also might buy his wife, or his wives. In these marriages, or in one of them, for the details are wanting, the ceremonial form of capture was preserved.

We cannot form any general knowledge as to endogamy or exogamy in the Malay territory; but we may conjecture that in the larger and thickly peopled islands, the customs were, and still are, very various. The Kalangs of Java were endogamous; and with them, before obtaining a girl in marriage, the man had to prove that he belonged to his own family. In the Malay archipelago, marriage, like everything else, is quite the reverse of uniform; for here we find several races of men and several orders of civilisation mixed up together.

IX.

Marriage among the Natives of India.

In India, the Aryan invasion, relatively of a recent date, did not destroy the inferior races, for the most part negroid, who before occupied the country. The people were only driven further back into the country, up into the mountains, from whence comes their denomination Paharias (or mountaineers), often contracted into Parias. But these ethnical helots, though their debased position, from the Brahmah point of view, has kept them beyond the pale of the civilisation of their conquerors, have in a great measure still preserved their ancient manners; and their matrimonial customs are exceedingly curious.

We have already spoken of the monogamy of the Veddahs in Ceylon, which must, no doubt, be attributed to the extreme intellectual inferiority which obliges them to live in couples apart. It follows, naturally, that these poor people are endogamous, but with certain restrictions. According to Bailey, they will readily marry a younger sister, though never an elder sister, nor an aunt.

From Ceylon as far north as Tibet a considerable number of the ethnical habits, the worn-out remains of the old races, practise polyandry, which in many cases would seem to be a diminished and legalised form of primitive promiscuity. This latter custom is still common in certain tribes, notably among the Sontals, among whom, as was the practice in ancient Peru, all the marriages are performed at the same time, once in every year; and they are always preceded by six days of promiscuous living. Besides this custom, we may mention the trial marriages in Ceylon, temporary unions lasting only for a fortnight, and which may afterwards be annulled or confirmed at pleasure.

Among the tribes living in the commune of Chittagong marriage is no more than an animal union; it is also, for the man, a convenient way of getting his dinner cooked, for it obliges the woman to labour.

Among the Reddies in India a girl from sixteen to twenty years of age is married to a boy of five or six. The wife then becomes the real wife of the boy's uncle or cousin, or of the father of the reputed husband. But the latter is considered to be the legal father of the children of his pretended wife. And when this constitutional husband has grown up, his lawful wife is already aged; and he, in return for the trick that was played upon him, does as was formerly done by him.

Among the Nairs of Malabar, among the Todas on the Neilgherry hills, among the Yerkalas of Southern India, among the Cingalese of Ceylon, we find different forms of polyandry in vogue; they are very different, but we may consider them as only so many forms of promiscuity.

The black women who belong to the superior caste of the natives of Malabar have usually five or six husbands, but they are entitled to marry ten or twelve. It is sometimes allowed to them to cohabit with any number of men, certain restrictions being made only as to tribe and caste. When the number of husbands is limited, the woman will cohabit with them all, each in their turn, for ten or twelve days. In these singular households a good understanding is generally preserved. It is also lawful for each man to belong to several matrimonial associations. This kind of conjugal intercourse is no doubt a sort of marriage, but it is marriage in its simplest form. With the Todas a girl, when married to a man, becomes the wife of all the younger brothers of the husband as they in turn reach the virile age; and reciprocally, they become the husbands of all the younger sisters of the first married wife as soon as they become nubile. The first child of these unions, incestuous as they appear to be to us, is attributed to the elder brother, the second to the second brother, and so on. A similar custom prevails among the Yerkalas; with them the maternal uncle may claim as wives for his sons the two elder daughters of his sister.

Polyandry is frequent among the Cingalese in Ceylon, especially with people whose means are sufficient, and the common husbands are nearly always brothers. The family, not the individual, is married, and to them the children belong as joint tenants or tenants in common. With the Teyars in India the women are possessed in common by the brothers, the uncles, and the nephews.

This polyandry in the old Indian races co-exists sometimes together with exogamy and endogamy. The Todas have five social classes, and marriage between the classes is forbidden. In many tribes a man may not marry a girl of his own colour. On the other hand, among the Kurds and among certain tribes in Central India, the form of marriage is a feigned capture.

We also find polyandry in the north of India, among the Mongolian mountaineers of the Himalaya, of Tibet, and also among the ancient European races, who are probably of Asiatic origin. The curious verses in the Menu Bramah Code authorising the brother to make fruitful his sister-in-law, if she be childless, may be considered as the remains of a primitive form of polyandry.

In terminating this ethnographical study on marriage we shall have to consider if polyandry has everywhere and always been the successor to promiscuity.

X.

Marriage in Indo-China, in Burmah, and in Tibet.

Among the Mongoloid Cochin-Chinese, the Cambogians, and the Burmese, the manners are very loose, and marriage can barely be said to exist.

In Burmah the nuptial ceremony consists simply in an exchange of promises, which become sanctioned by tasting a leaf of the tea-plant steeped in oil. People will leave each other upon the most frivolous pretext, and in the most careless manner. The husband has also the right to buy as many concubines as he pleases.

At the time of Marco Polo, no woman in Cochin-China might marry unless the king had first seen her and exercised his right of prohibition if he chose to do so; but in this case he was supposed to give her a marriage dowry. At the time of which we are speaking the reigning monarch had three hundred and eighty-six children.

In the thirteenth century, according to a narrative related by A. Rémusat, the king of Cambogia had five lawful wives, of which one was the chief wife; he also had several thousand concubines. The people, timidly subservient to the despotic will, considered it to be their duty to send their daughters in to the service of the palace, if they were in any way good-looking. And as is even now the custom in Japan, girls in Cambogia frequently did not marry until they had for a certain time led a licentious life, and they were in nowise considered to be dishonoured by it.

Neither the Cambogian husband or wife prided themselves in the least upon conjugal fidelity. According to a Chinese traveller who furnishes us with these particulars, the women would not consent to sleep alone for more than ten consecutive nights. The husbands were free to buy as many concubines as they chose.

In this licentious country girls were married as it was found practicable, when they were from seven to nine years old. The poor people only waited until their daughters had reached the age of eleven, on account of the singular custom of *Tchin-than*, or the legal and religious defloration, as to which we must say a few words. This strange ceremony was performed once in every year. The day was appointed by a public functionary, and parents who had children to marry were bound to make known the fact. They were then entitled to claim the services of a priest of Fo (a Buddhist) or of a Tao-sse priest. The holy man would ordinarily accept presents that were offered to him in return. The poor parents were naturally served the last, and sometimes their daughters were obliged to wait a few years longer than the rich girls, unless some pious-minded person paid for them the expense of the ceremony. On the appointed day the priest was led into the house in the evening in great pomp, amid feasting and rejoicing; he was again led out on the following morning, accompanied by men with palanquins, parasols, and drums and music. When the ceremony was concluded presents were again offered to the priest to buy back the girl, who otherwise would not have been allowed to marry; she would have been held to belong to the bonze who had condescended to deflower her.

On the Tibet side of the Himalaya girls may, before marriage, dispose of themselves as they please, and their reputation is in no way tarnished. Polygamy is not forbidden, but polyandry chiefly predominates. It is curious that in the very centre of Buddhism, in a country where religion touches the people deeply, where the practices of worship are joined to all the acts of their daily life, that marriage should be nothing more than a purely civil contract, as to which the priests have nothing to do. The whole ceremony consists of a mutual engagement taken before witnesses. Divorce, too, is optional if the husband and wife desire it.

In the opinion of the lamas, who, because of their profession avoid all communication with women, in the opinion of functionaries of high rank, and even of many Tibetans, marriage is considered to be odious and shameful. St. Paul, too, as we know, shared almost the same opinion. In Tibet only the poor people lower themselves by coming together for the purpose of reproducing their kind, and they co-operate amongst themselves to try and diminish the burden. The associates are generally brothers, and the eldest chooses one wife for them all. The children coming from these polyandrous marriages will sometimes call the eldest brother their "father," sometimes they will all share in bearing the title. Fraternal polygamy may also exist concurrently with polyandry; a young man who marries an old woman is therefore at the same time the husband also of the youngest sister.

The associate husbands usually live together in the house of their common wife, for it is she who possesses, and from her that the property descends to the children, and they, too, are held to be her property. In spite of her polyandry, the Tibetan woman is not immoral after her own manner of life. She is very hardworking; she spins, she digs, and she endeavours to earn for herself the title of companion by pleasing all her husbands equally. All travellers agree in saying that these joint households are usually very peaceful, and in nowise troubled by jealousies. The people did not even understand V. Jacquemont when he asked if the preference of the wife for one of the husbands did not give rise to conjugal quarrels. Adultery is sometimes common, and sometimes rare, and it is but lightly punished; we may conclude therefore that morality can hardly be judged by any determined standard.

XI.

Marriage among the Mongolians and the Mongoloids of Northern Asia.

The Tibetans and the Bhots of the Himalaya are already a civilised people, and their polyandry, very different to our ideas of legal marriage, is a regular and established form of union between the sexes. Manners far more barbarous exist among the Mongoloids in the north of Asia, for these people are still savages. In Kamtschatka, the man who wishes to gain a woman's hand must begin by making himself the servant of her parents, and this service will often last for a considerable time. When he is accepted he is bound to make an attack upon the modesty of his future wife. This is a sort of public ceremony. She wears her nightdress, her drawers, and bandages, and she is also protected by other women of the iourta. The man rushes upon her, the girl's father saying to him: "Touch her if you can." But she is strongly guarded, and several assaults are usually necessary. The man's victory is not proclaimed until after an intimate touching, which the woman herself recognises by crying out "*Ni, ni!*" This trial is imperative; but the parents may disallow it, and then the pretender finds that he has served all his time in vain.

The marriage customs of the Kamtschadales and the Tunguzians are certainly very singular, even in the form of marriage by capture. With them monogamy and polygamy are both customary, according to individual tastes. Marriage, too, may be cancelled at pleasure, and may be contracted between any relatives except between father and daughter, or between mother and son.

This brutal and fragile form of union, more or less softened, may be found among nearly all the Mongolian races in Asia.

These races, who have all at one time lived in a nomad state—some of them still continue to do so—and who formerly were so warlike, have all a preference for marriage by capture. The Tunguzians and the Turkomans still practise it in reality; the former kidnap the wives of their neighbours, the latter ravish the girls belonging to their own tribes, not including prisoners taken in war. Excepting in this latter case, the act of violence is compensated by a payment made in camels, or horses, etc. These ancient manners are however falling into disuse; and the regular form of marriage is by purchase, though the husband and wife, when they are young, are not consulted. They both belong to their parents, who marry them as early as they can; and after much haggling the contract is concluded for so many oxen, or sheep, or for so much linen, butter, flour, or brandy, that the future husband is considered bound to pay. In the time of Marco Polo the Mongolians used fictitiously to marry their children who had died when infants. They immediately burned the marriage document, to send it off into the next world to the deceased husband and wife, and the two families then considered themselves related one to the other.

When the marriage contract was arranged the ceremony was performed by a simulated capture. Among the Turkomans the girl, in her bridal dress, flies away upon an impetuous horse, carrying with her on the pommel of the saddle a kid or a lamb that has just been killed; her betrothed and the young people who were present at the wedding gallop after her. Among the Mongolians, those properly so called, it is necessary only to break open the door of the newly-married woman, and to seat her upon horseback, in spite of the feigned resistance of the parents, of her friends, and especially of the women.

The Mongolians have but one lawful wife; it is she who rules over the household, and her children only have the right of inheritance. But the husband may, if he pleases, buy a great many "little wives," who, however, must be submissive to the first wife. This disguised form of polygamy can only be within the means of the rich, for in Mongolia the men are much more numerous than the women. We may find here a strong reason for the celibacy of the lamas.

It is said that the women will make up for their deficiency in numbers by great freedom in their manners, both before and after marriage. Adultery, however, is legally punished. The man pays in cattle; the penalty is exacted by the princes, as the crime is regarded as a social offence. The woman is severely chastised by her husband, he may even put her to death. But so much severity is probably rare, for Prjévalsky tells us that the women hardly take the trouble to conceal their illicit passions.

Nothing can be more fragile than the conjugal tie in Mongolia. Divorce is optional, and, like marriage, it is purely an individual action; no civil or religious authority interferes in the matter. If a husband wishes to repudiate his wife, he is not bound to allege any reason; he simply loses the price he has paid. The marriage was only a commercial contract, and he cancels it. If the woman of her own accord returns to her parents, they are bound to send her back as often as three times to her husband; but if she returns a fourth time the divorce is held to be obligatory, and the parents are obliged to return some of the cattle that had been paid by the purchaser.

In the same way that the Mongolian marriage is but an attenuation of the Kamtschadaline marriage, marriage in China is also, in its turn, no more than a milder form of marriage in Mongolia.

XII.

Marriage in China and in Japan.

As the evolution in the majority of the races whom we have been lately considering is but little, if at all, known, we have not been able to point out the changes which have taken place in the connubial rites in each ethnical group; but we hope to do so more fully in speaking of the great civilised nations in Asia and in Europe. If we may believe their traditions, it would appear that promiscuity was originally the rule in China, and that they owe the institution of marriage to Fo-Hi, their first sovereign. Judging from the custom, common in China as well as with other people, of running away with the betrothed girl the first time that she crosses the threshold of her future spouse, we may be inclined to think that marriage by capture was once customary with the Chinese. The Chinese also were probably at one time a polygamous people, for with them, as well as with the Mongolians, legal monogamy coexists with the practice of buying "little wives," who are bound to be submissive to the first wife. In these families, half lawful and half unlawful, the "chief wife" is the reputed mother of all the children, they give her this title during her lifetime, and after her death they wear mourning for her, though not for their real mother. And here, as elsewhere, polygamy can only be practised by the nobles. Public opinion condemns the purchase of "little wives" in the middle classes, except when the lawful wife has been childless for ten or twelve years. The Chinese look upon sterility with opprobrium, it may even furnish just excuse for repudiation. Manners such as these show clearly enough that in the Celestial Empire the woman holds a very humble position. Her subjection is excessive; as a daughter she is a slave to her parents, as a wife to her husband, as a widow to her sons. The young Chinese girl has no idea that she may be consulted as to the choice of her husband. The matrimonial negotiations are often concluded by their fathers and mothers, or, in default, by their grand-parents, or by their nearest relatives, during the infancy of both children, sometimes even before they are born, if it happen afterwards that they are of different sex. Here, as in Mongolia, the girl is bought, and a portion of the price is paid when the contract is signed. If the betrothed husband dies before the marriage, public opinion will compel the girl to pledge herself to celibacy. When they are married, women in easy circumstances live as recluses, shut up in their own apartments. In case of the woman's adultery the husband has the right to sell her, or to cause her to be sold judicially. If she is seen *flagrante delicto*, the husband may kill the two guilty persons. This ferocious custom seems to be usual nearly all over the world, and in this respect the Europeans, who affect to look down upon the Chinese, have in fact no need to envy them. On the other hand, the Chinese wife ought to worship her husband; and her suicide, if she be left a widow, is considered as a most praiseworthy action, engraven upon the tablets of honour, and accomplished sometimes before thousands of spectators.

In China divorce is allowed when mutually agreed upon; but the husband may also claim it if his wife be sterile, if she be immoral, if she is disrespectful to his father or mother, if she is inclined to slander, or to theft, if she be jealous by nature, or habitually indisposed.

We shall have mentioned the principal features in the Chinese system of marriage if we add that it is exogamic. There are scarcely a hundred family names throughout all China, and marriage between persons of the same name is not allowed. We may therefore reasonably argue that in this custom there is still remaining a traditional vestige of the ancient form of marriage by capture, for it would seem that here at least we find an instance in a direct line of a connubial type, from its first and primitive form down to its legal institution regulated in all its details. And in fact over all this vast region we may perceive that marriage by capture is gradually tending towards monogamy, which, though fictitious in Mongolia and in China, is real in the Loo-Choo islands; and here the people speak with horror of the Chinese polygamy, though they still keep their own wives in a state of slavery.

Ancient Japan, which received from China all its civilisation, maintained, until the end of the feudal times, the Chinese system of marriage, which was despotically ordered by the parents, and which might also be cancelled by them. By degrees, however, the woman's subjection grew less strong: a Japanese girl has now a voice in the matter of choosing for herself a husband. The woman's adultery always gives to the husband the right to kill her together with her accomplice; but he may not kill one without killing the other. We may easily understand, that in Japan as in every other country in the world, there are two weights and two measures. In the case of divorce the Japanese wife goes back to her own family, but she is not allowed to have her children.

On the other hand, monogamy is not more firmly established in Japan than in China. It is not very rigorously imposed upon the man, on account of the liberty of manners allowed to the girls of the lower orders before their marriage. We know that in Japan poor parents will gladly hire out their daughters in large towns for a service of prostitution, lasting for several years. And some of these girls, who are so instructed in the art of pleasing, are very highly honoured; they appear in the religious processions, and, after their deaths, statues are erected to them in the temples. In a sort of way we may compare them to the celebrated Greek *ἑταῖραι*, the companions or the concubines of men in Greece. The profession of a prostitute was not at all looked down upon in Japan, nor did it in any way prevent women from marrying at the expiration of the term for which they had been hired. In Japan, as in China, the husband might bring into his house an almost unlimited number of concubines; but this, however, was a polygamic luxury only within the means of rich men.

If we go back as far as records exist to establish facts, the great Mongolian race, taken in its entirety, will show us a real evolution in marriage customs. We find nearly every form of marriage, from the brutal rape, still practised by the Kamtschadales, the Tunguisians, and the Samoyedes, down to the very relaxed monogamy of the Chinese and Japanese. In the interval we may notice the polygamy of the nomad tribes. During all these ages, the condition of the woman was always gradually improving. At first, she was captured as a beast snatches his prey; she is now bought and sold. She was originally bound to be absolutely submissive to her parents, she was a part of the family property; but by degrees her liberty increased, and she is now treated as a human being.

The curious organisation of prostitution in Japan may perhaps authorise us to think that, in far bygone times, there was a period of hetairism antecedent to marriage. This evolution is certainly a most important sociological fact, for the Mongolian race alone represents numerically at least one-third of the human species. We must not, however, suppose this idea to be a law common to the human kind, for humanity is a very homogeneous medley; and in past ages the different ethnical groups have moulded themselves in many ways according to the exigences of the struggle for life. For in trying to formulate sociological laws, rigorous and precise as the laws of physics and of chemistry, we are perhaps only indulging in fanciful notions which we may find at last to have been no more than a pleasant dream.

XIII.

Marriage among the White Races in Asia.

Pictet and several other distinguished Indian scholars have thought, after careful examination of the Vedic texts, that they could extract from them accurate knowledge as to the primitive Aryan marriages. We shall not follow them in their tentative adventures, for accurate knowledge on the subject is what we most want, and also that in which we are most deficient.

We are furnished with more precise ideas from the poem of the Mahabharata, which speaks of promiscuity as a very ancient custom, and in itself unblamable. Some remains of these primitive manners seem to have subsisted in India for very many years. In Goa, in Pondicherry, and in certain of the valleys of the Ganges, girls were obliged to present themselves in the Juggernaut temples, and vestiges of similar customs have been preserved even down to modern times. At Malabar, when the king was married the first three nights belonged to the high-priest, who received also fifty gold pieces in exchange for service given. This strange custom is probably no other than the Cambogian *Tchin-than*, of which we have already spoken. Every woman, too, who was tired of her husband, every widow weary of her widowhood, were free to dispose of their persons at pleasure, provided that they offered up a sacrifice in one of the temples at Tulavâ. And even nowadays troops of courtesans are attached to the different large Hindoo temples, for the pecuniary benefit of which they carry on their profession, and they are in no way stigmatised in public repute. Until quite recent times they were the only women in India to whom any education was given. In the days of Buddha the grand-mistress of the courtesans was very highly respected in the town of Vesali, and even Çaky-mouni, that divine incarnation, did not disdain to take up his abode in her house. If these facts are not sufficient to prove the existence of promiscuity at some ancient period, they will show at least that there was a great looseness in the moral life.

The Manu Code does in fact prescribe a marriage, apparently monogamic. This marriage is indissoluble, and the husband and wife owe to each other mutual fidelity. In case of widowhood the man may remarry; but it would be very reprehensible for the widow to do so. Men may always marry a woman in a caste lower than their own, but it is strictly forbidden to them to marry a woman in a superior caste.

The law enacts frightful penalties against adultery. The king ought to cause the body of the unfaithful Brahmin woman to be devoured by dogs in the middle of a public place; and her accomplice, if he is not a Brahmin man, ought to be stretched upon a bed of iron heated over fire. As regards the Brahmin husband, his digressions are punished very lightly; God, they say, protects his own people.

The Hindoo marriage is exogamic. A man may not marry a woman having the same name as his own; nor may he marry a relation nearer than the sixth degree.

As it very frequently happens, this exogamy coexists with extreme subjection on the part of the woman. The husband ought to treat his companion as a child; he ought to make her obedient to his will. He may allow her a few innocent pleasures; but she ought to obey him implicitly. Outside the marriage state the woman lives under the authority of her male relations. A woman may be cast-off if she is ill-tempered by nature, if she is a drunkard, if she has been childless for eight years, or has not for eleven years given to her husband a male child. The husband may once, or even twice, make his wife fruitful by his brother or by a relation; and widows may be treated in the same way, for they must "pay the debts of their ancestor."

The Code makes no mention of the suttee, a custom apparently due to more modern refinement; but it considers woman as a dangerous and malevolent creature. "It is in the nature of the feminine sex to endeavour to corrupt man. A man ought not to remain in a lonely place with his sister, with his mother, with his daughter, etc." *

Even in our own time, free will has no place in the Indian marriages; families of young people marry the husband and wife together just as it may best suit their own purpose. Polygamy is tolerated among people of high position.

The widowers remarry as soon as possible, for in India an unmarried man is partially excluded from society. As marriages are contracted without consulting the girl, and her parents think more about the material advantages than about any other consideration, one may often see Brahmin sexagenarians marry, or rather buy, a child of six or seven years old. Speaking generally, the Hindoo

* "Code de Manu," liv. ii. 215.

woman is an honest slave, who has not the right to eat with her husband.

In the north of India, the marriage customs are more primitive, more brutal. The Kasias of the district of Almorah—who by the way are very rigid Hindoos—still maintain the savage exogamy, the real form of capture. The woman so kidnapped is considered as a property which the man may knock about as he pleases. He may make her labour as though she were a mule.

In the region of the Himalaya, near to the sources of the Jumna, in Nepaul, the Aryan Hindoos have adopted the Tibetan form of polyandry. Women are there considered as merchandise, and are bought and sold as such. Fraser tells us that a countrywoman would cost from ten to twelve rupees: a sum pleasant enough to receive, but often painful to be obliged to pay. Girls are sold readily enough; the brothers in a family would hire a common wife, and they would make no difficulty about letting her out to strangers. There, as in every country where polyandry is customary, the women are in nowise scandalised, nor can it be said that this custom is always prejudicial to the general morality. For instance, among the polyandrous people at the sources of the Jumna river an innocent form of falsehood is held in horror; and the polyandrous people of Nepaul are the best cultivators in the country. There is no discussion as to the issue of their marriages; the firstborn is the property of the elder brother, and so on down the list.

In Afghanistan also the condition of the women is very low indeed. The Afghans are Mahommedans, and therefore polygamous; they buy their wives, whom they have the right to cast off and to hire to their guests.

For them the levirat is held to be a duty. The widow continues to be a thing possessed, as she was during the lifetime of her husband; and in case of second marriage the relations of the new husband pay the value of the wife to the relations of the first one.

Among the other branches of the white Asiatic race, among the Persians and the Semites, the sexual unions are still far from the ideal form of monogamy. In the last century the Persians used to contract temporary marriages. A conjugal lease was entered into for a certain stated time, at the end of which the contracting parties were again free. If it so happened that the woman was pregnant at the time, the temporary husband was bound to provide her with maintenance for twelve months. The child, when born, belonged to the father or to the mother, according to the sex. Chardin tells us that in his time there was great liberty as regards sexual intercourse. Marriage was polygamous, as in all Mahometan countries, and was contracted without the consent of the parents, all the children were considered legitimate, whether their mother was wife, concubine, or slave.

Among the populations of Asia Minor, where the Irân, Semitic, and Caucasian races are all more or less mixed together, we may also notice characteristics of very singular manners, especially in the manners of bygone ages. It would appear that promiscuity was formerly practised more or less in these countries. Strabo speaks of the Parthians, who considered that a woman should change her husband when she had given him two or three children. In Babylon there existed a law founded upon an oracle, that every woman should at least once in her life come into the temple of the goddess Mylitta and prostitute herself to strangers for hire, no matter how small the payment might be. Afterwards, but only for a short time, they would live chastely. In the same way at Cyprus, young girls were bound on certain days to go on the seashore, and by prostituting themselves offer up their virginity to Venus. In the Balearic islands, primitively peopled to all appearance by the Phoenician colonists, the newly-married women, the first night after their marriage, belonged to all the priests who were present.

In Lydia and in Armenia the priestesses were by special privilege allowed to be polyandrous; and in certain cantons of Media it was honourable for a woman to have at least five husbands.

Among the Semites, properly so called, a people so much given up to pleasures and sensual excess, as may be seen from their religion and their history, we hear of polygamy existing after the most primitive fashion. Solomon had seven hundred wives and three hundred concubines; and the majority of the Hebrew people imitated his example as far as it was possible for them to do so.

The Arabs, too, used to imitate their Israelitish cousins, both before and after the introduction of Islamism. But they differed from the Babylonians, inasmuch as they regarded the virginity of their wives as a very important consideration. Among the Jews, every woman accused by her husband of marrying him without being a virgin was liable to be stoned, unless her parents could prove to the elders the falseness of the accusation.

And nowadays in Arabia the same fault may be followed by immediate repudiation; and in Yemen the husband may kill his guilty wife. But this does not at all prevent the rich men in the holy city of Mecca from maintaining by the side of their lawful wives concubines, who are generally Abyssinian women.

There are other Arabs, who it is true do not inhabit their own country, who are utterly careless as to the fidelity of their wives. These are the Hassanyeh Arabs of the White Nile. Here the man will buy his wife for certain stated days in the week, and will pay for her so many head of cattle in proportion to the number of days of his purchase. During the days not included in the contract the wife may dispose of her person as she pleases.

We have now said more than enough to show that as regards conjugal manners among the white races in Asia, not only is there no innate nobility of thought, but these customs are very closely allied in many points to the inferior races of humanity. In Europe only do we find that man has succeeded in introducing more dignity of manner into his marriage ceremony and married life, and also to raise woman from the low state of subjection into which she is thrown in every other part of the earth. But this task, as yet very far from complete, was not accomplished in a single day.

XIV.

The Græco-Roman Marriage.

The Greek civilisation, which we may regard as the leavening of the human intellect, has been, like every other civilisation, grafted on to a state of primitive savagery. Tradition asserts that before Cecrops (about the 17th century before Jesus Christ) the

Greeks used to live promiscuously. Children used to know only their mother, and they used to take her name. Such is the legend, but we find it confirmed by other traces which have been perpetuated into historical times.

Lycurgus authorised husbands to lend their wives to those men whom they thought worthy of the honour, so that they might bear children. The aged husband, according to the same legislator, did a praiseworthy action in looking out for a good and virtuous man for his young wife. In the full bloom of the Athenian civilisation Plato blamed Minos and Lycurgus for not having authorised women to be common; and in his "Republic" he maintains that women ought to pass on from hand to hand. Socrates, following the advice of his master, lent his wife Xantippe to Alcibiades; and we know the high degree of consideration that was paid to certain courtesans in Athens.

At first the Hellenic marriage was coarse; the girl was bought either by presents or by services rendered to the father. From an early date the marriages were monogamic, but concubines were tolerated. It was not a disgrace to be the son of a concubine, for such, we know, was the case with Ulysses; but these illegitimate children could not inherit from their father. In Greece, as in Rome, the primitive form of marriage was not enough to establish a line of descent; that was based uniquely upon the declaration made in each case by the father.

Traces of an ancient form of marriage by capture subsisted for a long time in Sparta. The young man was held bound to run away with his betrothed, "who must not be little or weak, but a big girl, strong, and already able to bear children."* And for a certain time the newly-married man was not allowed to see his wife, except by stealth.

By degrees, instead of selling their daughters, fathers used to give them a dowry, and girls without some marriage portion were slighted; but in more remote ages, in Greece and in Rome, it was allowable to the young girl to earn her dowry by trafficking with her person. The dowry paid by the parents was determined at

* Plutarch, "Life of Lycurgus."

first by wages as a guarantee given in the presence of witnesses, then by a public act; it was hypothecated upon the husband's property.

In spite of her dowry the Greek woman was considered only as a thing. Her father might marry her without consulting her wishes. When she inherited, in default of male heir, she was considered as part of the inheritance, and was bound to marry the nearest male relative, or the oldest of the relations, who would otherwise have received the legacy. If at the time of her inheritance she was already lawfully married, her anterior marriage was annulled. The father might bequeath his daughter with the inheritance; and the husband had the right to bequeath his wife to a friend: this indeed did happen to Demosthenes.

The married woman, taken in the act of adultery, might be put to death by her husband—but after deliberation before witnesses.

The connubial customs of the Latins were very similar to those of the Greeks. Among the Samnites the nobles used every year to call together the young men, class them by order of merit, and then allow them to choose a young girl, one after the other, and in their hierarchical order. Traces of marriage by capture existed in Rome at the time of the emperors; they might be seen themselves in the custom of lifting the girl from the ground and in placing a dagger in her hair. The young girl was not allowed to marry herself. She was made the subject of a contract, very often in her childhood, but she did not become a lawful wife until she was thirteen years old. The father, who had the right to marry his daughter without her consent, had also the right to annul her marriage, and this exorbitant right was not attenuated until the time of Antoninus. In the early ages of Rome, the woman was not considered as part of the man's household, except as his slave; this, too, was the case with his children, for the emancipated son did not inherit. She was at first owned as a thing; the virtuous Cato lent his wife Marcia to his friend Hortensius, and took her back when Hortensius died. The Roman husband had the right to beat his wife; for, according to the expression of Monica, the mother of St. Augustine, the Roman marriage was only a contract for slavery.

For a long period the woman continued to be bought, and marriage *per coemptionem* always subsisted. If the betrothed girl belonged to a patrician family the sale was disguised by the ceremony of confarreation, which consisted of partaking with her future husband, before the witnesses, a cake given by the pontiff of Jupiter. For in Rome the marriage ceremony, the *justæ noces*, was long considered as a privilege belonging only to the patricians, and a religious consecration was held to be imperative. But once married, by coemption or by confarreation, the woman, her body, and her property, belonged to her husband; she was "in his hand."

The custom of dowry, however, modified this barbarous form of marriage. It assured to women an independence, which they very often abused.

In the dotal marriage the girl remained in the paternal family; she inherited her father's property, and she herself had the management of it. Generally she used to confide this charge either to the care of a special slave—the dotal slave—or to an agent, who became her confidant. This "free marriage" became the established form in the upper classes under the Empire; and the Latin writers were not sparing in their harsh criticism upon the arrogance and the misconduct of the women so married. These free and libertine manners were certainly very different from the savage customs practised in the early days of primitive Rome; the adulterous woman was then brought before the domestic tribunal, and executed by the parents themselves as they thought fit: "*cognati necanto uti volent*," says the law of the Twelve Tables.

From this rapid glance over the days of classical antiquity, it would appear that marriage became very slowly established as an institution; that its first form was barbarous; but that by degrees the customs grew to be more humane; and that woman, at first a slave, and liable to be bought and sold, to be bequeathed or to be lent, gradually acquired a considerable independence. We may add that from a very early period the Græco-Roman marriage was monogamic, but concubines were not forbidden to men. Concubines in Rome had a legal status, and Commodus publicly kept a harem, in which three hundred women were maintained.

XV.

European Marriage outside Greece and Rome.

Manners similar to those in the very early days of Greece and Rome might have been seen in the other Aryan races in Europe. These other nations were equally coarse, and did not in reality become civilised until Roman civilisation had become widely spread. We may therefore allow ourselves to smile at the fantastic schemes of those theorists who, in applying a sort of linguistic alchemy to the Vedic texts, would demonstrate the existence, according to the Aryan hypothesis, of a pure, noble, and monogamic marriage.

Marriage by capture was practised by the primitive Slavs. It was also for a long time customary in Russia, in Lithuania, in Poland, and in certain parts of Prussia. Young men would carry off their lovers, and afterwards enter into treaty for them with their parents. Quite recently, too, in Wales, a sort of sham fight was performed whenever a marriage took place. Among the Slavs, the Scandinavians, the Franks, and the Germans, marriage was no more than a sale of the young girl. The husband was bound to pay the *mundium*, that is to say, to buy the right of property from the father. By degrees, in place of this simple form of sale, the Germans substituted a dowry, to be paid to the woman *in perpetuo*, to the value of the husband's property. The bridegroom paid first the *osculum*, the price of the first kiss, then the "morning gift," the *morgengabe*. How many words have been wasted in vainly endeavouring to poetise these wild customs, which consisted simply in paying to the girl herself the value of her own property! But the custom of betrothal, also German, is more rational, and indicates a certain nobility of moral feeling. From a very early period the betrothal was taken to signify a real engagement between the parties, which could not be broken off without very serious reasons. But it was not all at once that these different peoples arrived at the sacramental form of monogamy.

Cæsar gives us instances of polyandry among the ancient Bretons. The primitive Slavs were for a long time polygamous; and one of their kings, Vladimir, before he was baptised, kept no less than eight hundred concubines in three different places. And even nowadays, in the Russian *Mir*, a peculiar form of incestuous concubinage will often be found to exist. The head of the family, the paterfamilias, will marry boys of eight or ten years old to girls of twenty-five or thirty, and will constitute himself the paramour of his daughter-in-law until the legal husband shall have arrived at the age of puberty.

In every barbarous European society woman has been considered, as elsewhere, a thing possessed. With the Saxons, the Burgundians, the Germans, and others, the widow, as soon as her eldest son had passed his fifteenth birthday, was placed under his care; without his consent she could not remarry, nor enter a convent under penalty of losing all her property. Under the feudal system, every female vassal of a royal fief was obliged, before marrying, to obtain the triple consent of her father, of her lord, and of the king. The lord could even sometimes compel her to marry the husband that he selected, as soon as she had reached the age of twelve. As Du Cange and Boëtius have shown, the young girl in a way owed the use of her body to the lord, from whence came the ignoble right of *marquette*, which in later times was commuted into a fine.

After she was married the woman was the slave of her husband. "Every husband may beat his wife if she will not obey his orders, or if she curses him, or when she gives him the lie, provided that he beat her moderately, so that death does not ensue." *

Among the Scandinavians, with whom the German laws were preserved longer than with any other people, divorce was optional for the husband; he might repudiate his wife at pleasure.

Adultery on the part of the woman was everywhere severely punished. The Germans used to make the guilty woman walk naked through the village. In some of the Celtic tribes the husband used to test the legitimacy of his newborn child by

* Beaumanoir, Titre 57.

letting him float on a river upon a shield. If the baby was drowned, the signification drawn was that the woman had broken the conjugal pact, and that she ought to be put to death. As late as the Middle Ages, the adulterous woman was shut up in a convent for the rest of her life; and in case of flagrant crime, the husband might put her to death, claiming, too, if necessary, his son's assistance. Such was the canon law; the makers of the code, it would appear, did not even dream of punishing adultery on the part of the husband. And still we hear of the woman's emancipation being effected by Christianity!

Among the Circassians, who have certainly preserved many of the ancient European customs, marriage was, until the days of Klaproth, permitted only between people of the same class: noble with noble, rustic with rustic. The husband might repudiate his new wife if she was not a virgin; in that case she went back to her parents, whose property she was, and they either sold her or put her to death. The adulterous wife was treated in the same way; but before casting her off the outraged husband shaved her hair and slit her ears. The right of divorce always rested with the husband.

We may see, therefore, that the white race, which now occupies the foremost position in the progressive advancement of civilisation, had, like the other races, a very humble origin from a moral point of view. We know, too, that no ethnical group of the white race is now living in a state of absolute savagery; but during thousands and thousands of years our primitive ancestors did not rise to a higher moral and social level than is now actually the case among the present existing inferior races. We may be permitted, therefore, not to despair of the one class of men, and we are also not justified in glorifying the other too highly.

XVI.

Evolution of Marriage.

We have now come to the end of our long enumeration. The ideas that we may have learnt are very incoherent; they are especially insufficient. We may, nevertheless, glean from them some general notions which we will now endeavour to summarise very briefly.

In the inferior degrees of civilisation, among the most primitive hordes, there has not as yet been invented any ceremony which deserves the name of marriage. Union, or rather intercourse between the sexes, takes place merely as desire prompts the people. One law only is known, the weaker has to give way to the stronger. In these human flocks man does not pride himself upon chastity, upon the sense of shame, nor upon humanity. Woman is possessed as though she were a head of cattle, and is kept in constant drudgery. Promiscuity will be found to exist more or less, but it is not the only custom. Very often the strongest man will arrogate to himself the property of one or more wives, either in capturing them from the neighbouring tribes, or in buying captured women from his companions, or in taking women out of his own tribe, merely because he chooses to consider himself the lion. The woman so possessed, being an object of booty, belongs absolutely to him who has captured her. She is his, as though she were his dog or his armour. He has the right to dispose of her, to lend her, to sell her, to beat her, to kill her, to make use of her, and to abuse her. She is his wife according to the fullest meaning that the possessive adjective can convey; to touch her without his authorisation is to commit treason against him, it is making an assault upon his property. This possession of a wife by capture may exist with endogamic promiscuity, and we are not at all entitled to consider it a necessary and general phase in the evolution of marriage.

When a woman is considered as an object, as a property, she is divided between different men, and the idea is not repugnant to them. If this form of sexual union is found to be convenient and advantageous, if in a sterile country where subsistence is scarce it prevents an over-increase of the population, as is the case in Thibet, the custom is continued without any scruples of conscience. It has at any rate the moral effect of restricting animal promiscuity.

The idea of possessing captured woman was very tempting.

Those poor creatures were deprived of all support from the tribe to which their master belonged, for he might abuse his captives as he pleased, and do with them what he liked; but the tribe from whom the woman was stolen would often reclaim their property; they would show their vengeance by exacting reprisals. From a very early period, therefore, the exogamic rape was legalised, the proprietors were indemnified by endogamic rape. To give effect to this, kidnapping was legitimised, and it became a sort of friendly transaction. Rape gradually became more and more an established ceremony, carrying with it the captor's right. The deed was proclaimed lawful after it had been done; the treaty was signed when the war was over. There was no law regulating the habits observed in their marriages. Each little ethnical group proceeded after its own fashion. In one group marriage was exogamic, in another it was endogamic; but from the time when these conventional customs were established the marriage ceremony may be said to have existed. The unions were more often polygamous, sometimes polyandrous, rarely monogamous.

What gave rise to the custom of monogamy? First, in a general way, it came from necessity; for wherever the number of women was not in excess of the men, the possession of several could only be enjoyed by the rich or the powerful. Other causes also aided: rivalry; the conflicts between covetous desires, for every man exacted his rights; and when the constitution of the family was established according to a more settled form, the line of descent became more regular, it naturally fell in more with, and gradually more and more conformed to, the system of monogamic marriage. But man did not comply readily. For a long time the husband considered monogamy as a legal fiction. Nearly everywhere he was allowed to maintain by the side of his lawful wife, concubines, "little wives," often slaves. During the inferior phases of civilisation, jealousy was forbidden to women. As time went on, monogamic marriages were more rigorously observed, especially in the Aryan race; but the infractions of the connubial tie were always lightly considered on the husband's side, and from this we may conclude that our unswerving monogamy is still distasteful to the greater part of the human race, especially to those of the male sex. But this form of marriage has been adopted in America, in Asia, in Europe, among the most civilised ethnical groups in each race, and we may now look upon it as superior to any other form of marriage as yet known. But we need not consider it as the Ultima Thule in the evolution of connubial ceremonies.

As monogamy becomes more generally established, the lot of the woman gradually improves; from being a thing possessed she rises to the rank of an individual person. During a very long period she is married, without her choice being at all consulted, and sometimes when she is still but a child. Her family, her own parents, contract for her; first they barter her for a few presents, or for money, as soon as money can be said to exist. Very slowly she acquires a certain independance. Sometimes her future husband will pay to her what he considers to be her value—this will constitute her dowry; sometimes her parents settle a dowry upon her, and this remains her own property. Once raised to the dignity of proprietor, the woman is much more respected, though her submissive condition be still more or less hard. For a very long time she might be repudiated at will, and for the slightest possible reasons. Adultery on her part, and on her part only, was severely punished. In our European marriage, the barbarous and feudal customs, the legal traditions of ancient Rome, and the Christian ideas have succeeded in effecting a lame compromise, and woman is now neither slave nor servant. She is simply a minor. The law makes of the conjugal union, at least in Catholic countries, an association that death alone can sever.

Will this for ever so continue? Evidently not. In the evolution of societies the last stage is never reached. Already we see that divorce, admitted, or on the eve of being admitted, in many countries in Europe, has shaken the fiction of an indissoluble monogamic tie.

As we have already seen, no form of marriage is absolutely necessary; mankind has had experience of a good many kinds. New innovations will surely be made. But in what direction? We can only answer that it will assuredly be in that direction which, socially, is the most useful. Utility no doubt changes with the constitution of societies, and societies themselves are so very different. Wherever the State will refuse to interest itself in the bringing up of children, a more rigorous form of monogamy will be necessary. The family must there be soundly constituted; for it is then only close to its own hearth, using its own resources, that fresh generations can find shelter, protection, and education.

On the other hand, wherever individual interests will go on gradually establishing themselves upon a sounder basis, the State will also gradually tend to leave to the family the care of bringing up its future subjects. By degrees society will trouble itself less with regulating the laws of marriage, and will think more of forming new generations. The care of childhood will become one of all-important interest; sexual unions themselves will tend to be more and more considered as acts of private life. The State will seek to bring up children, to instruct them well and thoroughly; and will ever consider itself more and more responsible for the performance of this important task. There will then be no reason why a much larger latitude should not be allowed in connubial contracts. Those interested will be allowed to come together as they please, to form these contracts as they would form others, maintaining only those few very general rules which experience has established.

We need not attempt to glorify "the sanctuary of the family." A man must do his best to blind himself, if he does not wish to see what sort of sanctuary this is in most families, how the child is tortured both in body and soul. We are justified in believing, contrary to the opinion of Mr. Herbert Spencer, that, in certain societies at least, the part which the family now holds in every-day life will ever gradually tend to diminish. We cannot predict how this great transformation will be effected in our social organism. On matters so deep rooted we cannot argue upon *à priori* principles. Sociological evolutions operate very slowly. In the first place, individual property must in a great measure become a usufruct, in order to give the State the means of disposing of such resources as would be necessary to defray the new expenses which would crowd upon her, in the hypothesis that we are now considering. But before such a condition can be effected, altruism must altogether gain the upper hand over egoism, the moral standard must be very considerably raised. This will not assuredly be the work of a single day. Progress, however, is fatal; for every society is an organism in evolution, which ethnical competition is always urging on towards a more perfect state.

CHAPTER II.

THE FAMILY.

I.

The Animal Family.

If we wish to study at all seriously the subject of family life, we must first begin by forgetting all the commonplaces that have been both said and written about it. There is perhaps no question on which a verbiage of rhetoric has flowed more largely. The family is a social fact like any other. We may investigate its rise and its development, indicate its good and its bad qualities, look for its origin in the animal kingdom; we may point out the reason for its existence, and even show that it is not absolutely necessary for the preservation of societies of men. Let us first study it in the animal kingdom.

In order that an animal kind of any sort should maintain its existence, it is above all things imperative that it should engender young ones, and that a sufficient number of these young should live. This indispensable condition may be obtained in many ways. As a general rule, the seed or offspring is more numerous in proportion as the animal is inferior, is less intelligent, and also in proportion as the adults think less of their descendants. We know that among many kinds of fish the female lays her eggs by hundreds of thousands, and that she then does not trouble herself about them. Of this seed so left to chance the greater part will perish, but a sufficient quantity will always remain to insure reproduction. The family here does not yet exist even in the most rudimentary state. We see its early commencements in some few reptiles. Certain female crocodiles will show a sort of solicitude for their eggs; they try to hide them; they will sometimes carry their young hatched ones to the water. The female crocodile in the Guayaquil river will hide her eggs in the sand, will come back at hatching-time, she will carefully break the eggs, and take her young ones upon her back down to the water. The male will follow her, but he is actuated by a very different care: those of his little ones that fall on to the ground during the journey he devours. For in the majority of animal kinds, solicitude for the young is first awakened in the female.

Among many kinds of birds, however, the male partakes of this tenderness with his female, especially in the monogamic kinds, where a temporary family may be said to exist. For ordinarily among animals the affection of the parents, even that of the mother, entirely vanishes as soon as the young are able to shift for themselves.

Among the mammalia, where a certain time is always necessary for the bringing up of the young, the female undertakes the charge; sometimes she is obliged even to protect her little ones against the ferocity of her male. Among the greater part of the vertebrae parental love is rare, or it is weak. As regards filial love, it is quite an exceptional thing; but instances of it have been observed in some of the most intelligent of the mammalia, for instance in the elephant. A young elephant has been seen caressing and protecting its mother, who succumbed at last to the arrows of her pursuers.

The large monkeys, notably the chimpanzees, live also after a rudimentary family manner. Progenitors and offspring are associates for a greater or less length of time; and ordinarily the troop will obey their adult male leader. He will keep the power in his hands as long as he is the strongest, and as long also as the younger ones do not think of affranchising themselves from his authority, either by forsaking him or by killing him.

It most frequently happens among mammalia that the female is the centre in the animal family, and that the younger ones will collect themselves around her. When the male remains in the association, it is much more out of attachment to the female than to the little ones. The "matriarcat"—where the mother is considered as the head of the family—that we find so frequently among the inferior human races, is also, in germ, in the animal kingdom.

But the family is far from being indispensable to animal societies. The bringing up and the education is with them the principal object, and this may be effected in different ways. In the most complex of all animal societies, in that of ants, so superior to many human societies, the family is suppressed. The care of reproducing the species is left to a special caste, and the progenitors have neither care nor thought with regard to their offspring.

In the human kind the institution of the family seems to indicate a phase of social development. Sometimes it nearly fails altogether, at another time it hardly rises above the family of certain superior mammalia. In the more or less civilised of the ethnical groups, the family is constituted in very different ways, as we shall see in our short review of the human species.

II.

The Family in Melanesia.

The family of the chimpanzee seems to exist still in all its wildness among certain savages wandering about in the forests in the interior of Borneo; these creatures are no doubt the remains of the negroid population, who primitively used to occupy the Malay archipelagoes. These aborigines prowl about the woods like wild beasts. The male will run away with the female, and they will couple themselves together in the thickets. As soon as the children are capable of finding their food for themselves, their parents let them go their own way. The family, if we may so call it, passes the night under a large tree. The children are hung on to the branches in a sort of net, and a large fire is lit near to the tree to keep away the wild beasts. The clothing of these people is confined to a strip of bark. And in the same way, in the Andaman islands, the man and the woman separate as soon as the infant is weaned; and henceforward, the father, who, by the way, is not always very readily ascertained, takes no further heed of the mother. We know that among savage people the weaning of the child does not take place until comparatively late; but even then, the uncoupling of man and woman is very similar to that of animals.

In Australia the family, in the European sense of the word, does not exist; we hardly find it even in the most rudimentary state. Parentage on the mother's side does exist, but the family knows no father, and the parental authority will very often be exercised by the uncle. As the marriage is exogamic, and the children are recognised to belong to their mother's tribe, they are, in case of war, obliged to follow her and to fight against their father, who, in public opinion, is not considered as their parent. It is true that here and there a tie between the father and the son is beginning to be formed. Sometimes, when the eldest son of a man has received a name, his father will adopt the same name; sometimes the son will succeed to the father's when the father has been a celebrated chieftain.

A similar system of relationship is also in vogue in Fiji. There the father and son are not considered as relations, but the nephew has the right to take as much as he pleases of his uncle's property. For among the Fijians, as among many other people, the terms "parents, children, brothers, and sisters" indicate the sequence of generations, the classes, together with the relative positions of the tribe, much more fully than they indicate the degree of consanguinity.

This shapeless family cannot be considered as a primordial institution, as "the cell" of societies. For between the structure of societies and that of animal organism there is no real similitude. The comparison between the histological elements of an animal and those of individuals or families constituting a human society is only an artifice of rhetoric. It may furnish a few metaphors or oratorical developments, but it cannot do more.

III.

The Family in Africa.

Filiation, through women, that which sociologists have called the "matriarcat," is manifestly the most inferior and the most animal form of relationship, and therefore it is very common in all primitive societies. It is general in Africa wherever negroes are predominant, but we must except the Bushmen, for they have no word to distinguished a girl from a married woman; with them neither family ties nor any sort of relationship is as yet known.

Nearly everywhere else the matriarcat prevails; the sons generally inherit only their mother's property; very often they are no man's sons, but a man's nephews, the sons of his sister, who succeed her or inherit from her. Among the Bechuana Kafirs, the power of the chief passes at his death to his brother, and in default to his maternal nephew.

Among the Kimbundas the children belong to the maternal uncle, and he has the right to sell them. The husband has no authority over them, and only considers as his sons the children of his wife-slaves. The relationship descending through women is in vogue also in Senegal, in Loango, and in Congo. On the Guinea coast the children strictly follow their mother's condition; if she is a slave they also are slaves, even though their father be the king. With the Cumbri the filiation and the inheritance descend through the mother: the son of a Cumbri man and of a strange woman is not a Cumbri. Caillié says, that in Central Africa the sovereignty is not transmitted to the son; it is ordinarily the son of the sister who succeeds.

Even with the Tuaricks the infant is the child only of his mother; he inherits her social position, and, like her, is noble or slave. Also, the collected property acquired by all the members of the family go to the eldest son of the eldest sister; the children inherit only those articles of strictly individual property which have been acquired by the father. Inheritance from the male is however gradually establishing itself, or is tending in that direction, among the African negroes wherever they have succeeded in forming themselves into any sort of complex society. Such has been the case in the kingdom of Dahomey. Nevertheless in Madagascar the matriarcat is still customary. "It is the womb that colours the child."

In the north and north-east countries of Africa, where Islamism or Christianity has penetrated the most fully, the patriarcat is by degrees supplanting the matriarcat; but with the Nubians the power of the chief still descends to his nephew.

Among the ancient Egyptians the patriarcat was in sway without any admixture. All the children of a man were equal, whether their mothers were slaves or free, whether they were lawful wives or not. Such is also now the case in Abyssinia, where, as we have already seen, the connubial union is so fragile that eviration, an accident frequent in the Abyssinian wars, breaks it, and makes the wife of the mutilated man pass over into the bed of his brother-in-law.

In short, in Africa, the family, in the European sense of the word, is not yet constituted; at least not among the real negro races. The interests of the family are nearly everywhere subordinate to those of the tribe. Sons and nephews are confounded; the latter will generally gain the advantage over the former. Under such a system the affection of the father for his children is naturally very slight. Sometimes the husband will submit to the community of the coverture to strengthen his ties of relationship towards the children of his wife. Adoption is everywhere easy; the real son and the adopted son are scarcely distinguishable. Adoption is considered as a true relationship if, as is the case in Madagascar, the father and son are sprinkled with each other's blood, and also if they each drink a few drops; or if, according to a custom in Abyssinia, still more significant, the son touches his father's breast with his own lips, and binds himself by oath to conduct himself as though he were a son.

IV.

The Family in America.

All over South America paternity hardly exists, or if at all, maternal filiation is everywhere predominant. Exogamy is general, and it follows the rule of feminine genealogy. Such is the custom among the Arawaks, among the Indians of Guiana, and in other places. In many tribes coverture is practised, and from this it would seem that an effort was being made in favour of paternal filiation. The Cayuvavas abstain from all work during the menstruation of their wives. The Guaranis will fast when their daughters become nubile, and when their wives are pregnant, and especially during the accouchement. And while their wives are pregnant they do not risk their lives in hunting wild beasts. Among the Chiriguanos, a tribe of the Guaranis, the woman continues her ordinary occupations immediately after the birth of the child; but her husband lies down for several days in his hammock; he fasts, he avoids every change of air, and becomes the object of tender solicitude. Coverture is observed also among the Abipones. This custom, so widely spread among the primitive races all over the earth, is equal to adoption. In this way a man establishes his paternity; he tries to institute paternal filiation in place of maternal filiation, formerly adopted, and still so prevalent, that among the Indians of Brazil the man was nearly altogether wanting in affection for his own children.

Exogamy and uterine filiation are still the rule in North America. It is the mother who gives the name; it is from her that property descends, and the rights of consanguinity are determined. The children belong always to their mother's tribe. The terms employed by the Indians of North America to designate the degrees of parentage, would seem to indicate a primitive state; for there the brothers and sisters lived together in promiscuity, and consequently a man's children and his nephews were not distinguishable. Then as the brothers and sisters could not marry each other, the brothers took to themselves wives in common, whereas all the sisters belonged to the same man. The woman consequently called the sons of their brothers by the name of "nephews," and inversely, they called the sons of their sisters by the name of "sons." The relationship among the Micmacs of North America is also still governed in this manner. Among some few tribes, after many restrictions, the maternal name is given to the children; but this is the privilege only of the chief, and of the rich people; the poorer classes still belong always to their mother's family, and the children continue to take her name exclusively. This is the practice among the Tlinkithes, in Russian America.

Here and there, too, we find the custom of coverture in North America, notably among the Choctahs.

As we have just remarked, the terms used by the Red Skins to designate the different degrees of relationship appear to indicate a restricted and familiar promiscuity; but we should perhaps do well to make allowance for the confusion inherent in the minds and in the language of savages, and to distrust indications hastily drawn. For instance, the Esquimau will call his father-in-law, or his fathers-in-law, by the name of father, even when there is no difference of age between them. The mind of the primitive man, like that of a child, notices things only when they are startling, when they are on a large scale.

Our knowledge is very incomplete as to the constitution of the family in the large states of Central America before the time of the Spanish conquest, but it appears that these ancient institutions were native born in the country. In spite of many legends that have been invented, we may consider that they originally grew from the stock itself of the American races, of which they represent the highest state of culture. As filiation from the woman's side is the rule over all the American continent, we may expect to find traces of it in ancient Central America. In fact, in Peru the matriarcat was general, and the patriarcat was only just beginning to show itself. It was instituted among the Incas, whose male descendants alone had the right of showing their origin, and whose sons inherited; but according to Gomara, among the mass of the people the inheritance descended to the nephews, not to the sons.

We must of course understand by this the inheritance of certain movable objects, for among the Peruvians land was considered as common property.

In Mexico the evolution of the family was in a more advanced state: the patriarcat was established. In fact it is always the paternal personality that predominates; it is the father who dictates to his children the rules which govern their conduct, it is he who establishes the moral precepts which have been handed down to us. The mothers warn their daughters that they ought to be submissive to their husbands, that they ought to obey them and endeavour to please them.

In America, therefore, we may watch the gradual formation of the family. Among people of an inferior stage promiscuity still exists, but it is gradually becoming less frequent. By degrees the uterine family was established, to give way, in process of time, to the paternal family. We shall again find this remarkable evolution in other parts of the globe.

V.

The Family in Polynesia.

It appears now to be beyond all doubt that many primitive societies used originally to live promiscuously, and that the family became constituted only in a very gradual manner. We might naturally expect to find traces of this coarse state among the voluptuous Polynesian islanders, with whom marriage was so fragile, and among whom an idea of shame was unknown. We may add that all things being equal, promiscuity has more chance of establishing itself in these islands, where life is necessarily much restricted, where it is impossible for the individual to isolate himself, and consequently where erotic temptations spring up at every moment. The Polynesians scarcely try to restrain themselves. On the contrary, in the Society islands, twenty married persons might be seen living in common in the same house, and all lying down upon the same mat. The family did not exist, except perhaps in its most elementary condition. The Sandwich Islanders, who had special words to express an adopted son, the parents of a son-in-law, etc., had no word signifying "cousin, uncle or aunt, nephew or niece, son or daughter, father or mother." In the Hawaian family nomenclature the relations were classed into five sections: grand parents, parents, brothers and sisters, children, and grandchildren. All the members of one of these sections were, between themselves, considered as brothers and sisters. The child would call his mother, or the sisters of his mother, by the name of "female parent," and "male parent" applied equally to the father, to the uncles, and even to distant relations. The word used to signify "child" meant, in reality, a little one. The Hawaian father was not the parent of his child. Adoption was rendered extremely easy; a man would give himself a father or sons almost *ad infinitum*. But in the last century this primitive promiscuity began to diminish. Brothers would generally own their wives in common, sisters had also their joint husbands, but their husbands could not also be their brothers. In Cook's days the uterine family was beginning to establish itself for the chiefs; their rank and dignities were transmitted through the female line.

In Tonga island maternal filiation was well established. The father was not considered as parent of his own son; the rank was transmitted through the women, and she sometimes would reign. But in these last few years the masculine filiation has begun to take the place of the feminine.

In the Society islands the masculine filiation was adopted for the chiefs, and even with some exaggeration; for as a matter of law the firstborn son succeeded from the moment of his birth, and his father henceforward was reduced to performing the functions of the regent, and was considered bound to pay homage to his son, and could not stay in his presence without uncovering himself as far down as the waist.

We have said how easily adoption was practised in the Sandwich islands. It was a general custom, and even by its abuse it shows how little importance was attached to the idea of filiation. In the Marquesas islands it was not uncommon to see elderly persons being adopted by children. Animals even were adopted. A chief adopted a dog, to whom he offered ten pigs and some precious ornaments.

The dog was carried about by a *kikino*, and at every meal he had his stated place beside his adopted father. In the Tonga islands no distinction was made between a real mother and an adopted mother.

From what has been said, we see that in Polynesia the family is still in embryo. And in many other races we find the same social conditions to be very similar.

VI.

The Family in the Mongolian Race.

In the Mariana islands there is no relationship between the father and the son. In many localities in Sumatra the father took the name of his firstborn, and in the district of Batta the title of chief was transmitted to his sister's son. We have already seen that, according to an ancient custom of Malay marriage, the father was considered as the property of his wife's family and might be driven out at pleasure. The establishment of a masculine filiation is necessarily incompatible with customs such as these.

In Burmah there is no word to distinguish between father and uncle, mother and aunt, son and nephew.

The Cambogians were more advanced even in the seventeenth century, for, according to a Chinese author, they had different denominations to designate the father and the uncle.

A great confusion still prevails in the family nomenclature among the rude Karians scattered over the Tenasserim, Burmah, and the kingdom of Siam. In their language the children of cousins are called nephews, the children of nephews are regarded as grandchildren, and the brothers and sisters of a grandfather call themselves respectively grandfather and grandmother. And by a singularity which we can attribute only to the likeness between the mental and social development, the family nomenclature of the Karians and the Esquimaux is almost identical.

We find remains of promiscuity among nearly all the pure Mongolian groups. In Bhotan, in Tibet, where polyandry is still practised—that is to say promiscuity in a restricted form—it is more often than not impossible to determine the paternity of a child; the matriarcat therefore prevails. Among the ancient Mongolians, the family must have been constituted after a most confused system, for Baber, the founder of the Mongolian empire in Delhi, in his Memoirs, speaks of one of his lieutenants having a whole tribe of maternal uncles.

Masculine filiation has been known in China for a long time; but the language still shows traces of the ancient social state in which brothers possessed their wives in common. A Chinaman still calls by the name of "son" the sons of his brother, but he speaks of his sister's sons as his "nephews."

In Japan, filiation is subordinate to the indivisibility and the inalienability of the patrimony. The inheritance passes to the firstborn, boy or girl. The eldest child is forbidden to go away from the property, and the spouse must take the name of the heir or of the heiress. The filiation is therefore sometimes masculine, and sometimes feminine. But the maternal uncle still bears the name of "second little father;" the paternal aunt will call herself "little mother;" the paternal uncle is a "little father;" the maternal aunt is a "little mother," and so forth.

As we have already observed, it does not follow that the imperfection in the family nomenclature should necessarily imply confusion between the members of the family. We know that this confusion does exist, and has existed everywhere; and we know that the Japanese are too intelligent to be suspected of not being able to distinguish by means of language that which they distinguish amongst themselves in their everyday life.

We may consider as certain that, among the Mongolians or the Mongoloids, the family has constituted itself, or is constituting itself very slowly; that a more or less restricted form of promiscuity has existed among the greater part of their ethnical groups, and that there are traces of polyandry still remaining; and finally, that in the heart of these races, filiation has established itself sometimes in the female line, but more often in the masculine.

VII.

The Family among the Aborigines of India.

In many of the aboriginal tribes in India the family is hardly yet formed; it is as yet only in process of formation. Among the Nairs, who, as we have seen, still live in a state of restricted and regulated promiscuity, there is no relationship existing between the father and the son, for the very simple reason that the son cannot recognise who his father is. The Nairs look upon their uterine nephews as their children, and to them their property descends. But we must except funded property; that is transmitted through the woman, and never goes out of the maternal clan.

Also among the Cingalese of Ceylon there is no parentage between the father and son. The tribe is expected to marry its own members; children belong to it as does the land, which remains always indivisible. These are the Malay customs, and this fact may assist us in determining the origin of the Cingalese people.

A similar system prevails with the Kasias, with the Kochs, where there is also no relationship between the father and the son.

A tribe in the south of India, named Macua, have instituted two sorts of marriage: one which cannot be dissolved without infidelity on the part of the woman; the other is a sort of free marriage, according to which the children ought to follow their mother in case of separation.

The fact is that, among most primitive people, children are looked upon simply as property, lucrative or onerous as the case may be. It is after this unpleasant idea that filiation is regulated in the law of polyandrous marriages. Certain polyandrous women in Nepaul assign their property in the firstborn son to the eldest of their husbands, who generally are brothers; the second child will go to the second brother, and so on in proportion. Again, in other polyandrous and Buddhist tribes inhabiting that part of Turkestan between the Oxus and the Hindoo-Ko, all the children belonged to the eldest of the mothers. This is a most curious instance; it is a case of male filiation, based altogether upon social conventions without the smallest care as to consanguinity, which all over the world man always thinks the less about in proportion as he is in a lower state of civilisation.

VIII.

The Family in the White Asiatic Races.

Unless we wish to run contrary to all common sense we must put aside many sociological hallucinations which have been laboriously extracted, with the aid of the forceps, from the linguistic interpretation of the Vedic texts. The white race, as little as any other, has risen up from nothing, already civilised, clothed in its most noble and intellectual attributes. Like the other human types, the white man has come from a very low state, and his evolution, like theirs, has been very gradual. For him, as well as for his fathers of different colours, it has not been a slight task to sketch out, to make perfect, to draw with accurate precision his many family relationships.

The race which once composed the Veddahs, whose origin is totally unknown to us, had already, it would seem, devised the patriarchal system. But with them the father was not only considered to be the generator, he was regarded as the proprietor, (*pitâ-ganitâ*, *pater-genitor*) and this had no doubt been his position originally. The Vedic Aryans also had their denominations to designate "the brother of the father," and the "son of the brother of the father." They have other words to signify "the father, the mother, the brother, the sister of the wife, the sisters and the brothers of the wife, the wives of these brothers." From all this we may conclude that the Vedic family were tolerably well constituted. But it had not been so always, for in the Manu Code we find remains of a coarser and more remote family state.

According to the Manu legislator, the infant of the girl-mother who has been clandestinely delivered of a child belongs to the man whom this girl marries; the child of the pregnant woman who marries without declaring her pregnancy belongs to the husband.

All the brothers of the father and the mother are considered the fathers of the son of either of them; all the wives of that one same husband are considered to be the mothers of the male child of any one wife. When a man has no child of his own, he may engage his brother or his relatives to make his wife fruitful. The lawfully born son, the son engendered by the authorised relation, the son adopted, or the son given, the son born clandestinely whose father is unknown, the son rejected by his natural parents, are all six relations and heirs of the family. The Menu Code is formed altogether upon the patriarcat system, and the filiation is wholly masculine. "The woman," it says, "is considered by the law as a field, and the man as the seed." Of however low extraction the woman be, if she is lawfully married she acquires the same status as her husband; it is the same also with the son. The right of priority by birth exists only in the male line. The eldest son takes the greatest part of the inheritance; the girls do not inherit, but the brothers ought to give them the quarter of their share.

An antique period of promiscuity is more clearly proved to have existed in Babylon, in Asia Minor, and in other places, as we may see from the worship of Mylitta, Anaitis, and Aphrodite. We have already seen that in order to obtain the right to marry—that is to belong only to one man—the women were bound first to declare themselves hetærai, in order to indemnify the community.

Among the ancient nomad people of Cyrene, and certain of the Arab tribes of whom Strabo speaks, women were assignated to all the members of the same family, then as moral progress gradually gave birth to social progress, the family by degrees came to be established. It is certain that among some nations in ancient Asia, filiation through the women—the matriarcat—was anterior to the patriarcat. The Lycians, Herodotus tells us, took their mother's name; they derived their genealogy through their mothers and their maternal ancestors; the children of a woman of noble birth and of a slave-father were considered as nobles.

We may suppose, but it is difficult to determine accurately, that there was among the Arabs a period of promiscuity followed by the matriarcat. In the Koran, which describes the manners of the Arabs at the time of Mahomet, male filiation is clearly established. "Women," says the holy book, "are the field of man." The husband ought to assign a dowry to his wives. After the death of the father, a son inherits the portion of two girls. The father's name is borne by the sons. We find that course manners were customary, because of the injunctions enacted against them. The Koran forbids sons to marry women who have been their father's wives; but it is specially said that the law should not be retroactive.

The degrees of relationship are clearly established in the Koran: it is forbidden to a man to marry his mother, his daughters, his sisters, his paternal or his maternal aunts, or his foster-sisters; for it was thought that in being nursed from the same breasts a relationship was formed. We find the same notion in Scotland and in other countries.

These manners, more or less altered by local customs of a more ancient date, have been introduced into Semitic and other nations where Islamism has taken root.

We may say then on the whole the patriarcat has now become customary with the white races in Asia, after they have gone through a phase, probably a very long one, in which the family relationships were terribly confused. And in the same way the so-called Aryan European nations have passed through a similar evolution, as may be seen by the numerous documents of which we have already spoken when treating on the subject of marriage.

IX.

The Family in Europe in Barbarous Ages.

The community of women has doubtless held its sway in many tribes or hordes at one time stationed in Europe. Diodorus Siculus relates that, almost in his times, it was considered in the Balearic islands that a newly-married woman should first belong to the parents and the friends of the husband.

Promiscuity necessarily determines that filiation shall be upon the mother's side; the man afterwards begins to take thought of his genealogical line, and endeavours to constitute his family. We can discover signs and traces showing this ancient social custom to have been at one time prevalent in Europe.

The coverture, which, according to Strabo, used to exist among the Iberians, is still in vogue in some of the Biscay and Guipuscoa valleys; for the manners of the Iberians have been partially preserved among the contemporary Basque people, their descendants. They considered that the family domain should remain indivisible and inalienable, under the care of the firstborn child; when it happened that the firstborn was a girl, her husband came to live with her and took her name, which was transmitted to the children, exactly as in Japan.

Among the Zaporogue Cossacks the relationship and genealogy was established through the female line.

Tacitus tells us that, among the Germans, the maternal uncles have as much affection for their nephews as the fathers have for their own children, and often more. It is probable that more or less confusion in the family relationships has existed in every country and during every period, when the land was regarded as common property, cultivated by all the people in the clan, who often all lived together under the same roof. Such was the case in ancient Germany; such, too, is still the case in the family communities in Croatia, on the confines of Austria, and in those of Lombardy.

There are so many small republics, in which family feeling runs very high, because it rests upon the keenest and closest interests. Nestor, the ancient Slav historian, glorifies the strength of this sentiment among his own people. He says that a Slav would violate the holiest of nature's laws were he to emancipate himself from the family ties. When everything is held in common, egoism is thought to be a crime. We should for the most part look in vain for anything similar in our family as it is now constituted. Our family seems to be the result of the crumbling up of the ancient communities, of which we must now endeavour to discover the early traces.

I.

The Family in Greece and Rome.

If we may believe in Greek traditions, marriage was unknown among the Hellenes before the time of Cecrops; from which we may infer that it was preceded by a period of promiscuity and confusion as to family relationships. Matriarcat was the result, as may be seen from a passage in Varro, quoted by St Augustine; and according to which, in the very early days of Athens, children used to take their name from their mother. Male filiation did not become established until afterwards: it was already an ancient custom in the Homeric age. Sons used then to take their father's name; girls did not inherit except in default of a male child, and they were considered as a property, of which the father had a right to dispose (see the preceding chapter on marriage). The Greeks pushed the theory of male filiation to an absurd extent. In the third part of his "Orestes," in "The Furies," Æschylus shows us at full length this strange conception of consanguinity. According to this theory the mother is but the repository—it is the father who gives life. And at last Orestes is absolved from matricide, because he was no relation to his mother.

In Rome also the family was constituted very slowly after a long period of confusion. But in Rome from the earliest times, masculine filiation, agnatic relationship, was established among the rudest clans. These people grouped themselves together, and so founded the nucleus of the great Roman nation. But at first there were few patricians who were able to say who their father was; the mass of the people did not go through the solemn marriage ceremony, the *justæ nuptiæ*. This mass was composed of plebeians, who lived in a quasi-promiscuous state, *more ferarum*, and knew of no legal paternity. The family of patricians was established upon a basis much more social in its nature than consanguinity—that of property; for the word *familia* was held to mean the collective body of slaves of the owner. Marriage alone was not sufficient to establish agnation, a masculine affiliation; a declaration, and the recognition of the child by his father were considered necessary. Children by the same father but by different mothers were agnate; but no legal relationship existed between children of the same mother and of different fathers. Until the time of Nerva adoption was symbolised by a feigned confinement; and no difference was made between the adopted and the consanguineous son. The latter even ceased to be regarded as part of the family, and therefore could not inherit, as soon as he was emancipated; that is to say he ceased to be his father's slave. In the earliest days of Rome everything was based on property; the soil belonged to the family, and the father had not the right to dispose of it by will. To be a member of the family community was the first and most important right; that constituted a quality which might be lost or acquired.

It was only slowly, and very gradually, that the idea of filiation, of consanguinity, came to overrule that of co-proprietorship, but the two were always closely connected; and nowadays, in our codes of law which have been based upon the Roman codes—that is in countries where the Roman law has been more or less adopted—the degree of relationship confers a strict and proportionate right of inheritance.

XI.

Evolution of the Family.

After the facts enumerated in the foregoing pages we may now follow the evolution of the family from its origin down to our own time, and we may even hazard a few conjectures as to its future.

In the most distant ages, when man began to recognise that he was different from the animals round about him, a sort of simian family must have existed everywhere in the human kind. Our primitive ancestors used to wander about through the forests in small groups, in which there would be the father, or rather the male, his wife or wives, and the children, forming collectively a temporary association under the paternal authority. The Veddahs, in the jungles of Ceylon, still continue to live in this way. As the intelligence of ourems human progenitors began by degrees to develop itself, and also the instincts towards sociability to make themselves felt, people came together in hordes, composed of several families: for everywhere union is strength. In these rudimentary societies, made up of very unintelligent beings, totally devoid of all moral delicacy, promiscuity naturally became the rule. All the women belonged to all the men, but especially to men of a certain age, to those gifted with experience and with strength. And now, in many Australian tribes, the elderly men have a possessive right over the women of their own group. In societies so established it is impossible that children should have a father: they belong to the community.

The family gradually emerged from this state of promiscuity; it is probable that this was mainly the woman's work. For it is a law, common among all mammalia, that the female should have for her young stronger affectionate instincts than is possible in the male kind. In the horde, the children had no fathers, but they had mothers, who for several years nursed them and gave them suck, and then afterwards, and by degrees, allowed them to go their own way. Feminine filiation, too, began to establish itself, and to become customary. Children inherited movable property from their mother; man's inheritance went to his uterine nephews, to the children of the paternal aunt.

As the moral and intellectual strength of our ancestors developed their genesic instincts became less animal. In their heart there was also some feeling of love; they came to attach themselves to one female rather than to another. Hence arose a certain predilection for the children of the favourite wife. They also began to think of their genealogy and of their descendants. They wished to possess for themselves one or more wives, generally bought, or else captured after a brutal fashion. They considered their wives, and the children of their wives, as their own absolute property. A great liberty of sexual intercourse was still common amongst all people in the tribe; but the women belonging to a man, as though they were movable objects or domestic animals, were still more or less respected. These women might designate the father of their children, who continued for a long time to bear their mothers' names, and were after considered as belonging to the tribe. At last, male filiation succeeded in establishing itself, but only in those ethnical groups where there was a somewhat complex civilisation, where marriage had become recognised as an institution, serving as the basis on which the family, clearly defined, was at last founded.

Such appears to have been the evolution of the family, independently of race, in all the ethnical groups which have succeeded in emancipating themselves from primitive savagery.

At the same time that filiation, first feminine and afterwards masculine, was establishing itself, man was thinking about his collateral relations; he was noting the different degrees of consanguinity, and applying to them different denominations, which he was able to determine with greater precision in proportion to the intelligence of his race and the richness of his language. The divisioning of property at last followed the degrees of consanguinity, man began to divide the capital that was once possessed by the tribe; and the family community was substituted in the place of the ethnical group.

Small consanguineous societies then grew up; clans arose; people having the same interests came to live together under the same roof. In these clans family sentiments ran high, and finally overcame the more general interests of the tribe or of the nation. In this social phase man's shelter, his place of refuge, was in his family; everything was made subordinate to its welfare, and family egoism was held to be a high virtue. It could not indeed be otherwise, for there the large community, the state, the nation, took little heed of the individual, who lived and maintained himself as best he might.

But in humanity the moral and intellectual evolution is never idle. The work of fusion, of consolidation, followed the fractioning into classes and families. Simultaneously with family interests, the general interests made themselves felt; the State put down her hand; and under the laws which she established individuals claimed for themselves more personal liberty. In process of time these legal trammels were no longer necessary for the common welfare, and their observance became gradually less rigorously exacted.

Such is the condition now in Europe, and in other countries which have civilised themselves after our mode. The family still exists, and inheritance is always divided according to consanguineous relations; but, on the other hand, the general interests impersonated in the state demand to be ever more and more respected. No doubt that the state is always endeavouring to direct the education of children to better advantage. It is ever trying to put restrictions upon individual property. Succession duties and deeds of sale are continually increasing in price. The public expenditure has to be provided for, and is every day becoming more onerous.

In a word, the family tie, morally and legally, is growing more slack. The authority of parents over their children is lessening. It moves in inverse ratio to the obligations, always augmenting, that the great social community imposes upon the individual.

If this movement continues, what will become of the family?

But here we must draw a line of distinction. As knowledge extends in human societies, consanguinity, both direct and collateral, will be more and more important; in proportion as we become more familiar with the hereditary laws, the vocabulary of our relationships will also increase. There is a strong social interest in knowing as far as possible the genealogy of an individual, for each man, in his virtues and in his failings, represents a whole line of ancestors. That is the scientific side of the family; but the sentimental side will evolve in the opposite direction. Family feeling will ever grow less and less. It will, by degrees, give way to the wider form of altruism, to the increasing care of the general interest. Who can deny that in the great majority of cases, family life is not for most children a deplorable schooling! It emaciates the body, it perverts the heart, and it warps the mind.

As social progress slowly advances, the state will gradually tend to substitute its authority for the blind and unwholesome family influence. It is impossible for us to say how this change may be brought about, or how far it may go. It may be effected in various ways. Here, as everywhere, or perhaps here more than everywhere, we must trust to shrewd observation, and to most prudent experience. The skein of sociological facts is infinitely complex; and not until after many trials and much groping can we hope to unravel it.

CHAPTER III.

PROPERTY.

I.

The Origin of Property.

As soon as any creature, animal or man, is capable of feeling pleasure or pain, as soon as he can recollect the value of impressions and is more or less capable of foresight, he endeavours to put on one side that which is displeasing to him, to possess that which he likes; he has in him then the desire for ownership. Nothing is more egoistical; but nothing is more natural, or more necessary in the struggle for existence.

Ants consider as joint property the galleries which they have constructed, the avenues leading into these galleries, the plant-lice which they milk, and for whom they have built stables; they claim even the possession of all the surrounding territory. Certain flesh-eating animals have also their hunting-grounds, which, in case of need, they will protect against fresh intruders. Our domestic dogs have the feeling of personal property to a very great extent, and they show it in a hundred different ways. Our children, who begin by being inferior animals, acquire very early a strong desire for ownership; the naïve egoism with which this shows itself was observed and noted by Pascal.

The desire for ownership is universal; we find it in every human society. It has been, it is, and will always be clothed in very many different forms, and some of these we now propose to examine. How many hollow declamations would never have come to light if the theorists of the rights of property had begun by studying its rise and its evolution! Property in itself has nothing of an execrable, nor of a sacred character. Like every large sociological fact, there is, and always has been, a reason for its existence; like them, it is destined to become changeable in proportion as the hearts and minds of human beings grow larger, in proportion as the feeling of justice becomes more delicate, and social joint responsibility grows narrower. Now, under penalty of death, ethnical competition imposes upon every group of men the obligation to progress indefinitely, to make better and better use of their talents; that is to say, to create for themselves an organisation upon a larger and more equitable basis. Past ages have given birth to the present, and from the present, in their turn, future ages will be born. Consequently, to enable us to form sound ideas and probable conjectures as to the rights of property, our most profitable study will be to retrace the different customs among the various human races.

II.

Property in Melanesia.

Among the Melanesians, property—that is, of course, territorial or landed property—is common in some ethnical groups, and individual in others.

Each tribe or horde of the Tasmanians used to possess their own hunting-ground, clearly marked out, and no one was allowed to trespass upon it under penalty of death. They considered the infraction of their right as a mortal injury done to them, for it was taking away from them their means of subsistence. The property belonged to all the members of the community, without distinction; the idea of landed property, held individually, was unknown to them.

With the Australians it was different. In certain tribes, individual property was customary; and poaching was everywhere considered a capital misdemeanour. Each male possessed a small limited portion of the land belonging to the tribe; he had the right to sell it, to exchange it, or even to subdivide it during his lifetime among his sons. It is curious to find among one of the most inferior races of humanity the idea of individual and alienable property; that is to say, property as it exists amongst the most civilised nations. And we need not suspect that agriculture has any influence, for in Australia it was not even known.

The scarcity of large animals in Australia may perhaps have

been one of the influences tending to promote the condition of individual landed property. The kangaroo and the emu were almost the only animals who gave long chase. Reptiles, the small opossum, the larvæ of insects, roots, gum, fish from the rivers, shell-fish on the seashore, used to constitute the ordinary daily fare. There was therefore no reason against stationing each individual man upon a large strip of the vast territory belonging to the tribe; though the tribe always reserved to themselves the supreme command over the property. For instance, the owners of property were not allowed to pick the eatable gum from the trees during harvest time; certain restrictions were placed upon the hunting; children were not allowed to eat the flesh of the emu.

Individual property exists also in New Caledonia. There every man, noble or plebeian, holds a strip, more or less large, of cultivated fields; and this property is respected even by the chiefs. The conditions of life in New Caledonia, the total absence of all large mammalia, the practice of agriculture, all lent themselves, better than in Australia, to the parcelling out of property.

These facts are instructive. They prove to us that individual landed property is not in any way the stamp and the seal of a very advanced state of civilisation.

We find individual property also in Fiji, but it is held after a much less democratic fashion; for there the chiefs, always liable to be deposed by their nephews, are the only proprietors, and they profit by the labour of their slaves. We see, therefore, that the majority of the Melanesian people seem to have a precocious taste for the right of individual property.

III.

Property in Africa.

In Africa landed property is rarely held individually, except in Abyssinia and in those northern countries where the Mussulman form of worship is practised; and even here, as we shall see, it is held subject to many restrictions. But all over negro Africa landed property, when it has been instituted, belongs in principle to the community, or to the chief who represents it.

Among the Hottentots, a nomad and pastoral people, cattle are the real source of riches; property in land exists only for grazing purposes or for hunting, and the limits are nowhere clearly defined.

Among the Kafirs, an agricultural race, the arable land belongs to the tribe, but it is not held in common.

Every year the chief parcels out the land already cultivated, or that which is in process of cultivation, among the members of the tribe. Each family, once in possession of its own bit of ground, establishes itself there, isolates itself, and lives mainly upon the grain that it has sown, and which it afterwards grinds between two stones. Similar customs, according to M. Fleuriot de Langle, prevail on the Guinea coast among the Iolofs, where each year the chief of the village, assisted by a council of the elders, divides out the land to be cultivated, partitioning the lots according to the needs of each family. In Saccatoo no man is allowed to enclose or cultivate land until he has first obtained leave of the governor.

In the regions of Equatorial Africa visited by Du Chaillu, where savagery is even greater, it appears that anyone may cultivate the ground if he so pleases. Villages never remain standing for a long time; as soon as anyone dies of sickness, the habitations are set on fire, and the people go away to establish themselves elsewhere. Men have at their disposition much more arable ground than they know how to dispose of. Riches with them consists in having a great number of wives and slaves, whom they can compel to work as they please. A few privileged persons arrogate to themselves the produce of the arable land, making use of men and women as though they were domestic animals.

In societies that are more stable and better organised the daily usufruct becomes an inalienable property; the powerful associate together, to form one or more castes, dividing amongst themselves the produce of the ground, which is cultivated for them by the slaves. This was the agrarian system in ancient Egypt. The common man, the fellah of those days, used to cultivate the soil on the banks of the Nile, sometimes harnessing himself to the yoke, for the benefit of the royal family, for the priests, or for the warriors. A fifth part of the harvest so gathered was taken to fill the store-rooms, for the man attached to the glebe had also to be fed. The sacerdotal caste alone were exonerated from this impost; they received charity, but they gave nothing away.

According to Mussulman ideas, the soil belongs to the sovereign. That is the principle, but the practice is not always scrupulously observed. In modern Egypt the greater part of the land is *mirieh*; the proprietors enjoy only the produce of it, and cannot leave their landed property by will without authorisation from the chief of the state. A portion of the land, however, is *moulk*, and the owners may dispose of it as they please.

In Mussulman Algeria there are various agrarian systems. The Arabs, properly so called, recognise four kinds of properties: that belonging to the state, that belonging to the religious corporations, that belonging to communities or to tribes, and that belonging to individuals. In point of fact it is the tribe who takes care of the property. The part belonging to each family remains undivided to those who have the right to it; they cultivate it in common and divide the produce among themselves. Each co-proprietor may sell his share, but the other members of the family have the right of bringing the property back again into its original line of descent; they may recover the portion sold in repaying the money.

In Kabylie, essentially an agricultural country, individual property has become instituted after the European manner. The fields, small enough, are often enclosed, each property has its title-deeds, mentioning even the number of the trees, and in which the ownership is fully established.

In the same way, too, in the different oases, covered with palm trees, each tree is considered as an individual piece of property.

The family organisation of landed property is customary also in Christian Abyssinia, and we shall find it nearly everywhere in spite of the difference between the religion of the people and their civilisation. The Abyssinians have carefully marked out their family estates, they seldom alienate them out of the family; and the women, who on their marriage might place a stranger in the enjoyment of the common property, do not inherit until default of male heirs to the sixth or the seventh degree.

On the whole, therefore, the alienation and the individual holding of property in Africa is only exceptional. We shall see that nearly all over the world this method of appropriating the soil is the least usual, and the custom that has most lately come into practice.

IV.

Property in America.

Community is or has been the rule nearly everywhere among the American aborigines. In Terra del Fuego the idea of landed property has not yet been received. The sea is the principal store-cupboard from which people take their food; no Fuegian owns anything beyond his canoe and his few utensils.

Among those savage tribes of South America who regard agriculture merely as an accessory—for instance, the people who live on the banks of the Orinoco—the districts set apart for hunting and fishing are owned in common by each tribe; but any bit of land put into a rough state of cultivation becomes the personal property of those who have more or less cleared away the ground. The territory is so vast that the tribes have not yet thought of repressing these insignificant attempts at personal or family appropriation.

In Columbia the majority of the Indians are quite ignorant of the notion of landed property held individually, but they have got a very keen perception of the rights of property possessed by the tribe at large over their hunting-grounds. They watch over their game with a jealous eye; the crime of poaching may terminate in a bloody conflict.

It is just the same among the Red Skins in North America. The vast hunting or fishing grounds of each tribe are the undivided property of all the members of the association; their frontiers are more or less clearly marked, and trespass is a crime that will often result in war. The property inside the enclosure belongs to the tribe, it is its own little country; the right of hunting and shooting belongs to all the members of the association, and each one may possess as his own the game and the fish that he has taken. The clearing away of a strip of land will confer the right of individual property over the products of the soil; but this is only a usufructuary title, for the Indians do not allow land to be transmitted from one individual to another. Some of the Red Skins on the north-western shore attached such importance to their property that they demanded payment of Cook's sailors for wood and water.

The rules that govern property among the Esquimaux are more numerous, and they are also more singular. Nowhere does the community proclaim its right over property more strongly; for with them this right is exercised over movable objects which nearly everywhere else are the uncontested right of the individual, of him who has fashioned them.

The Esquimaux form so many small associations among themselves, they often live all under the same roof, and they carefully determine the limits of a small district that shall be used by them in common. Their laws in this respect are very curious.

Whales, walruses, bears, all the large animals, no matter how they have been captured, are regarded as common property; for except in very rare instances, they do not consider that one individual, unaided, can capture one of these big creatures.

Of every seal that is taken in a winter station small parts of meat and of fat are distributed among the associates of the same group.

If the borrower of any utensil or any instrument lose or damage the thing lent, the lender is not entitled to claim damages; for a man does not lend that of which he is in actual need.

An Esquimau has not the right to own more than two kayaks. If he has a third he is bound to lend it to an associate of his commune; for that which is not being used is regarded as having no possessor.

Individual property, therefore, is by these rules limited to a few arms and utensils, and these only in a very small number.

But those who prefer their own individual proprietorship may leave the district inhabited by the association, and outside the frontiers build for themselves a hut which is their own personal property. In this way they may hunt and fish as they please, but they do so at their own risk and peril.

Also, everyone in the community has the right to appropriate to himself any piece of floating wood on condition that he is strong enough to draw it alone on to the bank, quite clear of the water. A stone placed upon it is then held sufficient to guarantee the proprietorship.

It is not without some astonishment that we find in a race so ill developed in every other respect a system of association at once so ingenious and so equitable, and in which there is such a strong sentiment of joint human responsibility, with which is also connected a regard for individual independence. The majority of Europeans, so proud of their arts, of their science, and, in a word, of their civilisation, are surely, as regards their social aptitudes, very much inferior to the Esquimaux.

V.

Property in Peru and in Mexico.

Like the Esquimaux, of whom we have just spoken, the ancient Peruvians took the idea of communism for the basis of their society. But their communism was not republican and equalising; it was rather patriarchal and authoritative, leaving the labour to the common people, whom the governing classes directed as they thought most proper. This is the largest attempt at a centralised and despotic communism that has ever yet been realised. It may be worth while to describe it with some detail.

The territory of the Peruvian empire was divided into three parts: one for the sun, that is, the sacerdotal caste; another for the Inca; the third for the people.

The lands belonging to the sun produced an income devoted to the maintenance of the temples, to the celebration of sumptuous ceremonies, and to provide for the wants of a numerous clergy.

The luxury of the court and the large household attaching to it, the immense family of the Inca, absorbed all the revenues of the royal domains.

The rest of the land was divided among the people, and the partitioning of the lots used to take place once in each year. Nothing was left to individual caprice. Every Peruvian male was bound to marry at a certain age, and the district to which he belonged furnished him with a habitation and a strip of land sufficient for himself and his wife. At the birth of each child an additional small piece of land was added to that originally granted, and the whole would increase or diminish each year in proportion to the number existing in the family.

The *curacas*, or men in government employment, received a lot, of which the value was made proportionate to the importance of their office.

On the other hand, the people worked for everybody. The three kinds of property were cultivated by them, and after a certain established order. The land of the sun was first looked after; God and his ministers naturally had precedence given to them. Then came—and this is curious in a despotic state—the lands of the incapable, and those who had become maimed through injuries received in the public service, those of old men, of the sick, of widows, of orphans, of soldiers in active service. Then everyone was free to work for himself, but under the general obligation to give assistance to his neighbours. At last, in the third place, people took care of the lands belonging to the Inca. This last work was undertaken as though it were a public rejoicing; the population used to sing as they tilled the royal lands, and they were clothed always in gala costume.

All their undertakings were performed in the same way; to work the mines, to graze and look after the numerous flocks of lamas, to shear them, to weave the stuffs of wool or of cotton, to make the roads, etc. But each Peruvian owed to the state only a certain stated portion of his time. As soon as his task was finished, he was replaced by another man; he was also maintained by the state as long as the state had need of him

The greater portion of the harvest of the wool taken from the lamas, etc. was stored away in repositories and divided into three categories, corresponding to the three great social divisions; but the stores belonging to the sun were obliged in case of need to supply the deficits of the Inca; and when his repository was overflowing, the surplus went to the sick and the infirm. The stuffs were fabricated by the women, who understood perfectly well the art of spinning and weaving. With these stuffs the families used first to clothe themselves, and the over abundant matter was put into the stores of the Inca. Men were employed to watch over the distribution of the goods, and also over the execution of the work.

Everything was done as it was wanted; the different employments usually passed on from father to son.

Thanks to this system, famine was unknown in Peru. There was no mendicity, nor was there any private charity. No one had to fear abandonment; the community, as far as it was possible, provided for every case of need; old age, sickness, infirmity, accidents, etc.

In the Peruvian system we find realised, point by point, certain modern ideal plans, usually considered as impracticable utopias. We may add that for centuries this system gave to Peru every kind of prosperity compatible with a poorly-developed civilisation. But would it necessarily exclude all progress, as has been imagined? What would have ultimately resulted from it? The savage Spanish conquest barred the way to any decision of these questions; but it is an important sociological fact that the system was in a very large measure successful. The device, "Everyone for himself and God for all," upon which we act in our European societies, has its good sides; it stimulates personal activity; it excites us all to work and to invent; but how many sacrifices does it exact? And how many times in the inexorable struggle for life must not the better man succumb before the worse, honesty before meanness! How many times has an honourable and laborious life ended in abandonment and unhappy old age! In a well-organised society we ought to succeed in conciliating the joint responsibility of all the members with their own individual independence. The problem may be difficult, but perhaps it is not insolvable.

We do not find anything existing in Mexico at all comparable to the systematic authoritative communism in Peru. But we do find a social organization, having the welfare of the community for its basis, to have existed at the time of the foundation of the Aztec empire; and we may still discover traces in the present pueblos of New Mexico, evidently constructed after the plan of those *casas grandes* of five, six, and seven stories, which caused much astonishment to the conquerors. In the anthropological section of the Exhibition of 1878, there were models of these curious pyramidical constructions, in which each story was less wide and less deep than the story below. The whole edifice is divided into rooms, into cells, into which one descends by a hole bored through the ceiling. There is no staircase; the communication from one story to the other is by means of outside ladders. Each building forms a village in itself, of difficult access, governed by a chief elected yearly. Everyone agrees in bestowing eulogy upon the Indians who live in these pueblos. They are peaceful, hospitable, industrious, and intelligent; and, in spite of their common life, they seem to practise monogamy with comparative strictness.

The general organisation in ancient Mexico was very different; the emperor nearly always maintained a sovereign right over all property. It was a sort of feudal system, in which the emperor, as principal proprietor of all the soil, used to grant the feoffs, but of which the investiture had to be confirmed at each new accession. In return, those who held the land were bound to appear frequently at court to uphold him in case of need with their armed vassals, and to pay him an annual tribute. But personal property did exist over certain estates conquered or given as recompense for public services. The owners were forbidden to dispose of their property in favour of a plebeian. Certain of these possessions were transmissible only to the eldest sons, and, in default of heirs, they reverted to the Crown.

We find, therefore, in America, that the idea of landed property is unknown to the Fuegians, and among other native tribes it takes the form of hunting or fishing preserves, held in common, but it is sometimes permitted to individuals to appropriate to themselves small strips of land, if they will be at the trouble of clearing it. In Central America landed property is better organised; in Peru it is used for the benefit of the community; in Mexico it is held individually. In this latter country we may say that the establishment of landed property was preceded by a superior civilisation. Both the Mexican and the Peruvian societies were about on a par of intellectual development, but in moral development the Peruvian was very far in advance.

VI.

Property in Polynesia.

In Polynesia property was and is still held mainly in three different ways: by the tribe, for the benefit of the community; by the community, for the benefit of the family; and individual property.

In certain districts of New Zealand there are small societies living in a state of absolute communism, even with promiscuity. The stuffs and woven materials are all common. To cultivate the ground, to manufacture the cloth, to watch birds, to go fishing in their canoes, is man's division of the labour. The women collect the fern roots, they pick up the shell-fish and other crustaceous animals on the seashore, they prepare the food, and they make up the articles of clothing. A portion of the ground is allotted to the use of each family, and this portion is again subdivided into individual parts on the birth of each child.

In Easter island there used to be large houses like the Mexican pueblos, in which there was room for a hundred persons. Similar buildings also were seen in Ulietea. In the Marquesas islands, every native on a journey has the right to enter any hut, to thrust his hand into the popoi trough, to eat as much as he wants, and may then go away without being compelled to utter a word of thanks. In these same islands theft is considered as a venial offence, and it is rarely punished.

Elsewhere the notion of individual property was instituted, sometimes in all its fulness. At Tongatabou the houses of the chief, and the cabins of his servants or of his slaves, were constructed in the middle of a plantation enclosed by hedges. In the Sandwich islands a feudal organisation existed, based upon the right of conquest. The conquering chief divided the conquered district among his great vassals, and they subdivided their part between their sub-vassals, who were taxed at will by their master. A similar practice was common in Tahiti. The lords of the district conceded portions of land to their vassals, who enforced the common people, the *tatous*, to do all the hard labour. Individual property was thus constituted in all its severity. Each strip of land had its own particular proprietor. These people had even imagined the system which in French Brittany is called "le domaine congéable;" sometimes the trees would belong to one individual and the ground to another. By means of a singularity which we have already noticed, the bare property itself passed from the father to the son immediately upon the birth of the latter.

In Tahiti people had arrived at the extreme right of individual property; the dying man had a right to leave it by will; he dictated to his relations and friends his last wishes, and these were always held as sacred.

Amid such an organisation of property, theft was not tolerated as in the Marquesas islands; the guilty exposed themselves at least to a bastinado, and sometimes to death.

In Tahiti, as in many other countries, the institution of individual property, almost Roman in its form, coincided with a social state that was but very slightly advanced. But in Polynesia these general conditions hastened the evolution of the right of ownership. In the smaller islands, where the chase after the larger kinds of game is unknown, because none of the large mammalia are to be found there, man must depend for his food mainly upon the vegetable kingdom, or, in order to get animal food, he must catch fish, or resort to the domestic animals. We may therefore, without doing harm to anyone, restrict the right of landed possessions, and when man has arrived at the stage of agriculture, individual property is then soon established. The chief, however, always exercises the main right, and the principle of primitive community is not abolished.

VII.

Property in Malay, etc.

In the Pelew islands the organisation of property runs contrary to the effect naturally produced by the mode of living on the islands. The different races of people do not all obey the same causes in the same way; everything therefore is in a state of change and evolution.

The individual here owns as his own property only his house, his furniture, and his canoe; the king is the general proprietor of all the land, and he allows certain people to enjoy the usufruct. When the usufructuary ceases to enjoy this privilege, the strip of land from which his produce came reverts to the king, and every year there is a new partitioning of liberated lands.

In the Caroline islands, also, a relative community exists. Each district possesses a large public house, in which people meet, where they keep their boats, their weaving spindles, and other instruments useful to the community at large.

But more especially in Java we find an agrarian community practised in perfect fulness; the system in many provinces may be compared to the Russian *mir*.

The Javanese think that all the soil belongs to the Creator, to God, and therefore to God's representative on earth—the sovereign. The sovereign allows the enjoyment of the property either to his commune or to the individual who has made it valuable. The recipient and his descendants enjoy all the benefits of ownership as long as he fulfils all the conditions that have been prescribed by custom. The soil, consequently, is inalienable; the majority even in the commune cannot lay their fingers on it.

The Dutch colonial government can only impose the taxes and customs duties upon the village communities. The cultivation of rice, the cereal *par excellence* in the island, is very favourable to the formation and to the maintenance of the different communes. For the successful cultivation of rice in Java it is almost always necessary to create a system of irrigation, and for the proper execution of this plan association is indispensable. The ground that has been fertilised by the common labour becomes naturally the undivided property of the labourers. The fields that have been watered are divided out among the families interested, sometimes every year, sometimes every two, three, or five years; but in order to obtain a share in this association, a man must first possess a yoke, that is, a pair of buffaloes or of oxen. There are therefore poor people who have no share in the allotment.

The Javanese community (*dessa*) is governed by a chief elected annually; and to him is allotted a share more or less large in nearly every community. The houses and the gardens attaching are the only landed private property.

But here and there individual property is beginning to be established. In certain provinces, the woods and the waste lands are common; and in clearing a portion of this common waste land the individual may become proprietor, sometimes for several years, or sometimes for an indefinite time. The land so cultivated may be transmitted to the descendants of the first proprietor, and these descendants may enjoy the property so long as they continue to cultivate it. The sale of shares in this common land is forbidden to any stranger.

This system seems to be favourable to the growth of the population. In fact, it renders quite useless that which Malthus calls moral restraint. From time to time a family swarm will leave the community, and go to found a new village elsewhere. The population of Java in 1780 was 2,029,500 inhabitants, in 1808 it was 3,730,000, in 1826 it was 5,400,000, in 1863 it was 13,649,680, in 1872 it was 17,298,200.

These facts may recommend themselves to the attention of those statesmen who are entrusted with the legislation of certain countries in Europe, where the system of individual property seems to diminish rather than to increase the numbers of the human species.

VIII.

Property among the Mongolian Races.

Among the nomad and pastoral Mongolians, property is valued by the number of heads of cattle, and these are always more or less held in common. Even when these flocks are possessed by one large proprietor, every individual belonging to a group of tents is in a certain measure interested in the profit accruing from these animals; he has a right to a minimum, determined according to his position.

In the hereditary transmission of property, the Tartars have invented the right of seniority. As a man's sons arrive at the age of majority, they leave the paternal tent with the cattle that their father is good enough to give them. After this deduction has been made, the paternal goods go to the youngest son. This custom, as human and as rational as the right of seniority is wanting in humanity and rationality, exists also in certain districts of India; it has also existed in certain English counties, in Cornwall and in Wales, therefore in a Celtic country. The same practice was once in vogue in French America, where it was called *le droit du juveigneur*.

The communistic habits in Tartary have caused the feeling of joint responsibility to rise to such a point, that the inhabitants in a group of tents are bound to go out and look for animals lost by travellers who have camped in their neighbourhood, and even to replace them if the animals cannot be found. The character of people is everywhere gradually tending to model itself upon social institutions.

The long duration of the Chinese empire may furnish us with a very instructive picture of the right of property. According to the ancient chronicles, about 2305 years before Jesus Christ, China was already an agricultural country; it was divided into communes governing themselves, electing their own chiefs, and to whom was apportioned a suitable piece of land. The remainder of the soil was partitioned out among those who would cultivate it, to men from the age of twenty up to sixty. Then, as has everywhere happened, the shepherds encroached upon the flocks, the chiefs of provinces usurped the hereditary right, sovereigns began to concede feoffs, etc. However, until 254 years before Christ, the cultivating families divided the arable land proportionately according to the number of men working upon it. One lot in every nine was cultivated for the profit of the state. Then, by degrees, the rich men took possession of the land, and let it out in small farms to the cultivators whom they had ousted, taking for themselves the larger share in the division of the profits. But still, the emperor is in principle the proprietor of the soil all over the empire. The treasury still treats with the autonomous communes who elect their own chiefs; but in fact, the imperial right confines itself to expropriating for non-payment of the imposts, and to confiscation for all crimes against the state. In return, the government assists and protects the vast system of irrigation which first began in the northern provinces of China 600 years before the Christian era. Landed property has, therefore, become individualised in China by means of a long and steady reign of violence and usurpation.

In Japan the origin of individual property is also very brutal; it rests only upon the right of conquest. It was in this way that the first Mongolian occupants of Japan established their feudal system. The chiefs conceded lands to their companions, the possession of which ennobled them and their descendants, to whom the privilege was also transmitted. As these fiefs were considerable, the titular holders divided the property with their liegemen, who constituted a second order of nobility, and they, for payment, sub-let the ground, which they were entitled to do, to the peasants, by the sweat of whose brow everyone was clothed and fed.

But the family community may also be seen at the basis of all this feudal structure. The first-born, of whatever sex, inherits the property, and has not the right to quit it. The qualification of heir overcomes every other; and the heiress, when she marries, gives her name to her husband. These are evidently the last remains of primitive communism.

II.

Property in non-Mongolian Asia.

The system of community was, or perhaps is still practised by many of the aboriginal tribes of India. Among the Nairs, landed property is transmitted through the woman's side, and it never goes out of the maternal clan. With the Cingalese, in Ceylon, it is the duty of the family to marry people, the children are considered as children of the family; and the land is never divided among individuals. The Teehurs in Oude live together in large establishments; they partake of everything in common, and the marriage tie is merely nominal.

In treating of property, of the imposts, etc., the Manu Code speaks only of the villages, which are still in India so many politic and economic unities.

Before the English conquest the right of property did not imply that of alienation. The natives could not understand that lands might be seized and sold for the payment of a private debt. They did not know nor had they any idea of what was meant by testamentary disposition. They neither sold, nor let, nor bequeathed their lands by will. By degrees, in certain districts, they began to alienate their property; but the consent of their relations, their neighbours, and their joint proprietors was first necessary.

The ancient Hindoo village was an agricultural community, used in common by those who had the right to participate. At the end of the year the fruits and the harvest produce was divided, as we find reported by Alexander's lieutenant Nearchus.

In the Punjaub, the village is still an association of free men, who have, or who think they have, one common ancestor. During the last fifty years certain villages in the Madras presidency have submitted only in appearance to the individual imposts; but in reality they pay it in one lump sum, and then divide it out amongst the different members of the community. The village owns the forests and the uncultivated lands; the arable land is allotted between the different families and belongs to them. A lot of ground is granted to the currier, to the shoemaker, to the priest, to the secretary and treasurer. Each family obeys a patriarch, who enjoys a despotic power. The village recognises a chief, either elected or hereditary. The idea of individual property is, however, beginning to take root. In certain districts, by the mere fact of his birth, a son has a right to a portion of the paternal wealth.

Among the Afghans the system of community is better preserved. After each cinquennial or decennial period, according to the local custom, a new equal re-allotment of land is made among the families. The estates may be exchanged, but only among members of the tribe.

The Bhots, a polyandrous people, have already established the system of individual property. With them the fortunes of the different husbands are all joined together in the wife; her children inherit conjointly, and even generally by heirship, in advance. The parents usually, at the time of their marriage, leave their property to their children in equal parts, keeping for themselves only that which is absolutely necessary.

In speaking of the system of property in Mussulman Africa, we have said that there were properties common among all the members of a tribe, of a community, of a family. This seems to have been a very ancient custom among the Semites, for Diodorus (ch. xxxiv.) tells us of its existence along the coast of Arabia Felix. In the island of Panchaia, agricultural communities used to reward their members, each according to his work, giving to the best cultivator the largest share in the harvest products. We may note as an exceptional case this desire for just equity, which is rare among primitive communities.

Among the ancient Hebrews the land was owned collectively by the families. There were rich families and poor families. But every fifty years there came a restoring jubilee; the sales of alienated lands were then annulled, property was equalised, and the Jewish slaves regained their liberty. It was like a great wave washing over all the land, and drowning the strips of individual private property. This periodical annihilation of the right of property would seem to show that the idea of property held in common preceded that of property owned by the different families.

X.

Property in Greece and Rome.

In the classic antiquity of Greece and Rome property held in common was also the first established.

Sparta owned a vast communal domain of forests and mountains, and the revenue coming therefrom served to defray the expenses of the public feasts. Sparta being a conquered country, cultivated by the ancient owners (the Helotes) reduced to a state of slavery, Lycurgus soon succeeded in re-distributing it into equal portions. He established also the practice of common repasts, a custom also in use in Crete and in many parts of Greece.

Portions of the Lacedæmonian soil might be granted even to strangers, provided that they conformed to the laws of the country; but it was forbidden to anyone to alienate land when it was once possessed. Other usages much more communistic were also in vogue; it was permitted to the Lacedæmonians to make use of the horses, the dogs, and the utensils of their neighbours, if the articles were not in actual use.

The prohibition to sell land, to bequeath it by will, for a long time maintained equality among the Spartans. The use of testamentary power was not introduced into the republic until after the Peloponnesian war. The right of testamentary disposition, and the faculty of inheritance allowed to women, gradually engendered opulence and the accumulation of wealth in the hands of a small number of individuals. Women also received considerable sums of money as their dower portions. The father, or in his default the guardian, used to marry as they pleased the daughter heiress and her property. In process of time two-fifths of the Laconian territory became female property. With these inequalities of fortune violent enmities and social dissensions arose between the rich and the poor: the ordinary corollary under similar circumstances.

Similar causes produced the same effects in Athens, where the laws of Solon had individualised property much more fully. Landed property belonging to private persons was parcelled out with great care, the proper distance to be observed between one man's trees and those of his neighbours was rigidly observed. Finally, and this is a graver matter, the right of bequeathing property by will was allowed to anyone who had no male heir, and landed property could by marriage be transmitted to heiresses.

Nevertheless, Solon and his successors placed very heavy restrictions upon individual property; these were probably the remains of a more early state of communism. According to Solon's laws, the sale of a property entailed with it on the part of the seller the loss of his right of citizenship; an increased scale of charges was also placed upon landed property. The rich were obliged to fulfil certain very costly duties, and mutation charges of a hundredth part of the value was placed upon the sale of every immovable kind of property.

Besides the right of individual property, customs absolutely communist in their nature had been preserved; the public treasury gave dower portions to poor girls; corn was sold at very low prices, or was given away gratuitously to the indigent; every day there were theatrical representations, at which every citizen might be present without ever untying his purse.

The dogma of individual property had a very unstable foundation in the Athenian legislation. At this time the state was not what it afterwards became—an abstract personage, troubling itself little about individuals, except to punish their offences or to impose taxes upon them.

In Rome, as in Greece, the idea of individual property rose very slowly from the older theory of common property.

In the early times the right of property applied only to slaves, to cattle, to movable objects, to everything that might be tangibly taken away (*mancipatio*).

Landed property in Rome arose first from the village communities; this community afterwards divided itself into family community, into *gentes*. The family and the *gens* were continued upon the male side. The family, strictly speaking, was constituted by the group of *agnati*; that is, a genealogical series in which the degrees are known, and may all be counted.

The *gentiles* also came from a common but a legendary ancestor; the rings on the ancestral chain which united everyone to their common progenitor could not be counted, they were lost in numbers.

The *agnati* enjoyed a reciprocal hereditary right, because they had the same domestic worship, and their ancestors lay in the same tomb.

The *gentiles* did not inherit until default of the sons of the *agnati*, according to the law of the Twelve Tables.

In the family associations everything belonged to everybody; the Roman family was a community comprising men and things. The members were maintained by adoption as well as by consanguinity. The father was before all things the chief, the general administrator. He was called father even when he had no son; paternity was a question of law, not one of persons. The heir is no more than the continuing line of the deceased person; he was heir in spite of himself, for the honour of the defunct, for the lares, the hearth, the manes, and the hereditary sepulchre.

The patrimony was immovable, as was the hearth and the tomb, to which it was attached. The man merely passes; generations succeed each other, who in their turn all play the same parts, they maintain the same worship, they take care of the property. The heir inherited of his own right; he was *heres suus*.

The emancipated son is excluded from the inheritance; the adopted son inherits only from the adopting family; a son can never arrive at the age of majority in his father's lifetime.

Dating from the law of the Twelve Tables, individual property was established because the right of sale or the right of testamentary disposition were both allowed.

But the sale of property was still surrounded by religious formalities; the presence of a priest was considered necessary.

The power of exchanging property lay in the right of bequest and in the right of sale. But the right of inheriting what was one's own was always extant unless there was an express disinheritance. Now disinheritance was a sort of excommunication precluding the disinherited person from the domestic sacerdotal performances, and from the hereditary sepulchre, which ever continued to be inalienable.

But some laws which came into vogue at a later date limited the right of bequest, and ill effects coming from the restriction soon made themselves felt. The Voconian law forbade leaving to anyone a greater share of the property than was left to the natural heirs; the Falcidian law insured to the natural heirs a quarter of the inheritance; the Glician law compelled the testator to allege valid reasons for disinheriting his children; the Voconian law prevented more than a quarter of the patrimony being left to women.

Then, the right of conquest, which in the days of antiquity did not consider property, whether the claims were family or individual claims, for a long time constituted in Rome an *ager publicus*, distinct from the *heredia*.

At last the religious and moral idea, which in primitive Rome presided over the law of property, died out by slow degrees. All feeling of joint responsibility became extinguished. The right of bequest, the right of sale, the right of inheritance granted to women, together with that of possessing a dowry, the inventory benefice, etc., ended at last by making property a perfectly movable object, which might become accumulated in large quantities in the hands of one individual. The effects upon the manners of the people were soon felt; social position was tested more by a man's money than by his merit; avarice grew apace; the foxes outran the lions. Secret influence became a source of fruitful industry, and was one of the means most employed; riches were acquired *per fas et nefas*, and Pliny was well entitled to write: *Latifundia perdidere Italiam*. Large properties swallowed up small ones. In some provinces all the *ager publicus* was owned by certain families; half of Roman Africa belonged to six proprietors when Nero put them to death.

The same crumbling away of the primitive community, the same transformation of the right of property took place, as we shall now see, over all the rest of Europe.

II.

European Property outside Greece and Rome.

Before the Roman conquest the system of community existed more or less all over Europe. Diodorus Siculus says that the members of the Celtiberian tribe of the Vaccaei divided their land every year for the purpose of cultivation; but the harvest was common and everyone received his quotient part. The penalty of death was even enacted against anyone who should disobey these injunctions. And in his commentaries Julius Cæsar speaks of the family community which he found existing among the Aquitanians. According to Strabo, the Dalmatians, every eight years, used to make a new allotment of their ground.

Horace says that the Getæ, on the banks of the Danube, used to divide the lands among them every twelve months. In the time of Cæsar, the Germans, who were then very little given to agriculture, used never to sow the same field two years in succession; each year the magistrates assigned to the families their different parts In Gaul the communal domains were considerable at the time of the Roman empire, and the present existing communal property still shows traces of the old system.

In ancient Germany bequest by will was unknown. The eldest son inherited; but only the house and the enclosure adjoining. Sometimes inside this enclosure habitations were constructed for the younger brothers when they married; for this little lot of ground was the salic land, transmissible by inheritance to the male children and to the male next of kin, women being always excluded. A strong hedge was put round this private property; but all the rest of the territory belonged to the clan, of which the members maintained that they all came from the same ancestor. There were allotments; lots of arable ground were formed, each member chose his lot by chance, fate determining the site. All these lots, except that belonging to the chief, were equal in size.

The allodial ground, or the special family domain, was owned by the father and son in co-proprietorship.

The territory common among the German clan was called *mark*, or *allmend*; it comprised the arable land, the woods and rivers; each family enjoyed only the right of usufruct. In Gaul, Cæsar founded a similar organisation.

The *mark* was considered as a little country; it had its own altars and its own tribunal

It was by means of agriculture that the idea of individual property seems to have implanted itself in ancient Germany. Among tribes still barbarous in their manners, dispersed over a vast territory thick with forest land, agriculture is far from having the same importance that it acquires in more civilised societies. As the clearing of the land imposed a considerable amount of labour, each strip of land in Germany that was put under culture became hereditary property. Hence arose the inequalities of fortune. Then the right of bequest, which the Roman legislation introduced into the country, dealt another severe blow to community.

And at last the feudal system established itself over the greater part of Europe in the primitive clans, who were a barbarous people, but still more or less republican.

Feudality, existing nearly everywhere on the right of conquest, thought little of the rights previously acquired by the conquered communities. Still, at the basis of the feudal right we may discover the principle of community. The conqueror, the lord maintains his sway over the whole property; he concedes the feoff, the benefice, but only the usufructuary enjoyment or retribution for past and for future services. The benefice was at first only a life interest, and it entailed the performance of certain functions, most of which were military.

By degrees the life interest extended itself into an hereditary right, and to this was joined also the right of bequest. Large domains then grew up, specially apportioned to the service of the church. These were called lands in mortmain. They were different from the common property inasmuch as they were free from all taxes; and these lands increased to such an extent that already in the ninth century one-third part of Gaul belonged to the clergy. On the other hand the feudal nobility encroached incessantly upon the communal territory, simply on the principle of might is right. They first invaded the forests, then the cultivated lands. William the Bastard, Duke of Normandy, desolated twenty-six parishes in his duchy to make a forest covering thirty leagues of land. The Nantaise forest, stretching from Nantes to Clisson, and from Machecoul to Rincé, grew upon the ruins of numerous villages destroyed in order that the Duc de Retz might hunt on horseback from one of his castles to the other.

In England, where, by virtue of the Norman conquest, the sovereign was, and is still in principle, considered as proprietor of all the lands in the kingdom, the feudal concessions made at first for certain services rendered became actually so many independent properties, and the larger property has at last ended by completely expropriating the communal property, and by swallowing up the small property. This absorption was at first usurpation, but it continues now under the name of purchase; for in England the legal expenses of examination of title-deeds are so heavy that the large capitalists only are rich enough to buy the smaller properties. In a word, the combined effect of past and present abuses in England have had the result of allowing all the land in certain counties to pass into the hands of five or six persons. One half of England belongs to a hundred and fifty individuals, and the half of Scotland to ten or twelve proprietors.

In spite of these changes, the moral side of which we shall soon have to examine, the ancient system of village and family communities is far from having altogether disappeared from Europe.

In Lombardy there are still agricultural associations, of four or five households together, living in common in large farm buildings, under the direction of an elected chief, the *reggitore*, and of a woman manager, the *massaia*.

In Spain, the right of enclosure abolishing the right of commonage was granted only quite recently, and it gave rise to more than one rebellion among the people.

In England there are still existing co-operative agricultural associations, of which no member may sell his share but with the consent of the community and of the proprietor. In the middle ages, associations of the same kind between the serfs or the peasants were common enough. They were persons having civil rights, or they were allowed to hold their property free from molestation.

In Ireland, village communities with annual drawing of lots, and sometimes labour in common, lasted until the reign of James I. Sir Walter Scott found similar associations in the Orkney and the Shetland islands. The small Breton islands, Haedic and Houat, are still possessed and cultivated in common under the direction of the minister, assisted by a council of twelve old men.

It was not until 1793 that the division of the greater part of communal property was made in France by virtue of a decree of the convention. But in spite of decrees and of the legal code, the family community still exists in the French Basque provinces according to the ancient plan. The law has been eluded, and the family property is always transmitted to the eldest child—boy or girl. The family takes its name from the property, and according to custom the heiress gives her name to her husband. The produce coming from the property, the reputed inalienable property of the heir or of the heiress, is given up to the general interest of the family, to the education of the children, to marrying, and to the establishing of young people outside the immediate family circle.

And, also, the Germanic *mark* has not altogether disappeared, in spite of the incessant encroachments of individual property. It still prevails in the sandy region of Nesslande, where the common field still undergoes the triennial rotation, and is divided into three parts, in which is sown the winter rye, the summer rye, and buckwheat. It is only after the full deliberation of the parties interested that the times for cultivation, for sowing, and for reaping are allowed to take place.

In Switzerland the *marks* or *allmenden* are still numerous in the cantons of Schwytz, of St. Gall, of Glarus, and others. In these communities the right of usage or of usufruct belongs hereditarily to the descendants of families who have held this right from time immemorial. The *allmenden* comprise the arable lands, the forests, and pasturages. The arable lands situated near the villages are divided into strips, and for these people draw lots every ten, fifteen, or twenty years. Sometimes the right is a life interest. The widow and the heirs inherit the right of use until a new allotment is made. The shareholders usually come together once in the year to hear the statement of the accounts, and to arrange any current undertakings. The assembly elects its own president, their own officers, and no one can refuse to perform a function imposed upon him. A council of a few elected members regulates the cutting of the woods and the allotment of the land; it orders the execution of the smaller works, it determines the payment of fines, of compensation moneys; it represents the corporation in judicial cases, etc. The *allmenden* are everywhere admirably cultivated, though it is not individual property; and everywhere the usufructuaries are kept from extreme poverty. A portion of the common property is set aside for public services, for schools, for the church, and for the administration of charity.

But it is more especially in the Slav countries, in Croatia, in Slavonia, in the military Austrian empire, especially, too, beyond the Dnieper, in Russia, that the village community has been preserved in all its integrity. The Russian *mir* is the type of this system, and it now exists amongst thirty or thirty-five millions of Slav inhabitants.

Formerly, in the Slav communes, the work of cultivation was performed in common, and the produce was divided out among those who had the right to it. This plan still exists in a few districts in Servia, in Croatia, in Austrian Slavonia, and among a few groups of the Raskolniks. In Russia, the arable land of the commune submitted to the practice of triennial rotation is generally divided into three concentric zones, starting from the village; then each zone is subdivided into narrow strips, five or ten metres in width, and from two hundred to three hundred metres in length.

It used to be the custom to draw lots annually for these divisions; now the partitioning takes place at different intervals, from three to fifteen years. The prairie also undergoes an annual allotment; but the periodical redistribution never touches the house and garden possessed by each family in the village, for these are considered as hereditary property.

It is the commune who pays the imposts; it is, indeed, perfectly self-governing, it names itself its own *starosta*. The *mir* assembled all together determine upon the time for the haymaking, and what is to be done with the waste lands; they admit the new members, they give or refuse permission to anyone to change their domicile, to absent themselves permanently on payment of a fine. The assembly, against whose decision there is no appeal, is composed of all the chiefs of the family, even of women, in the case of widowhood or their husbands' absence.

In principle, every male inhabitant who has attained his majority has the right to an equal part of the common land.

In the *mir* the common utilitarian reasons and arguments will dominate over every other. The status of co-proprietor or of co-taxpayer is never annulled; there are certain rights which cannot be avoided.

The chiefs of the family in a general meeting may dissolve the *mir*, and institute individual property, but for that two-thirds of the votes would be necessary.

The same utilitarian character is predominant in the constitution of the family, which is the unity of the Russian commune. The chief of the house is called *khozain*, a word which signifies a general administrator, and yet it gives no idea of parental authority. When a household is dissolved, the members do not inherit; they all partake of the collective property. For instance, all the male members receive an equal share, but the married daughter, or the son who has already left the house, is excluded. If those left remaining are all minors, a near relation comes to live with them and becomes a joint proprietor.

In Servia, not until all the members of the family are dead can the last surviving member dispose of the property as he pleases.

This organisation of property is very interesting; it has its drawbacks as well as its advantages. On the one hand, it brings together into close intercourse the members of the families and of the communes, by means of the joint responsibility of interests and of duties; it prevents both great riches and great poverty; it abolishes to a great extent the plan of payment by wages; it renders null every poor-tax and every organised system of charity.

On the other hand—and this is a very great objection—it trammels individual liberty, it makes each of the members of the community more or less the slave of all. It discourages all individual initiative action, and therefore all progress.

Does the sum total of good resulting from property held in

common outweigh all its disadvantages? Ought we to regret having lost it? May we hope to re-establish the system again amongst us upon a better footing? We cannot answer these questions before we have given a general glance over the theory of the evolution of the right of property. This is what we now propose to do.

CHAPTER IV.

THE EVOLUTION OF PROPERTY.

I.

FROM the foregoing long exposition of facts we may see clearly that our present notions of the rights of property are of a relatively recent date. The theory which now regulates the owner's right of property, submitting to the absolute will of one individual a portion of the common soil, is almost an anomaly, if we look synthetically at the evolution of the human kind both as to present and to past times.

At the commencement of every society the primitive families, or groups of families, the clans, the tribes, ordinarily composed of people connected with each other by blood relationship, hardly thought of fixed property (*la propriété quiritaire*).

The different clusters of men formed so many small associations, all struggling, the one against the other, for their existence. From this competition there came a sort of collective appropriation of the ground. Each little group protecting, *unguibus et rostro*, the patch of land on which it found its daily food, a general cantonment naturally followed. Chance as it might, each little clan was obliged to select their own domicile in a given district marked out with more or less clearness. In this settlement, relatively very large, each tribe lived as they best could, hunting, fishing, picking up here and there a few fruits or eatable roots. This little country was very precious to them, for it was in fact their larder. They became much attached to it, and as far as possible did not permit the intrusion of strange guests, though they never missed the chance, when it came in their way, of encroaching upon their neighbours' territory. Owing to this struggle, every man trying to satisfy his own appetite, or his own ambition, strife and conflict was perpetually raging between the different hordes.

Individual appropriation of land was not often thought of; the hunting-ground was possessed in common by all the tribe. Such has always been the case where man has been obliged to pursue for long distances large animals in order to provide food to enable him to live. Where the large game was rare, as in Australia, or where there were none at all, as in the islands of the Pacific, the male members of the tribe used sometimes to divide the common territory into individual properties, stipulating at the same time that restrictions for the general interest be duly kept. This precocious establishment of individual landed property has, theoretically, a great importance. It proves that the personal appropriation of land is in no way connected with an advanced state of civilisation.

When the hunters in the primitive tribes were so far advanced as to have come to the pastoral life, the pasturage still remained indivisible as the forest had been, for migration among the herds of men was one condition of their existence.

Quite at its commencement—at its very birth—agriculture suggested the idea and the desire for individual property, but the enjoyment of this property was confined at first to the usufruct. The hunting-grounds and the pasture land were so vast in extent that when any one member of the tribe, more intelligent and more provident than the others, thought of clearing away and sowing, after some rough fashion, a small patch of the forest or prairie round about his hut, his labour was at first unnoticed by his neighbours. But he who by the sweat of his brow, or rather of his slaves or his wives, first cut down or burned trees, weeded and sowed the ground, was generally allowed to remain the undoubted owner of his own little harvest. But then the art of planting and dressing was quite unknown; and also all the village inhabitants would often change their place of abode, and the strip of land that had been cleared was soon abandoned by the first agricultural occupant, who went away to renew his attempts elsewhere.

At a later date, when agriculture, instead of being merely an accessory, became an important industry, exacting large spaces of ground and the labour of every hand, the arable soil was then as the pastoral land, and as the forest, possessed by everyone in common. The tribes held despotic sway over the property; the soil was served out by the village communities, who obeyed their chosen chief; and after the harvest the produce of their labours was divided amongst those who had the right to it.

By degrees the holders began to seek their own individual independence; the clan was thus gradually broken up into small groups. Instead of dividing the property in common, periodical allotments were made to each family, and they in turn formed themselves into a usufructuary society, obeying their chief, who usually governed after a very despotic manner. But even then the prairie and the forest usually continued to be owned in common. This form of possession has been preserved in the Slav *mir* and in the Swiss *allmend*.

But the formalities of the allotment, and the servitude to which each family thereby became subject, imposed a restraint which people were at last determined to abolish. In principle the clan still remained proprietor of the ground, but the enjoyment of each lot was granted to each family for an indefinite period of time. The families also still lived together, each one obeying their chief, their father, or their general administrator. We find in the early days of Rome the most perfect type of this phase of proprietorship. Absolute master both of things and of people, the paterfamilias had the right to kill his wife and to sell his sons. Priest and king in turn, it was he who represented the family in their domestic worship; and when after his death he was laid by the side of his ancestors in the common tomb, he was deified, and helped to swell the number of the household gods.

This family community is certainly the most noble form in

which the right of property has yet been clothed; but it had its rules and its limits. Girls were not allowed to inherit anything, and the father of the family had not the right of alienation nor of testamentary disposition.

And also the rigid organisation preserved in the Roman family became intolerable to individuals, at least in certain countries, amongst the most perfectible, the most progressive people: it was then that the theory of individual property arose. The father had first of all the right of bequest, and also of disinheriting this or that member of his family; then he came to have the faculty of alienating his patrimony and of dividing it; his wives inherited, they received large dower portions, and when they married they took into another family a portion or all the patrimony belonging to their own. The quiritary property was thus established, and the right of individual property was pushed to the extreme point. The Roman law said: "Dominium est jus utendi et abutendi re suâ." But even in those days the great civic community attested its ancient right in the form of a reserve: "quatenus juris ratio patitur."

The right to absolute proprietorship has become established only in Europe; but, thanks to the Roman conquest, it gradually became customary over all the portions of the continent, severing itself more and more widely from the older theory of holding property in common.

The Roman empire was followed by the feudal system, which, based on the right of conquest, and drawing with it the right of eviction, caused in principle the property in every country to pass into the hands of the conquering chief; and he, according to his good pleasure, gave out to his companions portions of the stolen territory. This concession was in theory no more than usufructuary. The master remained the true proprietor, and the feoff granted at first for life necessitated not only the payment of fines, but also the obligation to perform certain functions: to mete out justice, and to take up arms on behalf of the lord in all his wars. By degrees these temporary concessions became hereditary, excepting the formalities of investiture, and the feudal owner at last enjoyed nearly all the rights of the Roman proprietorship. Once more the instinct of individual independence had nearly succeeded in shaking off the yoke of dominant mastership. But the communes for a long time kept their vast domains, contesting each foot as it was taken away from them by the great feudal lords or by the clergy. The former took the land by force, and the latter slyly, seizing when they could a new inheritance, and thus appropriating to themselves new portions of the soil.

It was not, however, until the break up of the feudal system that individual property became wholly enfranchised, in law as well as in deed. Possession then ceased to be a feoff conceded, a tenure exacting the performance of certain duties. It raised itself by its own efforts, and was dependent upon no social office. Property became saleable and movable according to the needs and the caprices of the proprietor. Confiscation by the master was no longer an evil to be feared, as was formerly the case, nor was eviction to be dreaded from the conqueror. This last evil had at any rate the advantage of compelling the proprietor to reside upon his land. Then, in process of time, our present notions of individual property came into vogue. According to our present conceptions, each strip of ground forms a small empire of which the master is the absolute owner. To him belongs not only the superficial extent but the foundation; he may dispose of what is above ground and what is underneath in every piece of landed property, be it small or large, and this property represents a cone or a pyramid, having its summit in the centre of the globe. The right of bequest became more or less admitted. In every case, in default of will, the property itself outlived the owner, and after his death it went to his direct or to his indirect heirs, who to entitle them to the privilege of absolute ownership had done no more than take the trouble to be born. It is as we read in Beaumarchais' comedy: "Qu'avez-vous fait pour tant de biens! Vous vous êtes donné la peine de naître."

That is the point at which we have now arrived. In every country which enjoys the European system of civilisation, the right of property has ever been in a state of evolution, always tending to give a greater degree of independence to the individual owner; in other words, the evolution has always worked in favour of individual egotism. Who can say that the evolution is now complete, or that we have yet realised the highest ideal system in the disposition of our property? A progressive evolution is for every society one of the conditions of existence. The right of proprietorship cannot therefore remain stationary. Let us now endeavour to conjecture in what direction it will most probably turn.

II.

Our preliminary exposition has shown us that in "the principle of proprietorship" there is no inviolable claim, that it contains no sacred or holy right. The *jus utendi et abutendi* grew very slowly, and as a rule established itself only by dint of violence and cunning. But, in spite of its origin, is there any social interest in preserving it in its present state? Is it essential to the progress of societies? We cannot deny that it has rendered great service, for in spite of innumerable acts of injustice in smaller matters, it has guaranteed leisure and independence to a minority, which would otherwise have been paralysed by the strongly-bound ties of common property. The majority of the privileged persons have used their right only for their own personal interest; but some have, of their own initiative movement, endeavoured to advance the much-tossed-about car of progress. Many, too, though in thinking only of their own interests, have served the common cause, from the very fact that they were free agents. But, after all, the total result has only been one of difference. We may therefore be prepared to believe that, in future, individual property will have to undergo very serious changes; for as man becomes more civilised, he also acquires more strongly the love of justice, and he learns better how to turn his social resources to a good purpose. We may add that because of the ethnical competition each nation must advance, or else, in course of time, disappear. But, in order to make good progress, every society ought to utilise as far as possible all its means of strength; and for that end it must create between its members a joint responsibility, always tending to become more serious and more binding in its nature.

The defenders of our present system of property, and (who would have thought it?) Proudhon himself joins issue with them, cry out that property is a shield against tyranny, "the cuirass of personality."* But the care of protecting oneself against tyranny will vanish, as tyranny also becomes dispelled, and despotic governments are now gradually on the wane. Again, the "cuirass of property" can only guarantee the shielded person; for the others, those who have no defensive armour, are tied hands and feet, and are governed by superior force. If the theory of individual and hereditary property, such as we find it constituted in countries where Roman and feudal law prevails, may be considered as a good cuirass, it is also an excellent pillow on which we may sleep very soundly, without fear of molestation. How many men are there who have lived as idle parasites, and who, if they had been compelled to work in order to gain their bread, would have made good and industrious citizens? But because of their presumed descent (*Is pater est quem nuptiæ demonstrant*) they have never in the whole course of their lives recognised the necessity for doing a day's work. The weight of their cuirass has prevented them from coming into the field of action.

The enormity of the privileges resulting from our modern theory of the right of property has been condemned by many writers whom we cannot fairly accuse of demagogism. We need not enumerate them all, for we should have to go back to Plato, to Essenius, and to the early Christians. But in confining ourselves to writers of our own times, protestations abound upon all sides.

M. Laboulaye says: "The right of property is a social creation. . . . . Every time that society changes the conditions under which it has existed, every time it alters the rights of inheritance or the political privileges attached to the soil, it is doing that which it is entitled to do. No one can find fault in virtue of an anterior right, for before and outside of its organisation nothing can or could have existed. Society itself is both the source and the origin of the law." †

* P. J. Proudhon, "Théorie de la Propriété," p. 237.
† E. Laboulaye, "Histoire du Droit de Propriété."

J. Stuart Mill, after proposing to disallow the right of inheritance to collateral relations, and to confine the amount of direct inheritance to making a sufficient provision, goes on to say that he does not think it fair or right that there should be a state of society in which there is a "class" of men who do not work, in which there are human creatures who, not incapable, or who have not bought repose by anterior labour, are exempt from sharing the toils common to the whole human race.

J. Fichte predicts that property "will lose its exclusive private character, and will become a real public institution. It will not be sufficient to guarantee to everyone his property lawfully acquired; we must see that everyone gets his own, that he is paid in exchange for his lawful work. . . . Work is a duty incumbent upon everyone; the man who does not work injures others, and therefore deserves punishment." *

Mr. Herbert Spencer says: "Equity, therefore, does not permit property in land. For if one portion of the earth's surface may justly become the possession of an individual, and may be held for his sole use and benefit, as a thing to which he has an exclusive right, then *other* portions of the earth's surface may be so held; and eventually the *whole* of the earth's surface may be so held; and our planet may thus lapse altogether into private hands." †

"We must at last recognise," M. de Laveleye well says, "the first maxim of all justice: 'everyone according to his work,' so that property come really to mean the result of labour, and that the welfare of everyone be in proportion to the assistance that he has brought towards the work of production." ‡—"Property has now been deprived of all its social character; it is totally different to what it was originally; it is now no longer collective. A privilege without obligations, without trammels, without reserves, it seems to have no other object than that of assuring the welfare of the individual." §—"The net produce of land is absorbed now in indi-

* "System der Ethik," b. ii. para. 96, 97.

† H. Spencer, "Social Statics," ch. ix.

‡ E. de Laveleye, "De la Propriété," xli.

§ Ibid. xx.

vidual expenditure, which in itself does not in any way contribute to the progress of the nation." *

What does all this mean? Are we to return to the Slav *mir*, making of each citizen a guarded workman, attached to the globe? Certainly not. The ancient system of holding property has died out because of its tyrannical character, and the general civilisation has progressed because of the degree of liberty granted to each individual. But individual liberty cannot degenerate into an inherited privilege. A reaction is therefore probable. In fully maintaining individual liberty this reaction will probably bring us back, slowly and by means of a series of graduated measures, to the life-interest usufruct of the land, thus rewarding intelligence and useful work, and also the labour given. We are, of course, to understand that this reform, by reason of its radical character, can only be effected with the greatest precautions, and made to work in such a way that it will bear upon future generations very much more than upon our own contemporaries.

But then the family would be dissolved! There are many families who never inherit anything. We may add that the same tendency, fatal to individuation, which has already destroyed common property, will surely, by degrees, slacken more and more the family ties in proportion as there is utility in replacing them by social ties of a more general kind. But what can be more sad, according to family sentiment, than the unwholesome greed coming from the right of inheritance? And this sentiment, so strong in the Slav communities, is with us gradually becoming extinct. If it still has any real power, it is certainly a remnant of a moral fundament bequeathed to us by our ancestors, who lived under a more collective system. No doubt among our contemporaries the best of them do not say that the lives of our fathers, of our mothers, etc., are a barrier between themselves and luxury, or welfare, or a hundred other desirable objects; but the worst of them do say so, and they repeat it very constantly.

We may add that the care of individual property is certainly the

* E. de Laveleye, "De la Propriété," p. 362.

principal cause of the diminution in the number of births, both in France and elsewhere. The holder of a property does not wish that it should be too much divided after his death; he endeavours to leave to his children the means of living in noble idleness; and, contrary to the patriarchs of old, his glory is to have only a few children. We may mention in passing the calm egoists who wish to enjoy, alone by themselves, both their actual property and their "hopes." If we continue to follow this plan, statistics tell us that at the end of five or six centuries the French race will be extinct.

But the reducing of individual property to the enjoyment only for life would naturally impose fresh duties upon society at large. The state would have to watch the young generations much more closely, to give to any who were capable of receiving it a complete general and special education; it ought then, in many cases, to substitute itself in place of the family, to cultivate the character of the young as well as their intelligence. Pedagogy would then become the first of the sciences.

When the work of education is completed, an attempt should be made to open to everyone the career for which he is best fitted, to establish district banks, who would advance to those who showed sufficient moral and intellectual guarantees a capital large enough to start them in their first trial. Effort should also be made to reward real deserving merit, so that a whole long life of toil should not terminate in abandonment and in misery.

To establish such a system very large pecuniary resources would be indispensable; the reforming body itself might furnish them. J. Stuart Mill has proposed that no inheritance should exceed a modest maximum, and we may follow this same course still farther. By means of succession duties charged upon the transmission of hereditary property, the state is ever imposing a tax upon the rights of inheritance. The amount of money coming from these rights might be increased in the most legitimate manner by establishing the amount of payment, not according to the degree of kinship, but according to the value of the property inherited. If this were prudently graduated over a long series of years, the progression would finally bring us, and without any sudden shock, to the abolition of the right of inheritance. At the same time we should, by taking warning from the lessons taught by experience, have learnt to ward off the social evils resulting from this great reform, by the side of which all political machinations are but as child's-play.

CHAPTER V.

MORALITY.

As the human mind becomes enlarged and enlightened, it gradually corrects itself of the infatuation with which it was intoxicated, so to speak, during its early youth. The human race does not know very clearly where it is going, but it knows still less from whence it has come. The line,

> L'homme est un dieu tombé, qui se souvient des cieux,

now causes us to smile. We know that man—the *genus homo*—was a very poor creature when he first made his entrance on to this world's scene. It was not without much effort that he has freed himself from animality, and from his head to his feet this poor little god is still impregnated with it! It was upon this coarse foundation that the noblest human qualities have been grafted; and the origin of these qualities is still clouded over by a thick fog of metaphysics amassed and brought together during many thousands of years. Some few glorious specimens of this kind have come to their full bloom, and certain sentiments dazzle our eyes, though they are in fact nothing else than the simplification of conscient phenomena very common in the animal creature.

We have explained elsewhere how the nervous cells, acting as registers, preserve the trace of molecular vibrations of which they have been the seat, and how these rhythmical vibrations are aroused and reproduced in proportion as they are the more often brought into action. These signs are very often quite unconscient: it is in this way, for instance, that the cells of the molar spine show many collective kinds of movements. The phenomena are still more curious in certain invertebrata, who, in virtue of an education that they have inherited, which we popularly call instinct, perform unconsciously series of very complicated actions, and to which on first thought we should be inclined to attach a moral importance.

If the insects had professors of philosophy to whom they paid money wages, we should undoubtedly be enlightened with many lucubrations on the abnegation of the female of the *liparis chrysorrhea* enveloping carefully her own eggs in an impermeable tissue which the animal fabricates herself by pulling out the hairs from her own body for twenty-four or perhaps for forty-eight hours. The operation is a cruel one, and the bombyx dies because of it; but certainly without having the slightest notion of the object for which he seems to be devoting himself. He has never known his mother, and he will never see his own little ones. He mechanically obeys an automatic habit which a slow work of selection has imposed upon his kind.

Moral instincts are not formed otherwise in the human brain. Under the persistent pressure of various influences, moral habits establish themselves and become hereditary. So that in many thousand circumstances an inner voice, the voice of our ancestors, cries out imperiously within us, urging us to do this thing or to abstain from doing that. All this will happen in the absence of any reasoning power, and very often in spite of it. That is the moral sense, the feeling of duty, which we may rightly admire when we see it push a Winkelried to die for his country, but of which we may very easily discover the scheme in certain animals and even among the coarsest savages.

From this point of view nothing can be more interesting than to watch a dog to whom man has succeeded in teaching so many curious habits contrary to his nature.

Without having been trained, one of these animals will suddenly stand still the first time he sees game; another, under the same conditions, will bring the game to his master; a young shepherd's dog will begin to run, not into the middle of a flock of sheep, but round about them. Why do these young animals act in such a singular way, so contrary to the nature of their savage ancestors? Clearly because the commands of their human masters, through many generations, always reiterated in the same way to successive generations of the animal, and followed wisely by reward or by rebuke, have at last fixed themselves in the animal brain, have taken root there, and have implanted hereditary habits. An automatic association of nervous inclinations to obedience has been formed, all of which act upon and strengthen one another. When once the instinct has become implanted in the nervous cells it holds imperious sway; the animal yields with pleasure to the unreasoning impulse which calls him; it would be painful to him to resist.

But if we really wish to look to the bottom of things we may find a sort of morality, very inferior no doubt, for the animal has no thought which way his acts tend, and the mechanical impulse to which he yields has in fact sprung only from the stronger will of several generations of masters. But instances of this low morality are far from being rare in the human kind. Is it not merely the caprice of the master, that, in many despotic countries, gives to the actions of their subjects their moral value? In one place, everything that pleases the tyrant is regarded as good, everything that displeases him as evil, and after this system has lasted for a sufficient number of centuries it is the most abject slavery that forms the morality of the down-trodden people. In Burmah, in Siam, and in other countries, man's submission to his master is boundless. Every Siamese, H. Mouhot tells us, respectfully calls himself "the animal of the king." In the legal language of some of the Javanese smaller states, crimes and offences are considered as treachery towards the sovereign, not injuries to the individuals who have been injured. In all the Mussulman states in the middle part of Africa the caprice of the master is the supreme law, and every man degraded by him is condemned, ill-treated, and put outside the common pale.

In societies still quite primitive and anarchical, the directing influence of a despotic will is often replaced by public opinion;

that is to say, by social custom. This is the case among some of the Australian tribes, from which we might in fact construct a theory of the embryo state of human morality.

In Australia, the animal kingdom is poorly represented. Therefore, the poor aborigines consider that the flesh of the emu (a sort of cassowary) is most dainty food; and as roast emu is a rare dish, it is reserved for the old men, who, in the Australian tribes, have great privileges reserved to them. It is strictly forbidden to all young people to eat of this holy meat: to do so is an offence against morality. And this inhibition, which has been handed down through many generations, has created in the Australian mind a peculiar and quite instructive code of morality; for the Australian is not a man capable of much reasoning. The flesh is weak, and the pangs of hunger are very strong. It may happen, therefore, that a young man hunting a long way from his home, far off from where his tribe is encamped, secretly infringes the law—the emu law. But then, as happens also in other countries, when the desire is satisfied, the moral instinct awakens. The guilty man hears the inner voice of conscience cry: "You have eaten emu!" Troubled by remorse he returns to his own people, he sits down in silence apart from the others, and his countenance would be sufficient to reveal his crime if he did not first avow it, submitting himself to the punishment which he knows he has deserved.

It is to the traveller Sturt that we are indebted for this curious fact;* but many other instances of inchoate morality have been observed among the Australians. In this respect, nothing is more typical than the Australian *vendetta*. The same idea of morality which forbids the eating of the emu, will also in certain cases impose upon the individual an imperative duty of vengeance; but it is a blind vengeance, taking little heed of justice or of right. If a native has been injured by a white man, he will be satisfied if he revenges himself upon any white man. In the opinion of the Australian, there is no such thing as natural death; every disease is the result of some evil act, which ought to be avenged; it therefore imposes upon every man a long series of bloody actions, which

* Sturt, "Hist. Univ. des Voy." vol. xliii. 293.

he considers it his duty to perform. And as the Australian mind is very simple, his impulsive movements have no mutual restraint one upon the other, these moral obligations are very strongly felt. Dr. Lander relates that an Australian, whose wife had died of illness, declared that he must kill some other woman belonging to a far-away tribe, so that the spirit of the departed might rest in peace. He was thwarted in his desire, and was threatened to be put in prison. Then a serious conflict arose within him. Broken down by remorse, he became sad; he grew pale and wan, until one day, listening only to the voice of duty, he escaped. After a certain time he came back, holding his head up as though he were easy in his mind: he had acquitted himself of what was to him a sacred obligation.

Facts such as these lay bare the mechanism of primitive morality, which is perfectly observed in the animal and in the savage, until it becomes entangled with some process of reasoning. This is merely a question of education. Certain associations of sentiments and ideas have been slowly written down in the nervous conscient centres, and under the shock of impulse they will almost fatally unfold themselves.

It is in virtue of psychical facts of the same order that the elephant, broken loose, will after a sufficient domestication once more obey docilely the voice of his tamer. In London, a tamed elephant who had grown furious, and whose death had been resolved, was observed to obey mechanically the commands of his keeper, even under a whole battery of fire that was being showered upon him. For the brain of the elephant preserves very tenaciously any impression it has once received. The elephant is an animal capable of receiving education, and perceiving the effects of morality.

We find instances of this automatic morality in all humankind, in the most savage people as well as in the most refined. The Hottentot huntsman coming back to the kraal empty-handed will bear unmoved the reproaches of his famished wife; but all his stoicism will vanish if his housekeeper unties her only article of clothing, her apron of modesty, and with it slaps him on the face.

At this pitch of opprobrium the Hottentot can resist no longer; he puts on his hyena-skin cap, and rushes off like a madman, determined that, cost what may, he will kill or steal a head of game of some sort to wash out the affront. That is to him the moral, the point of honour,—such a powerful incentive in our better-regulated societies. For reasons of the same kind, a Bedouin Arab, Niebuhr tells us, thinks himself bound to pursue with cruel vengeance anyone who says to him: "Your bonnet, or your turban, is dirty. Arrange your bonnet; it is all awry." And in the same way our smart young gentlemen make it a point of duty, if one gives them the lie, to wash it out with blood,—though for the most part they themselves have no horror of falsehood.

During the early phases of social development, morality does not rise higher than the instinctive education that we find in every sporting dog. Man does not reason out his actions; he never weighs their value in the scale of utility; altruism has yet to be born; morality is still a mechanical habit. The Kamtschodalian considers that to violate a woman, whom he has found a long way from his own iourts, is quite a lawful action; but the same man would consider himself dishonoured if, when he had captured and safely landed a seal into his boat, he should then be weak enough to throw it overboard, should a storm arise.

It is to this total absence of control that we must lay the charge of coarseness in all primitive morality. At first man is obedient only to a stronger force. The Australian languages are devoid of words to express "justice, fault, crime." A Bushman—one of the very savage Hottentots—said one day: "A man commits a bad action when he runs off with my wife; I do a good action when I run away with the wife of another man." Burton says that in Eastern Africa a theft may distinguish a man, but an atrocious murder will make a hero of him. A negro, too, said to the same traveller: "What! must I die of hunger when my sister has children whom she can sell!" In New Archangel, four men, who were in love with the same girl and jealous one of the other, agreed at last to cut her in pieces with their knives. During the performance, Kotzebue tells us, they were singing aloud: "It was impossible for thee to live. Men looked at thee, and thou set their hearts aflame." Wallace reports that at Timor he has known public opinion justify two officers who poisoned the husbands of their mistresses, because these husbands were only half-caste men.

And if we go back to past ages, we shall discover traces of this savage morality among the superior races. The Manu Code inflicts upon a Brahmin the same degree of punishment for the murder of a Çoudra as though he were to kill a blue jay, a mongoose, a frog, a dog, a crocodile, a crow, or an owl,—for the Çoudra man is only a slave. In the same way, the Wergeld in ancient Germany placed human life upon a scale in the inverse ratio to a man's social importance. In China there are associations of men outside the pale of the law, in which moral ideal consists in giving and receiving blows with perfect impassibility, to kill others in cold blood, and not to fear death for themselves. In our so-called civilised Europe, do we not know too well that the most frightful human hecatombs become not only legitimate, but glorious, when they are justified by political passion?

How can man from this inferior degree of morality raise himself to such a fever-heat of virtue, of heroism, of abnegation, that we admire among the most noble specimens of our kind? It is not that the essence of moral qualities has altered, but that the standard has become higher. The social intelligence has grown slowly, and by degrees the moral impressions, stored away in the nervous centres, have become considerably enriched. Experience has taught us to test the value of individual acts from the general utility point of view. At the same time public opinion has become less confined, and the social verdict more severe. The number and the strength of moral customs have increased. Many of these, considered lightly by the uncultured conscience of the primitive man, have awakened a feeling of repugnance in the man truly civilised; to do them, or even to be a witness to their performance, would entail a certain amount of disturbance in the impressions fixed from time immemorial in the cerebral cells; hence, if the case were to happen, moral suffering and remorse would follow. On the other hand, education has aided, more or less powerfully, to ripen these inherited tendencies, and at the same time punishment has followed the infraction of certain admitted moral rules. It is the total sum of inherited tendencies that forms the most solid basis of individual morality. Education may develop these tendencies; it cannot supply them.

As is natural to expect, a higher degree of morality has been developed in those societies whose conditions of life are most stable, where the same system of culture has been pursued through a long chain of generations. An exception, however, may be found among men very little civilised, or even among certain intelligent animals. For instance, two Indian elephants fell into a ditch dug purposely with a view to capturing them. One of these succeeded in getting out of the ditch. Then, instead of taking flight immediately, he lent a charitable assistance to his less fortunate companion.

As a general rule, honesty, veracity, etc., in business transactions, is uncommon with the primitive man; but it is not unknown among all savage tribes. According to the abbé Domenech, some tribes of the Red Skins plant in the middle of their villages a post which they call "the tree of probity," and on to this all waif and stray objects are hung. In the same way the most honourable conduct is preserved in the transactions of the Esquimaux between themselves, but between themselves only, for they do not consider themselves bound by any law of morality with regard to the stranger. Wallace says that the Borneo Dyaks are most scrupulous in their notions of veracity. Mungo Park saw a negro woman in Senegambia follow her son, who was severely wounded by a musket-shot, lamenting her boy, and enumerating all his good qualities: "He never told a lie in all the course of his life."

On this matter there is, as regards the evolution of the moral sense, an important distinction to be made. It is the least elevated portion of morality, that which we may call commercial morality, which becomes the first developed. It is not rare to find probity in transactions, fidelity in a given word, co-exist with the greatest disregard for human life. For instance, the over-nice veracity of the Dyaks does not at all prevent them from running after people to kill them. The Sandwich Islanders, who made a sport of infanticide, used to observe religiously the taboo, and would confide themselves without hesitation to an enemy when he had sworn to them friendship over a crest of yellow feathers. The ferocious Turkoman will shed blood as though he were pouring out water, but he will leave with full and entire confidence, in the hands of a debtor, the written acceptance that the debtor has signed. "What will he do with it!" the creditor says. But the debtor has need of the acceptance, to remind him of the amount of the debt and the date of its falling due. The maintenance of societies, even the most rudimentary, is first the all-important matter; a certain amount of good faith in daily transactions, in the habitual current of life, especially in the middle of a social group, is absolutely necessary. Kindness, charity, humanity, respect for the feelings of others, are not born in man until a much later date.

As everything else, morality has evolved slowly, and it is only too evident that it is still far from having arrived at its highest point of development. The most advanced of human societies are still in a furious conflict of egoism, of cupidity, and of cruelty. The primitive instincts of the beast are not yet extinguished in all hearts, and the moral level is very little elevated even among people who call themselves civilised, for nobility of character is too often the cause of want of success in the struggle for life. How often does it happen that duplicity, meanness, greed, hard-heartedness are used as means to overcome the more delicate individuals who have freed themselves too quickly from the state of low morality, and who therefore are fighting with courteous weapons against unscrupulous rivals, to whom every weapon is good, provided that it is destructive. Facts in support of this sad truth abound both in history and in daily life. It would be waste of time to enumerate them. We have done enough if we have succeeded in drawing the broad outlines of the origin and the development of the moral sense. Few subjects are so well worthy of the meditations of our teachers, of our legislators, of all those upon whom is incumbent the delicate task of forming and moulding our individual characters.

CHAPTER VI.

THE CONSTITUTION OF SOCIETIES.

I.

Animal Sociation.

Man has long enough deceived himself with the idea that he was made in the image of the Divinity. It is now more than time to say and to repeat to this poor creature that he is animal in every fibre and in every particle of his existence. In the foregoing chapters we have several times pointed out the mental similitudes which connect man with his inferior brethren. Many tendencies, many aptitudes, on which man prides himself, may be seen more or less developed in the animal, and it would be only too easy to point out the same manifestations as regards social matters. Man is no doubt a sociable creature, and we may agree with Aristotle in defining him "a polite animal." But the definition does not apply only to man.

We need not say that we regard the word "society" in another sense than mere juxtaposition. The polypi, the madrepores, the ascidiæ, oysters, etc., live together in aggregations, but their grouping together is not a matter of sociability any more than the clustering of buds or leaves on a tree. We may say the same of the clouds of grasshoppers which from time to time devastate Algeria, or the immense columns of butterflies which in summer may sometimes be seen swarming in our own country. The idea of society implies that of active concourse; a social state can exist only where creatures endowed more or less with sensibility, with will, and with intelligence, pursue together the same object in common.

A glance over the animal world from this point of view may be interesting and instructive. As regards social aptitudes the palm is far from belonging to the mammalia, even to the mammalia most approaching to man; nor could we deem it to the inferior human groups. Many mammalia come together only temporarily during the season when prompted by their amorous desires. Deer will form small societies amongst themselves, but these societies are no more than mere family associations. Reindeer, buffaloes, wild horses, elephants, certain kinds of monkeys sometimes constitute large agglomerations, in which a sort of government of a hierarchical order may be found established. Flocks of wild reindeer are guided and protected by the old males, who in their turn act as sentinels while the others are resting; and they are ever careful to stop the more foremost of the herd, or to drive in the laggards. In the same way the tribes of elephants do not gambol and frolic unless they are under the guard of a few old males.

The chief of the hordes of cercopithici is careful also every now and then to climb up to the top of a tree to explore the surrounding country, and by guttural noises he communicates the result of his examination to his associates. Anthropomorphous monkeys collect together only in little groups, in polygamous families, living under the despotic authority of a male adult, who is served and obeyed until the young ones revolt and assassinate him. The gorillas so grouped together in little hordes have made themselves masters over a whole district; they arm themselves with sticks and stones, and drive from off the soil and away out of the country everything which is displeasing to them.

But how rude are these slight associations if we compare them to the firmly-established republics of bees and of ants! We have all heard of these societies, so numerous and so well-regulated, in which the nervous instinct, which makes men say and do so many foolish things, is made subordinate to the social interest, where the system of castes is fully practised, and where the division of labour is pushed so far. In the republic of bees there is in the first place a female whose only occupation is to furnish citizens to the state; then there are males or drone bees, and working bees or neuters. These last divide themselves into working nurses or working wax bees. Social foresight is pushed very far, they are careful to provide against eventualities by filling everywhere the closed cells with wax. If they wish to build a hive, the work is all divided out. Some of the female working bees furnish the necessary materials, others plan out the work, which is finished by another set of female workers. And during all this other females provide the labourers with food. If the female bee is hungry she has only to lower her trunk. At this signal the purveyor of food opens her honey bag and pours out a sufficient quantity. All this industry is performed freely and spontaneously; there is no despotism in these societies, and individual initiative has no other guide than the instinct of duty. The female bees so unsexed and deprived of the joys of maternity are not wanting in solicitude for the young. They take care of them, they bring them up, and the affectionate sentiments which subsist between the nurses and their children are later transformed into one of the strongest social ties.

Ants, too, have created for themselves a similar social organisation. With them, when the females have become pregnant, they themselves pull out their wings to devote themselves to the *foundation* of a new tribe. When this new tribe has become numerous, they will soon construct an underground habitation, in which everything will be ordained towards the perpetuation of their kind. They will build a city and defend it most courageously, and this city will have gates opened always in the daytime and closed at night. Among the termites there is only one fruitful female ant. There are also winged and idle males, and the neuter aptera, some of whom are busy in constructing the great phalanstery of the republic, while others form a warlike caste apart. We may also mention the ants who have domesticated the plant-lice, those who, like the *formica rubescens*, have confided to their slaves the care of constructing their nest, and to furnish their larvæ. These ants are so aristocratic that they cannot eat alone; they will die of hunger if their slaves are not ready to put their food into their mouths!

This last case is a social monstrosity. The ants, ignorant, we may suppose, of our human prejudices, have, in pushing matters to an extreme length, realised for themselves caricatures of our own aristocracies. But what good examples may we not find in these little republics, where family feeling is established, and care for the welfare of others is pushed to an extent of which men have hitherto been incapable!

The surest way to make us admire the cities built by bees and by ants is to consider the societies which human beings have formed for themselves.

II.

The Melanesian Societies.

In the Melanesian races we may see a long graduating scale of societies and governments, but the boundaries of the humblest people in this series are very little removed from animality.

The Tasmanians, grouped together in small hordes, each having their totems, used to live almost in an anarchical condition. In the time of war, but then only, each horde would flock round their temporary chief, and they would obey his orders until danger was actually upon them.

In these small societies there was no sort of division of the social labour. No aristocracy, no caste, no slave. For slavery will imply a certain degree of civilisation. In the earliest days of every society wars are pitiless; no quarter is shown to the conquered people.

But with the Tasmanians, an idea of law had begun to show itself. To infringe a man's right of property over his wife or wives was considered a fault deserving severe chastisement. The delinquent, fastened to a tree, had to expiate his crime by serving as a target for the javelins of his neighbours. He was allowed, however, to ward them off, perhaps with a shield.

The Tasmanian hordes must have been very small, for the Australians, who were more civilised, live together in very small agglomerations. Dampier saw them always in groups of twenty or thirty, never more, counting the men, the women, and the children. On the banks of the Murrumbidgee, Sturt met only fifty individuals over a space of one hundred and eighty miles. But in the forests on the banks of the Murray the country is more thickly peopled.

The social organisation of the Australians, which we know better than that of the now extinct Tasmanians, is, we believe, more complex. In their tribes the Australians have created for themselves a sort of aristocracy. In the same way that hordes of chimpanzees are despotically governed by the old males, the small Australian societies are governed by their old men, or by the most robust of their male members. A whole collection of rules and customs puts at the discretion of these privileged persons the life and the property of the weak, of the young, of the women. Everything that they like becomes their own as a matter of right: the best bits of food, the handsomest women, or whatever they may fancy. As we have already seen, the young are forbidden to eat of the emu. All these tyrannical customs are not supported by any family reasons; they merely suppose an agglomeration of people more or less large.

For the weaker ones servitude becomes an absolute necessity; to think even of escaping from it is considered a crime, often punished with death. But still there does exist a sort of social justice. For certain offences, the injured person has a right to inflict upon the wrongdoer so many cuts with a lance upon this or that part of his body. Sentiments of joint responsibility may also be observed, though they are naturally of a very barbarous kind. As they never consider the death of a man to have come by natural causes, it is with them a sacred duty to avenge the death on the real or supposed murderer, or upon one or more members of his tribe. It will make no difference if the deceased has been killed, or if he has died accidentally, or of sickness. In this latter case, his death is imputed to some wicked machinations of a member of a hostile tribe; and it therefore becomes a strong point of duty to avenge the death by killing a number of persons, proportionate in number to the importance of the deceased.

If we will recollect that in the small Australian societies the family and the laws of inheritance are governed by precise customs, we shall at once agree that these societies of men are higher than the societies of monkeys, though they are very far inferior to the societies of ants and of bees.

Among the Melanesians in the Fiji islands, the despotic power has become organised; it has been constructed after some system of rule. The Fijian chieftains, whose dignity is hereditary, enjoy plenary power. As they come near to them, the common people prostrate themselves; to speak to them they employ a whole vocabulary of servile words and expressions, such as "god," "root of war," etc. etc. The subjects hold their property only under the good pleasure of the chiefs, who dispose of it for their own use if they wish to do so. At the time of Dumont d'Urville's travels, half a century ago, the whole population of the archipelago was under the orders of a monarch, who owned more than a hundred wives, who claimed also, as his tribute, young girls, the teeth of whales, boats, the stuffs fabricated from the mulberry tree, mats, bananas, pigs, and anything else if he wanted it.

In Fiji judicial penalties are regulated according to the rank in society; the gravity of a crime varies with the social position of the culprit. A theft committed by one of the people is a much more serious offence than murder perpetrated by a chief. But the faults held worthy of punishment are not numerous. They are: theft, adultery, rape, magic, arson, the want of respect towards an important personage. The offences will group themselves, for the most part, under two main heads: offences against the master, and violation of property. This same remark will be found applicable to many other ethnical groups, and to many other races of men. We shall see, in point of fact, that in many barbarous societies men think much more of the rights of property than of the value of human life.

In New Caledonia society is better administered, it is also rather less servile. According to M. de Rochas, each New Caledonian tribe forms in itself a small feudal organism. At the bottom of the social scale we find the villeins, the labourers, who may be proprietors, and who may consider their own bodies to be their own property, more or less, on condition that they pay to their master certain fines. Above them there is an hereditary aristocracy, according to the right of male primogeniture. This aristocracy, composed of vassals and sub-vassals, obeys a much-respected suzerain; but in him is not vested the power of disposing of the life or the property of his nobles. He will, however, be less ceremonious with his villains; and a chief named Bouarate, who has left behind him a glorious name, thought it perfectly legitimate, every now and then, when dining at home, to eat one of his inferior subjects. Another chief who made use of his people as though they were a provision market, thought well to preserve human flesh by salting it.

In New Caledonia slavery does not yet exist, and the wars are therefore very bloody. But some small sparks of humanity are beginning to appear; if in the middle of a battle, for instance, a chief desires that any particular life may be spared, that individual will be respected.

The same prisoner may afterwards be adopted, if the chief gives his consent; henceforward he forms part of the tribe to which his adopted father belongs. This facility of adopting and of assimilation is common among the primitive races, in whom a patriotic sentiment is still very ill defined.

On the whole, the social organisation of the New Caledonians seems to be the most complex and also the most advanced of any of the Melanesian people. The primitive confusion has been followed by an hereditary hierarchy, and, despotic as this hierarchy may be, a certain individual right is already beginning to be recognised; also, to a small extent, humanity is sometimes practised. But we may ask if this social constitution, relatively advanced, is in fact the work of the Melanesians themselves. We do not find it elsewhere in Melanesia, and it is certain that New Caledonia has more than once received emigrants from Polynesia, who have introduced there the custom of the taboo. In any case, whether it be native or imported, the New Caledonian civilisation indicates a certain degree of perfectibility among the people who have adopted it.

III.

Societies in Southern Africa.

The black or the negroid races on the vast African continent are anthropologically very different. They certainly do not all come from the same origin, and many of their ethnical groups have evolved separately. Nevertheless some sort of sociological gradation may be noticed in Africa, starting from the south and going northward.

At the bottom of the scale we must place the Bushmen, people who wander about in families, in small groups, in human flocks, through the forests and plains of southern Africa. They are ignorant of agriculture, they have no domestic animals, they eat whatever they can kill or steal, crunching, when they can find them, a few roots, the larvæ of ants, and grasshoppers; they live in naturally-made grottoes, or in holes scooped out in the earth. They are not more advanced than the most humble types of the animal creation.

The higher kinds of the same race, the Hottentots, properly so called, have founded amongst themselves pastoral societies, and they are a people much more advanced. But strangers to agriculture, at least in their native state, they will sometimes descend to the bestial life of the Bushmen, if it happens to them to lose any of their cattle. Thompson has seen some of the members belonging to one of the most advanced of their tribes—the Koraquas—go through this sociological retrogression; for though progress is general in humanity, it does not always move regularly.

In the villages in the Hottentot kraals, no form of government has yet dawned upon the people. In time of peace, each clan scarcely know of other laws than a few of their own customs. But over each kraal there is a chief, whose authority in time of peace is almost nominal; his office is sometimes temporary, and sometimes hereditary. In some of the Koraqua kraals, the chief used to abdicate in favour of his son, when the son in fighting against his father had succeeded in ousting him; it was literally no more than the law of the stronger party. The Hottentots have no social hierarchy; an aristocracy and slavery are equally unknown. But a wide difference is made between the rich and the poor; the possession of a large flock gives to a man a great social influence. It is the aristocracy of cattle, not of money.

This same aristocracy may be seen also among the Kafirs, who are neighbours and rivals of the Hottentots; the Kafirs are both an agricultural and a pastoral people, and their social state is of a more complex kind. "The organisation of each Kafir tribe is a coarse feudal hierarchy, governed by an absolute monarch, whose authority becomes mitigated by deputations of the people. Every man is the chief of his own family; he is the undoubted master of his wives whom he has bought; he is the master of his children, until the son is old enough to share with him the parental authority. Each father of a family ordinarily holds his property under a suzerain, near to the *cotla* (the Kafir forum), on which he builds his house. And this suzerain owes his obedience to the chief of the tribe. He is the master supreme. It is he who divides the land according to the individual wants, for in Kaffraria landed property is held in common. It is he who is at the head of every hunting or war expedition, which he orders and conducts as he pleases. But still this kingling, who sometimes rules over towns of 8000 or 10,000 inhabitants, does not usually take upon himself to decide a matter of much importance, without first convoking the *pitsho*, or the national assembly. In this assembly the orators are allowed the greatest liberty of speech; the king must not allow himself to be angered at anything that is said. But, as a consolation, he is not obliged to take into account any opposition that is made to him.

"The Kafir kings, proprietors of large flocks, are necessarily rich men, for they are the great purveyors of the tribe, and they do not impose a regular impost except a tax upon the game killed, and upon the booty taken in war. We may add the fines that they inflict upon guilty persons, but these are relatively rare; the Kafir code is purely traditional and is very elementary, it punishes only a very small number of crimes or offences. Theft is punished by chastisements or by death. Adultery is reprimanded as though it were theft; the manners in this respect are far from being severe. On the other hand, human life is very little respected. The husband may kill his wife for the most futile reasons. In Kafir towns and villages murder will produce little or no sensation among the people. Everyone defends himself as he can, and he takes his revenge after his own fashion."*

* Ch. Letourneau, article "Cafres" ("Dict. Encyclop. des Sc. Médicales," t. xi.).

Power and social rank are hereditary. In some tribes the servility is excessive, for the inferior ought to salute his chief, saying to him: "You are my chief, and I am your dog." The Kafir aristocracy consider wealth as the basis of their existence; for among the Matchlapis those who own a sufficient number of cattle to maintain a family have the right to the title of chief.

In some tribes, notably the Bachapins and the Bechuanas, there is also a servile class who hunt with the dogs for the profit of their masters, and are obliged to bring to them the skins of all the animals killed. Sumptuary laws forbid members of this class to make use of the same skins for their clothing that the rich men use—jackal skin, for instance; their quality of slave is hereditary, as are also the dignities among the rich.

In a word, the Kafirs have not got beyond their primitive anarchy except to organise a state of slavery. Many other races, as we shall see, have followed the same steps; even this is progress, it is a barbarous moving onward towards a better future.

We may also remark that in these coarse societies the law first concerns itself with the care of property; but it pays very little attention to any regard for human life. This is a significant fact; it is also a very common one, and we may connect it with what we have already said in speaking of morality. Pity, charity, justice, etc., are implanted in the heart of man, but they are the fruits of a later season.

The Kafir society, with its hierarchy, its castes, its monarch, has already taken rigid forms very similar to those which we find among the majority of the human groups at a certain time in their evolution.

Among the Gaboon tribes we may study this social condition in its primitive state. In this region each tribe divides itself into clans, they again subdivide themselves into small villages, more or less nomad, and each having their independent chief. The power of these chieftains is generally hereditary. Ordinarily it passes from one brother to another, as among the Kafirs; but the elders have the right of passing their veto; they may also exclude the rightful heir, and elect another.

To this same council of old men also belongs the right of deciding if the community ought to emigrate; if they ought or not to declare war. The king decides only upon matters of daily practice.

Slavery is in vogue among these people, and we may see clearly the reason. Punishment there is inflicted in a very simple manner:—a man is condemned to slavery or to death. Of all rights, the right of property is the most strongly respected. The adulterer, that is, he who has violated feminine property, is sold as a slave; so also are thieves and insolvent debtors. These men pay their creditors in the shape of service done. Sorcerers, and children of whom people wish to rid themselves, share the same fate. Prisoners whose lives have been spared are also sold. But for slavery, all these disinherited persons in the Gaboon tribes would be put to death. A slave trade is everywhere carried on between the tribes, on the seashore and also on board the slave-ships. But in the interior of the country, as well as on the sea-coast, a slave is the monetary unity. When slaves are kept in the village, it is intended that they shall reproduce their own kind; their descendants are servile by the right of birth. Slaves in this category are treated with some consideration; they are not sold out of the country. A sort of feeling of humanity has already arisen in their favour, for they have the right to take refuge in a neighbouring village, and to choose there a new master, who is bound by public opinion to take charge of them.

IV.

Societies in the Centre of Africa.

Certain negro groups whose social condition we have just sketched are tending more or less clearly to institute slavery, castes, and the theory of absolute power. This organisation, still in an embryo and confused state among the black men in Gaboon, is spreading and becoming implanted in the portion of Africa north of the equator. We find there, all across the continent, a large zone that we may call the servile zone, and which stretches from Senegambia and from Guinea as far as Abyssinia.

Over all this vast region, occupied by the best specimens of the black races of men, by populations in which Berber or Semitic blood is more or less mixed, monarchical despotism is exercised almost without control.

In Ashanti the king, who has 3333 wives, a mystical number rigorously determined, inherits all the gold belonging to his subjects. In his purchases he makes use of weights a third heavier than those used by the rest of the nation. He is surrounded by children, who carry for him his fetich arrows, and they have the right of pillage over all the common people. Whenever this demi-god spits, children, who carry with them elephants' tails, carefully wipe up the royal saliva, or cover it over with sand. When he sneezes, all his assistants draw two of their fingers across their foreheads and across their chest; this is equivalent to asking for a benediction.

The Bambarrans of Kaarta are a little less servile. At his coronation the king has to undergo a sort of investiture. He is lifted up on a large strip of oxhide by the representatives of the caste of the blacksmiths, and he then listens to the following allocution: "Before accepting the power you must know four things:—First, that you are our master and that all our heads belong to you; secondly, that you must treat us as your fathers have treated us; thirdly, that you must make the laws respected and protect the nation; fourthly, that you must win the favour that you receive by signalising yourself in a warlike expedition."

This ceremony, however, does not at all prevent the monarch from enjoying absolute power, or even from choosing his place of residence where he pleases.

Among the Mandingos, and among them only, do we find the despotic power at all mitigated. Before declaring war, before concluding peace, or deciding upon a matter of any importance, every Mandingo king ought to take the advice of a council composed of the elders and of the principal members of his small nation.

The Timmanis have also their deliberative assemblies, that which in the greater part of negro Africa is called *palavers*. But these assemblies are purely formal; the orators as they speak are careful to watch for the expression of the king's face, and to make their opinions conform to his.

Among the Fellatahs, in Soudan, despotism is absolute; the king will concede, possess, or sell, at his own pleasure the government of his provinces to whomsoever he pleases.

At Katunga, in the valley of the Niger, when the king holds an audience, the eunuchs, the courtiers, and all the assistants ought, before sitting down, to prostrate themselves lying with their stomachs on the ground, and be bare of clothing down to the waist. Then they must drag themselves along the ground towards their master, kissing the earth as they do so, and rolling their heads in the dust. They compete one against the other as to who shall most besmear himself.

The same unlimited despotism exists in Borgou, where the king is the judge over every matter of business in dispute, and he decides as his caprice directs.

At Kiama, still in the valley of the Nile, the king, when he is mounted on his superb charger, and followed by a numerous armed escort, has close to his side six young girl slaves holding in their right hands three light javelins; they wear no other clothing but a band of cotton around their heads and a string of glass-ware around their waist.

The King of Boussa thought that the European system of monogamy might be good enough for the common people, but for monarchs he considered that it was very impertinent.

We are indebted to Captain Speke for some curious details in the great monarchy of M'tesa near the large lake on the Upper Nile. Here we find absolute power almost in the form of a caricature. Whenever the monarch holds an audience the nobles and the high dignitaries place themselves round about him, crouched up or kneeling, and they remain perfectly silent in this form of adoration. Any default in a ceremonial form is punished instantly with death. The courtiers or the ambassadors, when they have to report upon a mission, approach the king grovelling in the dust. The monarch expedites business very promptly by ordering confiscation of property, fustigation, and capital punishment, while young women, quite naked, are handing about cups of banana wine. Every royal decision, no matter what it be, must be considered as an act of grace, and he whom it concerns must roll in the dust, grovel with his belly on the ground, and utter cries of rejoicing. At the same time the great king receives the fines which he has levied, the presents that are brought to him either in cattle or in women. The latter are usually young girls respectfully offered by their fathers and destined to replace the odalisks in the harem whom the monarch causes to be slaughtered when they have ceased to please him.

Farther northward, but still in the basin of the Upper Nile, the Niam-Niam kings every now and then will rage and howl like wild beasts over one of their subjects; they will cut off his head with their scimitar only to shew that this human animal was their property. The Monbuttous consider that every object touched by the king becomes sacred, no once is henceforward allowed to touch it. In the Mahomedan states of Fezzan, Sennaar, Darfur, the government is also the most absolute monarchy. In Sennaar, upon the accession of the king, all his collateral relations are put to death, all the land belongs to him, all the inhabitants of the country are his slaves.

So also in Abyssinia every man is born a slave of the sovereign. The monarch overrides all law, he is proprietor of all the soil, he decides absolutely upon every civil or ecclesiastical matter of business, he is at once pope and sultan.

In Madagascar, too, the sovereign exercises an unlimited power. He disposes at will of the life and property of his subjects.

But though he is the personification of the most complete despotism, the African monarch is not the only master. Under him there is ordinarily a long hierarchical scale of tyrants, one more brutal than the other. There are castes of aristocratical bullies who, after grovelling on their belly before the king, trample upon the neck of the slave. With some few unimportant differences, this is the sort of social organisation among all the servile people of whom we are now speaking.

In Ashanti a sort of council of state, composed of four members of the aristocracy and of the principal chiefs, have a consulting voice in the administration of the affairs of the country.

Among the Bambarrans of Kaarta there are three aristocratical castes: (1) the caste of the blacksmiths, of which the chief crowns the kings upon their accession, and has the right of justice in his caste; (2) the tanners, (3) the griots. The revenues of the state are composed of the booty taken in war, tribute-money paid by the neighbours, a tithe *ad valorem* duty put upon all merchandise transported by caravans, arbitrary taxes levied capriciously by the king, and tithes in kind which weigh upon the labour of the different castes or corporations. A similar system exists also in Bondou and among the Timmanis. In Mahomedan countries such as Bondou, or among the Fulahs, Musulman laws are more or less in force; but everywhere the imams who exact them are good courtiers.

The Mandingos, as well as the Bambarrans, have their professional castes: blacksmiths, and shoemakers, also orators and musicians; and in the first rank must be put the profession of the Koran.

Even in these barbarous societies, the organisation becomes special and complex. Despotism is conducted after a regular form; and the most civilised of these agglomerations, the Mandingos, for instance, are beginning to attach a value to artistic and intellectual aptitudes. Despotic monarchy is none the less violent; it exists to such an extent that, in Boudou, to kill a lion is an offence for which pardon must be asked of the chiefs; for this is considered as a want of respect to the sovereign. In Kiama men salute their superiors by lying down with their bellies touching the ground; the women kneel, their elbows on the ground and their hands open. In Kano, the governor lets out all the shops at so much a month, and he regulates at his pleasure the price of every article.

The government and those men in power think little about the people, except as to how far they can vex and oppress them; otherwise they allow them liberty to do what harm they like. The people in the towns are always fighting one against the other; they will at once jump at any pretext for pillage and robbery.

This social state, which we find almost everywhere throughout this African zone, is evidently a rough sketch of the feudal system; and in Abyssinia it has showed itself in the shape of a real feudality quite similar to that which has once existed in Europe. There are endless ties, rights, and duties among the Abyssinians, establishing a reciprocal joint responsibility, which each man prides himself in observing. The freed man, who is bound to no subjection, is outside the law; he is outside the social pale.

In Madagascar the nobles, before they were made obedient to the royal power, used to live as petty sovereigns, shut up in their fortresses, pillaging their neighbours, plundering travellers, as was the practice for so long a time with our own feudal barons. And like our old barons, the Madagascar nobility used to pay a tithe to the suzerain or to the king. On the other hand the villains were taxable, and the labourers were mulcted at pleasure.

Analogies such as these seem to convince us of the idea of a law of evolution.

But hitherto we have only spoken of the governing classes, or rather of the oppressors. Beneath these privileged persons there lies groaning a large mass of servile people, who have many duties to perform and who enjoy very few rights.

In the countries of which we are speaking, as in Gaboon, the soldiers are taken from the servile classes; and all those who are condemned for insolvability, or for any offence, are also considered servile. The newly-made slave has no surety, no guarantee; he may be killed or sold at will. The hereditary slave is somewhat more respected; and among the Mandingoes his master cannot put him to death until he has first held a *palaver* to decide upon his conduct.

It is everywhere incumbent upon slaves to cultivate the land, to look after the cattle, to perform every servile office. Sometimes they are allowed to have a habitation of their own, and there to carry on a trade, but only on condition that they will make over to their master all or a portion of their profits.

The number of slaves is enormous. At Boossa and in the neighbouring districts, four-fifths of the population were slaves according to R. and J. Lander, and three-fourths according to Mungo Park.

In Madagascar, where slavery has the same origin as elsewhere, and the class is made up of convicts and insolvent debtors, a whole servile hierarchy has been established, each class having their own different rights. The Hova slaves are the most favoured, then come the Malagasi of every sort, and lastly the black man.

We see therefore that from the Guinea coast to Madagascar, across all the vast African continent, the social state is in reality everywhere very similar; in certain regions only we find it more or less clearly defined. But to complete our short sketch, we must first say a few words as to the administration of justice.

It would appear that nothing is more confined than the moral horizon of the poorly-developed man. For this being, as yet so very inferior, every desire is lawful; for him, his own personality, his own small self fills the whole universe. During this phase of his evolution, man has not the faintest idea of liberty, of humanity, of right, or of justice. If our opinion on this point be true, we shall find superabundant instances in support of it in Africa.

In Ashanti, to kill a slave is quite an indifferent matter. But the murder of a high personage by another of equal rank entails upon the assassin the punishment of death; but the guilty man is allowed to kill himself. One of the king's sons ought never to be put to death, no matter how black be his crimes. Every chief, but every chief only, has the right to kill or to sell his wife if she is unfaithful to him. Cowardice, being a want of respect to the king, entails capital punishment. Every amorous intrigue with a woman belonging to the royal family is punished by castration. In Abyssinia, as in many other African countries, the accusation of sorcery involves torture, death, and other punishments. In a word, justice is no more than the vengeance of damageable or disagreeable acts done to those in power, and especially to the king, from whom everything proceeds.

Over all this region of Africa which we are now considering, the right of high and low justice belongs in principle to the king, who is almost quite free to decide any matter just as he may please. In the smaller states the master will judge for himself. With the Bambarrans of Kaarta the members of the royal family have only the right of middle and of low justice; high justice is the royal prerogative. Crimes entailing death, and therefore submitted to royal jurisdiction, are theft, murder, and adultery. But the blacksmiths and the royal family are not liable to capital punishment; they may be subject to confiscation of property, banishment, or flagellation. But the flagrant adultery of a blacksmith with the wife of a Massassi, or a member of the royal family, or with a woman of another caste than his own, is a crime punished by death. In Soudan the death inflicted will vary according to the caste to which a man belongs, or according to the religious faith he is supposed to profess. In Mahomedan countries the true believers are decapitated, the others are impaled or crucified.

Among the Mandingoes, relatively more intelligent, the organisation of justice has made a step in advance. The trial is public, in a local spot set apart for the purpose, similar to the *coffa* of the Kafirs. Witnesses are called, and a special functionary, an hereditary judge, passes sentence.

Among the Koussakes, murder only is punishable with death; and the guilty man may buy back his life by indemnifying the friends and parents of the deceased. The matter is considered private, and no one thinks of interfering with the general interests of society. This rude conception of justice is very common all over the middle part of Africa. In point of fact, there is never an offence committed; damage is done to the master or to an individual. So also in case of murder, the people who live in the Syouah oasis deliver up the guilty man to the relations of the victim, and they are free to kill him, to torture him, to give him his liberty, just as they may please.

In Darfur, the sultan will judge according to his own caprice. His will is law, and is controlled only by the representations of the imana. In the provinces, the master's authority is delegated to the functionaries, to the sub-tyrants whose commands are acts of arbitrary authority upon a smaller scale.

In Abyssinia, there is a class of legists who boast that they know the Pandects and the Institutes by tradition, brought to them from Byzantium; but the prince is above all the laws. It is to him that, in principle, application must be made for relief against all wrongs. His door, his windows are besieged with people crying, weeping, and lamenting, praying for justice, and if by any chance there are no oppressed people who really want to make their claims heard, others are hired to play their part.

In Madagascar, the functionaries administer justice in open air, but always in the name of the sovereign, and reserving for him the decision in graver matters. In Madagascar, also, as in other countries, as in Ashanti, for instance, judges, when they find themselves embarrassed, often resort to judicial proof by means of poison. This mode of judgment, the judgment of God, might be exacted either by the judges or by the parties, and the trial was made upon the accused or upon an animal; it was also practised in civil cases as well as in criminal.

The foregoing facts will show clearly enough that all over the middle part of Africa the ideas of justice and injustice are still in a most confused state. Against the slave, and especially the newly-made slave, everything is lawful; and among men who call themselves free, the oppression of the weaker by the stronger is the common rule. We may notice, however, that in each of these small states, the despotism of the master, hard as it may be, is still a sort of social safeguard. It will often happen that the death of a despot is the signal for a long string of acts of unbridled crime and violence; the bad instincts of men which fear alone could keep in check are then let loose. The oppressed man will become oppressor in turn when it is in his power.

A certain sort of social organisation has, however, established itself: individual property exists, but it is tolerated rather than lawfully constituted. The master has learnt not to kill the goose who lays for him golden eggs. In place of confiscating everything, he will content himself by capriciously imposing heavy taxes. At Kano, in Houssa, the governor takes for himself two-thirds of the dates and other fruits brought into the market.

The idea of offering protection to society in general has not yet begun to dawn, nor yet the notion of scrupulously weighing the degrees of guilt in offences and in crimes. Human life is not guaranteed; and very generally, as among the Bambarrans, theft is reputed as the greatest of crimes. For in many barbarous societies, and even also in some civilised societies, man thinks more of his property than of his life,—especially of the lives of others.

If we pass from these rude disordered societies to that of ancient Egypt, a country where the organisation was better administered, we shall see that the different functions were more clearly determined, the offices were strictly hereditary, the control and the administration of power were better defined. No doubt the old Egyptian race has assimilated to itself some of the Berber and Semitic manners; but at bottom it does not differ from the coarse states in negro Africa. It has done no more than regulate and bring to a high degree of perfection the tendencies and the social ideas common among all the African races. At a later date these customs became modified.

V.

The Social State in Ancient Egypt.

In ancient Egypt, as all through the middle part of Africa at the present time, there was at the basis of society a mass of servile people, who, reduced to the condition of domestic animals, performed all the hard agricultural and industrial labour. On the backs of these humble sons of the state were piled the aristocratic castes, ingeniously ranged, the most useless, as of right, occupying the highest places. A despotic monarch, in all his parasite splendour, sat at the top of the pyramid, the result of the labour and the suffering of a large nation.

We are not called upon now to describe minutely the social constitution of ancient Egypt, but we may recall the main characteristics.

There seem to have been four castes: the priests, the army, the agricultural, and the commercial people. Outside these divisions everyone else was a slave.

The superior castes, the warriors and the priests, possessed as their own a portion of the territory. It would seem that the priests had first instituted a pure theocracy, to which the military caste substituted afterwards a monarchy, the government of a royal family set up in the middle of their own body.

Social immovability was the general law in the state. No one went out of his caste, and every man followed his father's profession. The doctors even had their hereditary specialities. They were each, from father to son, obliged to concern themselves with the same kinds of diseases.

At the head of all society, the king and queen sat upon the throne, honoured with the title of gods, and surrounded by a numerous train of courtiers ranked in minute hierarchical succession. Orders of chivalry set off the situation of the courtiers. There were clasps and collars of honour, belonging of right to the courtiers in chief, to the "parents."

Passive obedience was the general law in the kingdom. The government knew everything, would do everything, directed everything. No initiative was allowed to individuals.

Nevertheless the constitution of the Egyptian state differed from the rude monarchies now existing in Africa in important matters, and the balance was altogether in their favour. Justice, which was usually administered by the sacerdotal caste, was carefully organised. The pros and cons of each case were pleaded, but only in writing; the judges delivered their verdict after they had consulted the books of Thôt. The desire that equity should be done was so great that dead bodies were judged before allowing them to be placed in the family sepulchre.

Various torments were inflicted in the future life, in hell, on the shades of the guilty persons. These shades were sometimes decapitated, sometimes cooked in hot caldrons; sometimes they dragged about their heart which was torn out. The Egyptian notion of hell was very like our Christian notion; it would seem that from them we borrowed our model.

There is one very honourable feature in the ancient Egyptian society, and one that shows a true sentiment of joint responsibility. The people held it to be an imperative duty to give succour to their fellow-creatures. A man was judged guilty of homicide if he did not proffer assistance to another in danger. To pursue with justice a guilty person was obligatory; he who did not fulfil this task was beaten with rods.

Nothing is more typical than this form of society; it preserves in its full bloom every primitive social tendency. As yet the individual does not exist. The structure of society is as though it was congealed; it does not suppose the possibility of progress, and is resolutely antagonistic to it. But the public conscience, though it made legitimate and held to be lawful oppression and privilege, was not opposed to the feeling of justice. Every illegal violence was forbidden, and mutual assistance was considered a duty. We may add that the insolvent debtor was not reduced into slavery; he paid his debts with his property, and not with his person.

Egypt seems to have realised the social ideal conceived by the African races. It now remains to us to describe those that have been imagined by other human races.

VI.

Societies in North and in South America.

Like the African continent, in America we shall see different phases of human societies.

The stupid Fuegians are at the bottom of the scale. According to Cook, their daily life is more like that of brute beasts than the life of any other people. Essentially wandering and vagabond, they go about in small hordes, changing their place of abode as soon as they have exhausted the supply of animal food, or more especially the shell-fish at any point on the sea-shore.

The Patagonians, the Araucanians, the Charruas, and, speaking generally, all the nomad tribes who ride about the pampas of South America, whom D'Orbigny has called the Pampas nations, have already made a commencement towards social organisation, though it is still very rude. As their means of subsistence are more certain than that of the Fuegians, they can group themselves together in larger numbers, they can constitute tribes. But in these tribes the government is of the very simplest kind. As the principal occupation of the people is fighting, the men, before undertaking an expedition, all sit in a circle to deliberate their plans; and they designate a chief who is charged temporarily with the care of the expedition. While they are under their tents there is no subordination among them, no submission from one to another, not even from the son to his father. As they live only upon the produce of the chase, of their razzias, and upon the flesh of their horses, they leave a district as soon as the pastoral supply of food is exhausted. Their principal passion is war. We find here an anarchical state in all its brutality. They have yet no idea of protection, nor of social justice. Everyone acts as he pleases, he defends himself as best he can, and he takes his vengeance after his own fashion. The Botocudos of Brazil are hardly more civilised than the Fuegians. Absolutely ignorant of agriculture, they wander about stark naked in little hordes, living upon game, which they devour raw.

The Guarayos have made a step towards social organisation. Each group of families has an hereditary chief, whose power is confined to giving advice in time of peace and directing the warlike operations. Two offences these people punish very severely. These are the two principal forms of violation of property—theft and adultery.

Among the Caribs and the Tupinambas, the chiefs have nothing to do with the administering of justice. Everyone for himself avenges the offences he has received, and if he does not do so, he is stigmatised by public opinion. The manners of the greater part of the natives in these vast regions are therefore almost bestial. We may therefore now explain the facility with which the Jesuits reduced to a state of animal domesticism the Indians of Paraguay. These poor creatures began their daily task at eight o'clock in the morning, under the eye of the corregidor. The men used either to till the ground, or to labour in the workshops; the women wove the cotton. The submission of the Indians to the curés was absolutely servile. The whip, administered without distinction of sex in an infantine fashion, was used for public faults; and sometimes, obeying the voice of their own conscience, the sinners would come themselves to solicit chastisement for the omission of duties they had forgotten to perform. The Paraguayas, devoid of every individual initiative, docilely submitted to this mechanical existence. Life gave them no pleasure, and they died without regret.

The Columbian savages, a more intelligent and more energetic people, choose for themselves chiefs, whom they honour as beings of a superior nature; but they first make them undergo a most severe training. In the first place, every candidate for the office has to submit to a rigorous fustigation, without showing any sign of pain. He is afterwards laid in a hammock, with his hands tied, and myriads of venomous ants are showered over him. Then underneath the hammock a fire is lighted, composed of herbs, so that the patient shall have the full benefit of the heat, and be thoroughly enveloped in smoke. Under all this treatment he must be completely passive. The least sign of impatience, or the slightest possible groan, will at once disqualify him for office.

We shall soon have to describe the more advanced civilisations in Central America; but to preserve the order of our plan, we must first mention the ill-developed societies of South America.

In a sociological point of view, the Californian natives may be compared with the Fuegians. Living in a state of complete anarchy, they know of no other right than brute force. Everyone acts as he pleases, without taking heed of his neighbour. Vice and crime are unpunished, or rather, public opinion does not recognise either vice or crime. Everyone must defend himself as he can. Such at least is the description given of them by the Jesuit Baegert, who lived among them for seventeen years.

Though the Indian Red Skins are much superior to the Californians, the organisation of many of their tribes was, and is still, as rudimentary as that of the Araucanians or the Patagonians. They have no form of government in time of peace; in time of war they obey the bravest of their warriors. Birth has nothing to do with this; physical and moral force only will give a man prestige. Important matters are also often eloquently debated by a council of old men. They have no idea of social justice; every man avenges himself as he can.

Such of these tribes who have laboured at all seriously at agriculture have materially improved their social condition. We may mention the Iroquois, of whom all the tribes have united against their white or red enemies. As the Tchéremiss have done elsewhere, the Creeks established among themselves a system of common agriculture. At certain fixed times everyone had to perform his portion of the work in the fields, and the harvest was divided among the different families.

Certain tribes had constituted among themselves a barbarous form of justice. Among the Comanches the adulterous woman was punished by having her nose slit; but it was the family of the injured husband who awarded the penalty.

The peaceful Esquimaux, whom Ross could not succeed in making understand what was meant by warfare, and who possessed no fighting instruments, sometimes obey a chief to whom they pay a certain fine, but usually form themselves into small free communities. They could not understand that there should be chiefs and officers among the men in Parry's expedition. In the little Greenland republics, all the citizens are equal. The community will not admit a new member, but with the general consent; every associate is bound to hunt seals and whales as long as his age will permit him, and he has no son to replace him.

This instinct of equality, these sentiments of joint responsibility, joined to a great softness in their manners, make of the Esquimaux a race apart, and their origin is probably very different from that of the Red Skins, their enemies and their tyrants.

Among the societies of Central America which will now claim our attention, ancient Peru will offer an interesting example of a communist society; and though it was much more complex than any association of Esquimaux, the Peruvian organisation was morally inferior to theirs, because it was founded upon despotic monarchy and the system of caste.

VII.

Societies in Central America.

We find among the American races anatomical resemblances and also mental analogies so striking as to justify us in supposing a common origin, and also to allow of the very unequal degrees in their social organisations making themselves subordinate and taking their place in the successive stages of a great evolution. The Fuegian and the Botocudo would be at the bottom of the scale; the people belonging to the ancient states of America would be at the top. But the large empires of Mexico and of Peru ought not to be considered as civilised islands in the middle of a great ocean of savagery. Round about them there were other societies less important, but still having a social organisation very far above that of the Red Skins or the Guarayos.

At the time of the conquest of Hernando Cortez, there were in New Mexico, where we may still see them, communities, phalansteries more or less,—the pueblo. Each pueblo is a small republic, composed of from forty to fifty families, inhabiting in common one enormous house, of difficult access, constructed in the shape of a colossal mound of huge stepping-stones, and of which each step forms one story of the house. Each story contains so many cells, and in each cell a family resides. One cannot mount or descend from one story to the other but by a ladder. The Indians who live in these pueblos are skilful agriculturists; they have learnt how to fertilise the soil by means of canals. Their manners are soft, and on the whole they are much superior to the Indian tribes of huntsmen.

Small monarchies used to exist elsewhere, but these were little more than a form of great empires of Central America upon a reduced scale.

The Natchez were governed by a grand chief who was brother to the sun. This demigod had the right of life and death. His wives and his slaves were bound to immolate themselves upon his tomb. Below him there was a line of hereditary nobility.

In Florida the caciques were regarded with servile respect.

The inhabitants of the Antilles islands used to obey chiefs by the right of birth—men who enjoyed an absolute power; they sprang from the gods, and they governed the elements.

In Bogota there was a numerous people already skilled in agriculture; they had built important towns and convenient houses, they had traditional laws repressing certain crimes, recognising as their absolute master a king, whom they approached with awe and trembling; and whenever the king went out, the people used to strew his path with flowers. This monarch imposed his own taxes, and he also received rich presents.

This is, upon a small scale, the same organisation as that of the Aztecs. The chroniclers of this empire do not make it date farther back than the end of our twelfth century; and we find there a theocratic and monarchical society very similar in its principal features to that of ancient Egypt.

The Mexican monarch was chosen from among the brethren of the defunct monarch, assisted by four delegates of the nobility; and he henceforward became a quasi-divinity, the representative at least of the gods upon the earth. He was always surrounded by a most complex ceremony. In his harem there were no less than 3000 concubines. His sumptuous repasts were served by handsome young girls. His body-guard, his postal couriers, his ministers of peace and war, were everywhere stationed along the roads in the empire.

Below the king, and always near to him, were a numerous aristocratical and hereditary class, who owned the soil and occupied the principal places about the court and in the administrative offices. To these men was confided the government of the provinces and of the towns.

The principal occupations of this aristocratic caste were governing and carrying on wars, which often had no other object than that of procuring so many prisoners, which the religion of the country imperiously exacted should be offered up as sacrifices. All the great social labour was performed by the serfs and by the slaves. The serfs were attached to the glebe; they could not go away from the little bit of ground which they had been commanded to cultivate; they were transmitted hereditarily from one owner to the other, together with the domain, as though they were domestic animals. Their lot differed from that of the slaves in only one particular: they could not be sacrificed to the gods.

As in Africa, there were in Mexico several categories of slaves. The majority were prisoners of war. Others had been condemned to servitude for various crimes. Some of them were voluntary slaves, for free men had the right of disposing of their liberty. Poor parents could also sell their children. In this case a contract of sale was drawn up before witnesses, in which was clearly specified the kind of service that might be exacted. The slave might have a family of his own, he might even own other slaves. The masters rarely sold their slaves without strong reason. They had the right to do so, nevertheless, and there were fairs at which this kind of traffic was regularly carried on. The slaves were dressed in their best clothes, and they were bound to sing, to dance, and to entice buyers by showing off their good qualities. The vicious slave was sold with a special collar round his neck, and if he repeated his offence a second time he was destined to be sacrificed—a most important matter in the Mexican society.

On the other hand, the children of the slave were free. In Mexico no one was born a slave.

Outside the rank of slaves and agricultural serfs there was a corporation of working men, and in Mexico each of these men fulfilled a certain office, to which was connected its feast days, and its protecting divinity, its patron. In these corporations the profession was hereditary; everyone was bound to follow the occupation exercised by his father.

A similar organisation prevailed in Mexico, properly so called, and in the tributary kingdoms of Tezcuco and Tlacopan. Even in Tlascala and in other republics bordering on the Aztec empire, there was the same social hierarchy. The monarch only was wanting.

The Aztecs seem to have had that sort of care for justice which is compatible among a despotic and barbarous people; but their legislation was not complex, and it was very severe. Theft and murder, even that of a slave, were punishable with death. The adulterer was stoned. In a young man, drunkenness entailed the penalty of death; in persons of ripe age, it was punished with degradation and confiscation of property.

The care of administering justice was put into the hands of special functionaries; each of these had his own separate jurisdiction. Inferior magistrates decided upon the smaller matters; they were elected by the people. In each province, a court, composed of three members, who were named for life, decided upon important matters. This court was subject to the revision of a supreme judge, named by the king and also for life, and from his judgment there was no further appeal.

A portion of the Crown lands was affected to the maintenance of the superior judges.

At Tezcuco these judges used to come together every eighty days, under the presidency of the king, to decide upon difficult or important causes which had been reserved for them.

The judge found guilty of bribery or of corrupt influence underwent the penalty of death.

The attribution of judicial functions to a special body of men will everywhere indicate a state of society free of barbarism, or at any rate endeavouring to become free. It shows a marked progress in the division of social labour. The organisation of justice among the Aztecs implies therefore a certain moral elevation, contrasting strongly with their ferocity and their sanguinary religion, which degraded them so much below the Peruvians, of whom we must now speak.

Though they were almost neighbours from a geographical point of view, separated only by two states, both of whom were fairly civilised, it would appear that each of the empires of Mexico and Peru was totally ignorant of the existence of the other. It is therefore all the more curious that in the social organisation of the two countries there should have been so many features common to both.

The monarchy of the Incas, of which we have already spoken,

offers to the sociologist a most interesting subject for study. We find there the largest communist society that has ever existed, and it seems to have realised in a great measure the utopian ideas of many an ancient and modern reformer. But still, in spite of its own peculiar characteristics, the organisation of ancient Peru is not very widely different from that of ancient Egypt or of Mexico. There also we find at the top of the social pyramid an absolute monarch, and under him the castes in their order. A large body of servile plebeians are made to labour, and by the sweat of their brow the rich men lived and were made happy.

In Peru the king was not merely, as in Mexico, the vicar of the gods, he was the son of the supreme god, the offspring of the sun; and "when he was recalled to the home of the sun, his father,"—or, in plain language, when he died—he was thought to be greater than a god. The divine monarch, so to say, was supposed to tower above and look down upon his people. High pontiff, the representative of the sun, he presided over the grand religious solemnities; generalissimo of the army, he levied and commanded his forces; absolute king, he imposed taxes, legislated, appointed or revoked all his functionaries and judges at will. He was, in fact, an earthly sun.

He was the eldest son of the coya, or the lawful queen, and was therefore brought up at the military school, as were all the descendants of the Inca grandees, the reputed descendants and the successors of the founder of the monarchy. Like them he was broken in to warlike and gymnastic exercises.

At the death of an Inca, his palaces, his places of residence, everything that had belonged to him, remained in the state in which he had left it; and his pompous funeral was always accompanied with the sacrifice of an infinite number of human lives. His servants, his concubines, his favourites, sometimes numbering a thousand souls, were immolated upon his tomb; for it was necessary that he should be properly waited upon in the other world before he came back to give life to his mummy, which was carefully embalmed and deposited in the grand temple of Cuzco, by the side of those of his predecessors.

After the king, but very much below him, came two privileged castes: the Incas, or descendants of the polygamic series of the kings, and the *curacas*, or caciques, of the conquered nations.

The noble Peruvians of royal blood wore a particular costume, and many of them used to live at the court. They only were intrusted with the high offices in the state; they only held military commands. The laws, very severe upon others, were not made for them. They bowed only to the master, before whom they always performed a slight act of reverence.

As we may expect, the greater part of the public property was intended for the maintenance of this caste of divine blood.

Underneath these men came the *curacas*, or the caciques of the conquered nations.

The Peruvian government generally allowed them to remain in possession of their offices, obliging them only to come from time to time into the capital, and that they should there educate their children.

The great mass of the people were governed much in the same way as a careful cultivator will bring up and look after his domestic animals. Every male inherited his father's profession; he was not allowed to choose another employment. By right of birth a man was either labourer, miner, artisan, or soldier. The population, divided into groups of 10, 50, 100, 500, and 1000 persons, each having its chief, was attached to the soil. The government officers treated the people kindly, as though they were a flock of sheep. Every man had his task set out for him beforehand, he was married; a portion of ground was given to him for his maintenance. His morality was watched; he was dressed; and in case of need assistance was given to him. In the empire of the Incas liberty and misery were equally unknown.

In Peru, the spiritual government was mixed up with the temporal government. Nothing was more natural, for the Inca himself was God. The mission, therefore, of every Inca was to extend the territory of his empire, and to wage a perpetual crusade against the unfaithful people. Peru was consequently often engaged in war against its neighbours. The conquered people were treated leniently; they were only made submissive to the laws of the conqueror, who endeavoured to assimilate them with his own people. There was in Peru none of the brutal tyranny similar to that which the Mexicans inflicted upon their vassals.

In virtue of the passive condition of the Peruvian people, a methodical and well-regulated centralisation was everywhere the rule. The kingdom was divided into four parts, into each of which there was a great highway starting from Cuzco, the capital, the *navel* of the monarchy; and this town was itself divided into four corresponding parts. Four viceroys governed the four provinces, which in their turn were subdivided into departments; and these dignitaries were obliged to live in Cuzco for a portion of the year. They formed a sort of council of state, under the command of the master.

A system of statistics, by means of *quipos*, was kept, showing a register of the births and deaths, and also of the revenue of the empire. With the aid of these documents, the government was enabled to impose taxes and other means for collecting the revenue, to authorise the public works, and whatever else was necessary.

The postal arrangements were well organised. There were relays of couriers carrying *quipos*, and stationed at equal distances from each other along all the roads. Small buildings were erected, offering shelter. There were also other constructions, upon a larger scale, intended either for the Inca, or for the public functionaries, or for the troops when on the march. In Peru there were no other travellers.

The laws were not numerous, and were very severe. Capital punishment was common. Such was the penalty for theft, for adultery, for murder, for blasphemy against the sun or against the Inca—they were both considered equally—for setting fire to a bridge, and for other crimes. If the inhabitants of a town or of a province revolted they were exterminated. To rebel against the child of the god-sun: what an abominable crime!

This description, incomplete as it is, may be sufficient to give some idea of the social economy of this singular country. Nowhere do we find has the idea of monarchical government been realised more ingeniously or more minutely. It was the great ideal dreamed of by the first founders of the empire; servitude was flourishing in all its glory, and it was exercised tenderly. A superhuman power conducted everything, ruled everything, foresaw everything. The subject was a simple machine, an automaton without initiative movement, bound to serve a superior caste, and also an all-powerful master.

VIII.

The Polynesian Societies.

The old empires of Mexico and Peru show to us the monarchical phase in all its splendour; it was the adult age of royalty. In Polynesia, where every archipelago formed a small independent society, and where the ethnical evolution has not everywhere progressed with the same quickness, we can in some way study monarchy in its embryo state, one stage after the other.

When the Polynesian society has arrived at its highest pitch of development, we find all the social degrees which used to exist in the monarchies of Central America: a caste of slaves, recruited from the prisoners taken in war, a popular class, an hereditary aristocracy, and a despot crowning all. But we do not find this gradation to be perfect in all the islands. The crowning of the edifice is often wanting. The chiefs of the tribes owe no tribute to a suzerain monarch.

The few inhabitants of Easter island were divided into small circles, each governed by a chief. Also in the Navigator islands the chiefs were very numerous, and they used to govern the people by means of the rod.

In Noukahiva each valley is the country of an independent tribe, having its own laws, its priests, and its chiefs. Here a monarchical system prevails among the tribes, and each one of these small ethnical groups is usually obedient to a powerful old man, who owns a great many cocoa-nut trees and bread-fruit trees.

The power of the chief is considerable. He has the right to claim for his own anything to which he may take a fancy wherever he goes; and whenever the people hear of the coming of the sovereign or of his lady, they immediately hide all their most precious objects. Before declaring war the Noukahiven chief ought to call together his nobles in council; but the more usual custom is to rush wildly upon a neighbouring tribe which has given offence to a woman, or who has run away with a victim and offered him up as a holocaust to the gods.

The chief, whose office is ordinarily hereditary, can command scarcely more than a few hundred individuals, for the tribes are not large. Under the chief, there are, not counting the slaves, two social classes: the nobles and the villains. The nobles, or the *akaikis*, hold their privilege by right of birth; but its ranks are also open to men valiant in war, to those who marry a *woman-chief*, or to those whom the chief has adopted. The inferior class, the *kikinos*, are ordinarily attendant upon the others both in peace and in war. The *kikinos* also eat with their superiors off the same plate, they sleep upon the same mat, their wives are often in common, and they may change their patron at pleasure.

The *akaikis* enjoy some important privileges. They may claim as their own any object that pleases them; they may impose a tithe upon the harvest, and an impost upon privileges of every kind. They may drive the *kikinos* off from their property; and they may proclaim the tabu. We shall speak again of this latter prerogative.

In the large islands of New Zealand we do not find that any large form of monarchy had been established. The islanders had grouped themselves together in small tribes, governed by a chief, whose authority they obeyed only in time of war. During peace this chief had no other right than that of living in a noble idleness, upon the produce of a tax of a tenth part imposed upon all sorts of produce. He enjoyed the privilege of being allowed to keep the heads of all the prisoners taken in war. As regards his own head, it was always eagerly sought after by the hostile tribes. The office of chief was often transmitted from one brother to another; it might also be acquired by great riches, by great valour, or by great sacerdotal influence. But to be chief, a man must first belong to the aristocratic class, to the caste of the *rangatiras*.

In many of the Polynesian archipelagoes, the people owed obedience to one master, who governed the whole feudal hierarchy. This was the case in Tahiti, in Tonga, in the Sandwich islands, in the Gambier islands, and in others. In these islands the sovereign power, nearly always hereditary, was transmitted sometimes through the direct and sometimes through the collateral heir. In Tahiti, the crown passed virtually to the infant of the king upon the day of his birth, and from that time his father was only king-regent.

Great marks of respect were due to the Tahitian monarch and to all the members of his family. When any one of their great personages passed, everyone was bound to uncover their shoulders.

At Tonga, to salute the monarch, people prostrated themselves before him; they put their heads under the sole of his august feet. The body of the king was not tattooed, as was that of his subjects. On no pretext was it allowed to anyone to stand behind the royal head. Whenever the monarch condescended to sleep, women used to assist him in this important occupation by smacking his thighs. The king of Tonga was the natural heir of his subjects; but ordinarily he inverted the inheritance in favour of his own son. But sometimes the royal power was transmitted from the deceased prince to his brothers, or in their default, to his sisters. Such was the usage in Tonga island at the time of Cook's voyage. At a later date (1845) the moribund king had the faculty of choosing his successor out of the royal family, provided that his will was ratified by a consent of the chiefs. Then the accession of the new prince was celebrated at Kava with much feasting and ceremony.

The power of the Tonga monarch had some of the same characteristics that we find in real monarchies. The life, liberty, and the property of the subjects were at the free disposition of the master.

Nevertheless, in the majority of these archipelagoes, the supreme dignity did not remain indefinitely in the same family. The most powerful chiefs were ever struggling to usurp it. In the Sandwich islands only do we find a real dynasty dating back from time immemorial.

In all of these Polynesian islands, whether they were or not governed by a monarch, there existed an aristocratic caste, generally hereditary and possessing excessive privileges.

The power of the principal chiefs was almost absolute. In the Sandwich islands the common people were bound to prostrate themselves before them, literally "lie down to sleep in their presence." In Tahiti, they had the power of life and death over the populace. A chief who had killed a villain got into a furious fit of anger when he was told that, for a peccadillo of such a nature, he would, in England, have been hanged. There were even culinary privileges. In Tahiti, the common people were not allowed to eat pork. In the Sandwich islands, when the wind carried the chief's canoe towards Cook's larger ships, it struck against and sank, with no sort of compunction, the canoes of the common people. It was above all things necessary to lie upon the ground, belly downwards, the moment the great master landed. And in the combats, people endeavoured mainly to strike the enemies' chief; for his death was generally followed by the defeat of all his party.

When they were young, the Polynesian chiefs were almost worshipped; people approached them with fear and trembling. And if, during their lifetime, they had done any action worthy of renown, they were, after death, really put into the same category as the gods. Their prestige was so great that, standing naked and unarmed in the middle of the people, they were obeyed if they gave the slightest sign. But this unlimited power was enjoyed only by the supreme chiefs, those who were of high birth, and who also owned sufficient land to feed a great number of people.

On the other hand, the plebeians were everywhere at the mercy of their great men; the *toutous* of Tahiti, the *kikinos* of the Marquesas islands, the *tuas* of Tonga, and others, could call nothing their own. Without being actually slaves, they yet constituted a servile class. They were the domestics and the soldiers of the chiefs; and at Tahiti it was usually from this class of men that the human sacrifices were chosen, of which the gods were so terribly fond. A man was born and died *toutou*; he could not change his condition.

There was also a hierarchy among the governing classes, and in the great archipelagoes there was an organisation that was almost feudal. At the top of the aristocratical edifice, in Tahiti, were the *Arii*, or princes, lords either of a whole island, or of a portion of it. After these came a sort of feudal baron. These men also had their vassals; and under these vassals there remained only the *toutous*. At Tahiti, however excessive the power of the *arii* might be, they were not allowed to confiscate the property of their inferiors; but they arbitrarily claimed a large portion of their harvest produce.

The assistance of the petty Tahitian nobility was indispensable to the high aristocracy to enable the great men to carry on their wars; the *arii* never embarked upon an expedition without first consulting their vassals. All the dignities, privileges, and titles of nobility were strictly hereditary, and in default of a male heir the Tahitians had recourse to the expedient of prince-consort or lord-consort, who was empowered only with the duty of raising a male heir.

At Tonga there was also the same sort of social order, but more frankly despotic. The king was the supreme master, and the nobles possessed their domains only as fiefs, which were ever at the king's disposition. The nobility and the titles were hereditary, and always commanded most servile respect. It was forbidden to touch the person of a chief or to enter his house. The members of the aristocracy had a special dialect for their own sole use; they were not in the least bound to respect either the persons or the property of the common people. The Tonga nobility was subdivided into four classes, one subordinate to the other. Some of the professions were strictly hereditary, specially the carpenters and the fishermen. Above this social edifice the royal power was predominant; it was despotic and subject to no control; this society, so very feudal, had even its knight-errants. They were young warriors, who always went armed with their javelin and their club, ever in search of warlike adventures, and sometimes going in quest of them as far as Fiji, boasting that they would eat the bodies they had killed.

In the Sandwich islands the aristocratic caste also subdivided itself into different classes, of which the two principal were: that of the district chiefs, who enjoyed an absolute authority; then the nobles who were simply proprietors, and who had no functions given to them. Below these men, who were noble by race, were their servants: the villains, the plebeians, the *tontous*, who had no rank, no right, no property.

At Tahiti, and in the Easter islands, whenever a chief said: "Hog, who does this tree belong to!" the owner would never say "To me;" he would answer: "To us two," or "To you and to me."

In this coarse state of society, founded only upon hereditary despotism, justice was the last thing thought of. No judicial function had been constituted. Sometimes the injured parties would address themselves to the chiefs; but the chiefs had other things to think about, and they cared little for protecting the weak against the tyranny of the strong. They did not often inflict chastisement but for injuries done to themselves, and then the punishment was terrible. The slightest offense, the least wrong, done either to themselves or to their favourites, were in their eyes unpardonable crimes. In reality there was no social justice in Polynesia. Everyone avenged himself as he pleased. "An eye for an eye, a tooth for a tooth," such is still the law of laws. But a very rude morality is beginning to be seen, and public opinion has sanctioned or reproved certain acts of vengeance. As in the majority of barbarous societies, theft and adultery are considered as the gravest offences; and the penalty was also simple and very severe. Theft, rape, and adultery were usually punished with death. In New Zealand the thief was decapitated, his head was hung on to a post in the form of a cross. In Tahiti the husband might kill his adulterous wife, but ordinarily he contented himself with beating her.

In Tahiti the murderer was attacked by the friends of the deceased. If he was conquered, his house, his furniture, his lands, became the property of the assailants, and vice versâ. In these conflicts the stronger despoiled the weaker. As regarded infanticide, no one thought anything about it; every man had a right to do what he liked with his children, either born or those about to be born. Theft was the greatest crime. Sometimes the culprit, condemned by the verdict of public opinion, was given over to the vengeance of the persons injured, and was also deprived of the right of defending himself.

At Tonga the judgment of God was sometimes invoked against the thief. The suspected man was made to bathe in certain places frequented by sharks, and if he was devoured, or even bitten, he was adjudged guilty.

At Noukahiva justice was less strict: to be able there to pass oneself off as an artful thief gave a man a sort of distinction. Murder was avenged by murder. A man, if he was famished, was by the law of public opinion allowed to eat his wife or his child. Adultery was considered as a crime only in princely families, and even there it might be held lawful under certain circumstances. At the time of Krusenstern's travels, at the commencement of this century, there was close to the wife of the potentate, whom this traveller calls the king of Noukahiva, a functionary called the "king's fire-lighter." The duty of this functionary was in the first place to be always near to the monarch to execute his orders, then to take his place in everything and for everything beside the queen, in case of his prolonged absence. In a country where such a custom exists, adultery cannot be regarded as a great crime. In all primitive societies adultery is assimilated to theft, and it becomes lawful when authorised by the proprietor or by custom.

In Polynesia, public morality, so indulgent or so heedless of many acts reprimanded or condemned in Europe, had taken a peculiar special form; it had adopted the tabu—such was the expression always used. By tabu we must understand a sort of prohibition that might be put upon anything by the priests, who in this respect acted nearly always concurrently with the chiefs; and they had made of the tabu a powerful instrument of despotism. The tabu was sometimes a very wise provision.

When a bad harvest of bread-fruit was feared, a tabu was placed upon the bananas and wild ignames and other substances; in this way a reserve supply of food was assured to the people. Pigs and hens were tabooed when they became scarce. Certain bay leaves were tabued for fishing purposes when the fish would not approach them. In New Zealand an iron pot of European origin was declared tabued, and therefore all cooked food was forbidden to the slaves. The idols, the morais, the burying-grounds, the bodies of the priests, the chiefs and their dwelling-places, sometimes whole districts, were tabued. Many tabus were strange, vexatious, and capricious. A tabu was sometimes placed over an individual, forbidding him to leave his house for a certain number of days, to light any fire, to eat after the rising and before the setting of the sun. Because of the tabu women were not allowed to touch men's food, not even that of their husbands, their brothers, or their children. They could not go into the morais. The woman was tabued just after her confinement. She was obliged to be in a cabin apart, and was not allowed to touch her food; women came to her who put the meals into her mouth. The tabued chiefs were fed in the same way, sometimes for months together, during which time they were bound to abstain from all intimate acquaintance with women. At Tonga the hands which had touched the feet of the king were tabued until they had been washed, and before this ablution a man could not make use of them. In New Zealand the backs and the heads of free men were tabued, they were therefore forbidden to carry burdens. In the Marquesas islands public disasters, sickness, etc., were generally considered as the consequence of the violation of some tabu. And therefore the violation of a tabu was everywhere considered as a crime, and was severely punished. In the Sandwich islands a woman who had dared to eat of pork on board a European vessel was sacrificed to the gods. Generally speaking, to violate a tabu was a capital crime, if it was even suspected a man was put to death. The revocation of a tabu necessitated a religious ceremony, for which human holocausts were indispensable, and as the care of designating these victims belonged to the priests, this prerogative conferred upon them the right of life and death over the common people. From the characteristic custom of the tabu we may conclude that a moral sense is beginning to awaken in the Polynesians. They had a keen sense of what was lawful and what was unlawful, but their infantine morality was wholly irrational, and was very often confounded with their religious obligations. No act was considered blamable until a priest had declared it to be so.

Such was the ancient social organisation in Polynesia, as it had spontaneously developed itself. It was a despotism, often brutal, but more often tempered by the thoughtless puerility of the race. In time, the Polynesians would, doubtless, have come to possess institutions more rational and more human; but the intrusion of the Europeans has cut short all progressive evolution in the country. It is at Tahiti, especially, that the deplorable effects of this change have been the most startling. There the English missionaries have succeeded in civilising the people after their own fashion. They have enforced their ascendency by making one-half of the population exterminate the other half. Then they founded a monarchy grotesquely theocratic. One missionary, the Rev. Mr. Nott, armed with a holy ampulla, anointed and consecrated the prince Pomare king of the half-devastated archipelago. Then, with the assistance of the secular power of this eldest son of the Church, the missionary endeavoured to impose upon the islanders the religion and morals of the English people.

Up to this time, love had been free, and was practised unreservedly in Tahiti. The voluptuous Polynesians lived for little else than enjoying the pleasures of sensual love; and they saw no harm in it, when legitimate property was respected. The missionaries considered that these delinquencies were grave offences, and many female sinners were condemned to give their assistance towards carrying out the public works; they had to labour at the construction of the roads, the bridges, etc. The great high-road which runs all round the island of Tahiti represents a very large amount of gallantry on the part of the Tahitian women. The most curious thing is, that each condemnation was preceded by a public debate on the matter before a tribunal, in which the offenders fully set forth, with a sort of cynical candour, all the circumstances of their fault. The Tahitian morality gained little; but the people have at last acquired the vice of hypocrisy, which was before unknown among them. They did not sin the less, but they endeavoured to hide their faults. Then, to extract confession, which grew to be less spontaneous, the pious legislators resorted to torture. A slip-knot put round the waist of the women, and drawn tight by two men, drew from them the confession of their faults. After which, their faces were indelibly marked with a special tattoo sign.

We know the result of this imbecile tyranny. The rudimentary morality of the race perished and was not replaced. The Tahitians borrowed from European civilisation nothing but its vices. In the smaller islands especially, where the natives were more or less free from the vigilance of the missionaries, in Raiatea, in Tahaa, in Bora-Bora, the principal occupation of the converted people was in distilling alcohol and intoxicating themselves. The number of islanders diminished with wonderful rapidity, and the queen island of the archipelago—the lovely Tahiti, the New Cytherea, of which Cook and Bougainville have left us such graceful pictures—counts only a few thousand inhabitants. *Ubi solitudinem faciunt, religionem appellant.*

IX.

The Malay and Indo-Chinese Societies.

A man must surely be a missionary before he can believe that the morality of a people can be changed suddenly and by dint of force. The mental condition of any race, their desires, their tendencies, show the very life of the people, the series of cerebral impressions resulting from their acts through a whole chain of generations. These characteristics, which have been gradually forming themselves from father to son during the course of centuries, cannot be readily effaced.

The Dutch colonists in the Malay archipelago, who think less of saving the souls of the natives than of their commercial interests in the country, have obtained results far more satisfactory than those of the fanatical English missionaries in Polynesia.

Without apparently changing anything in the social organisation of the Malays of Java, in Celebes island, and elsewhere, the Dutch have confined themselves to placing beside each native chief a resident, who is called "the elder brother" of the little prince, and he never gives orders but only "recommendations." And besides the resident there is a controller who visits the natives, hears their petitions, and inspects their plantations. New modes of cultivation have been introduced, and even new sorts of culture—coffee, for instance—and schools also have been established. It is a paternal and disguised form of despotism which seems to show excellent results. In some districts in Celebes island the aborigines, who were savages, now enjoy a relative civilisation. They are now well clothed, well housed, well fed, well brought up; and, instead of diminishing, their numbers are on the increase.

But we have now to describe the native societies of the Malay archipelago, and also the monarchies founded by this race on the continent, in Siam, and in Cochin-China. These Mongoloid people have all more or less freed themselves from primitive savagery, but they have not yet succeeded in getting beyond the phase of despotism.

In Celebes island there were quite recently savage tribes, each having their own particular dialect, and constantly at war one with the other. The hut of the chief, built upon pile-work, according to the old Malay custom, was also decorated with a few human heads. It was an act of duty to place upon his tomb trophies of this kind, but cut freshly for the occasion. As far as possible the heads of some of the enemy were offered up to the departed spirits of the master, but in their default slaves were sacrificed.

In all the tribes of the Malay archipelago the power of the chiefs is absolute. It is so in Lombok island, in Celebes, and in others. In this last-named island no one dare to stand upright in the presence of the rajahs.

There are slaves everywhere. At Sumbawa they are attached to the glebe and are sold with the soil. In Timor the islanders are perpetually fighting amongst themselves, and only to capture slaves from each other; these slaves are everywhere bought and sold as merchandise.

The laws are very severe, and sometimes remind us of those in

Polynesia. In Timor the custom of the *pomali* is very like that of the Polynesian tabu; to protect a garden from thieves it is considered sufficient protection if a few palm leaves, suitably arranged, are hoisted.

At Lombok, theft, which in most primitive countries is considered a great crime, is punished with death.

As in all barbarous societies, the Malays are very conservative; they are very strongly attached to their ancient customs. A tribe of Dyaks agreed to impose a fine upon anyone who, in cutting down a tree, should notch it in the European way; the tree must be gashed by perpendicular strokes only, according to the ancient custom always observed by their ancestors. In Malay, that which in a family has belonged to the ancestors is often regarded as possessing a peculiar value. In Sumbawa, a house which has sheltered several generations becomes quite sacred, and nothing is considered more precious than the stones which, because they have long been used as seats, have become polished. With tendencies such as these, we can easily understand that the ideal of despotic monarchy should have been realised.

In Cochin-China, in Siam, etc., wherever the Mongoloid races, of whom we are speaking, have grouped themselves together into semi-civilised states, they have created the most despotic forms of monarchy that can possibly be imagined. To lay a stick over the back of an inferior, or to expect the same treatment from a superior, was in Cochin-China considered as the most ordinary way of governing. Any idea of showing resistance to a despotic caprice was never conceived. The abject condition of the Siamese, which formerly astonished Finlayson, also, twelve years ago, was the cause of much wonder to the French traveller, H. Mouhot.

In Siam, the king, who is the absolute proprietor of the persons and the property of his subjects, has only the right of standing. In every degree of the Siamese hierarchy, everyone literally grovels before his superior, and exacts also that his inferiors shall grovel before him. At a dinner given in Bankok by a functionary of the fifth order, servants brought in the dishes walking as animals on their hands and feet. When the king gives audience he is placed upon a throne, in a sort of nest, twelve feet from the ground, curled up in the sacramental attitude of Buddha, while his assistants are prostrated before him, their faces on the carpet. "I, a hair; I, an animal," are the respectful formula which in Siam a man uses in speaking to his superior. If anyone passing the gates of the royal palace neglects to take off his hat, hard balls of clay are thrown at him by watchmen, who are made to stand there specially for the purpose. The king has a harem of six hundred women, perpetually renewed by the voluntary gifts of fathers of families.

While the high dignitaries of state, overladen with gilding and decorations, are walking on board the long boats that carry them up and down the Menam river, women, and children, and officers, prostrate themselves before them, and are bound to collect in golden vases the saliva of the masters.

In Siam, the imposts laid by the terrestrial god upon his respectful subjects are enormous. According to H. Mouhot, the treasury does not allow the peasant to keep for himself more than two and a half per cent. of his income; the master has the right to take everything.

People are forced to sell sugar, pepper, benzoin, and other things to the king, who resells them at his own price. To have the right of fishing in the watercourses, of distilling arack, the privilege must be bought from the Crown.

But some traces of progress may be affirmed, even among the Siamese. Finlayson relates to us how their way of looking at adultery has gradually become changed. At first, the injured husband might avenge himself as he pleased; he might kill, or receive payment for the injury done him. Then the law intervened, and accorded the right of killing the two delinquents, but only in the case that they were found *flagrante delicto*. The punishment too must be instantaneous. Afterwards it was thought that there was no capital crime, except for women belonging to the palace, and a fine was adjudged sufficient punishment.

If progress will penetrate among the human cattle of the Siamese monarchy, where, indeed, will it not make itself felt?

X.

The Social Organisation of the Nomad Mongolians and of the Thibetans.

To examine the whole human kind in order to ascertain the desires, the tendencies, the degree of development in each of its principal races is a most serious task, and we could not hope to arrive at any result at all satisfactory, except by limiting our work, by confining ourselves only to the characteristic facts, to the main features of the great whole. We are compelled, therefore, in this short review, to omit a thousand interesting details, to pass over, almost in silence, whole ethnical groups of people. We shall say nothing as to the government of Burmah, where, as in Siam, the most despotic monarchy prevails. The few preceding pages will have shown us clearly enough that the Malays, and other congeneous people on the Asiatic shore, have in fact emancipated themselves from savagery, properly so called, but only by organising the most abject form of servitude.

The great Mongolian race, the first in the world in point of numbers, and the second in moral and intellectual dignity, have learned the most various and the most ingenious social manners and customs. But all the different people composing this large group have not advanced with equal strides; and from the hindmost to the foremost in the struggle may be seen a most interesting and long progressive series.

According to Chinese traditions, the ancestors of the sons of heaven were in a very remote bygone age wretched savages, who had not passed the Age of Stone, and who used to wander about in small hordes at the foot of the Thibet mountains, much in the same state and condition as are nowadays the least civilised of the Kamtschadales. This is indeed probable, for traces of the Stone Age may still be seen almost everywhere in China, and all the northern half of Asia is inhabited by the nomadic Mongolians. It is only in Thibet, in Japan, and in China, that the yellow race has succeeded in founding societies worthy of being called civilised.

Somewhat owing to Chinese influence, many Tartar tribes have tried to organise themselves in a regular way; but westward among the Turkomans, and eastward in the basin of the Amoor river, at the end of the last century, the Mongolians were still living in a primitive state. These were the people with whom La Pérouse entered into relation on the shores of Asia, as far north as Saghalien island. They lived still under the patriarchal system. Every family had its own chief. The people were very soft-mannered, and they professed a great respect for old men. Their dogs, which they harnessed to their sledges in the winter, were their only domestic animals. They lived mainly upon fish, which the women dried and prepared. In race and in manners they were quite the same as the people in Saghalien island.

The nomads of Khorassan, or, perhaps, to speak more accurately, the Turkomans, are still without government, and live almost upon a footing of equality. They form small groups of thirty, a hundred, or two hundred families, each having for their own debonair director a white beard, whom they respect; but in the community he exercises only the functions of counsellor and arbitrator.

These men are constitutional chiefs, they are governors who are paid for their trouble, obedient like everyone around them to traditional customs, not grasping at exorbitant power, which, as a matter of fact, would not be tolerated. The Turkomans say: "We are a people without chiefs, and we do not want any!" An organised despotism is rare among nomadic primitive people of any race, and there are few people who show as strongly as the Turkomans their love of equality and of individual independence.

Among the more civilised Mongolians who, spiritually, are subservient to the grand lama of Lhassa, and temporally to the Chinese government, social equality has now quite disappeared. Mongolian princes who boast that they are the descendants of Gengis-Khan, and sometimes great lama dignitaries, are at the head of the different ethnical factions. Below these princes is the caste of the nobles, proprietors of the soil, and equally the descendants of Gengis-Khan. This caste subdivides itself into other lesser castes, whose rank and dignities are transmitted hereditarily from father to eldest son, or in default of lawful child to the nearest relation. Lower still, below the princes, the nobles, and the clergy, is a servile mass, who, though they smoke with their masters and live under the same tent, are nevertheless slaves. Every noble Mongolian has the right to impose upon his slave any injustice he pleases; he has over him the right of life and death. There is no way of escape from this tyranny, for in spite of the nomad life which prevails almost everywhere, the districts are clearly defined and flight is not possible.

Everyone, princes, nobles, and people, belongs to a squadron, to a regiment; everyone serves under some banner. The princes hold their office under the government at Pekin; from there they receive their salary, and every three or four years they have to go to the capital to pay their homage. Mongolia is one large camp, through which a population of some millions of shepherds are perpetually wandering. They all serve in some military grade, and they form the reserve of the Chinese army—a reserve composed only of cavalry. It is a warlike colony, and all the important affairs are decided at the Foreign Office in Pekin.

But Mongolia does not pay tribute to China. She governs herself, in spite of the presence of a Chinese functionary beside the Mongolian governor at Urga, the most important of the embryo Mongolian towns.

Tartar customs, drawn up into a code by the Chinese government, serve more or less as guiding rules to the princes, who, by the way, trouble themselves little as to any notions of equity. The tax upon the cattle, which they lay upon their subjects, weighs only upon the poor. The fixed ratio is one sheep for every twenty, two for every forty; never more, though a man had ever so large a flock.

In its social organisations, Mongolia is as different from all the other half-civilised countries which we have already mentioned. The majority is made to work for the minority; but this minority does not now exact exorbitant privileges, and enforce them brutally. A man is slave or master by right of birth. There, as everywhere, man is a hierarchical animal; to command is pleasant to him, and it is not very irksome to him to have to obey. And as man is an intelligent creature, he will justify his mode of action by a hundred reasons. Why should not a noble have the predominance, since he is descended from Gengis-Khan! Why should not the slave humble himself? He has in him no heroic blood. As man is tractable and susceptible to hereditary tendencies, the inclination to impose his will, or to be obedient to that of others, is generally born in him. That is, assuredly, the principal reason which, all over the world, has prevented, and still continues to prevent, the servile classes from wringing the necks of their oppressors.

The masters by the right of birth are not the only men whose foot weighs heavily on the poor Mongolian. He has also masters by divine right, and of these it is more difficult to free himself; for these masters take hold of men by the ideal side of their nature, by their imagination, and hence they tyrannise over every inner thought.

A whole world of lamas is scattered over the plains of Mongolia. They have peopled quantities of rich monasteries, into which the faithful souls pour their offerings. Let us say, however, in praise of the lamas, that, more generous than the Christian clergy of the early centuries, they admit the slave into their ranks, which for him is equivalent to emancipation.

Though they devote their lives to divine objects, the lama clergy are far from taking no interest in matters passing round about them. The high dignitaries of their church are also great lords, and the koutouktou of Urga, a sort of Buddhist cardinal, owns round about the town nearly a hundred and fifty thousand slaves, forming a whole social category among themselves. In Thibet itself, the spiritual power, as the sectarians of positivism say, has completely absorbed the temporal power. From this confusion, so much to be regretted from a Comtist point of view, has resulted a curious theocracy, in which the principal abuses of semi-barbarous societies have become common, but in nowise improved.

As Father Huc judiciously observes, the political organisation of Thibet, as governed by the lamas, very strongly resembles that of the old pontifical states. The fountain-head of this clerical society is the papal lama, in whose hands, in principle, resides all legislative, executive, and administrative power. This terrestrial god, as we know, is an incarnation of Buddha. He cannot die, he only transmigrates. In spite of his immortality the grand lama has not the gift of omnipresence, he therefore delegates his most important functions, the government of his provinces, to dignitaries of a second order, to the koutouktous, or cardinals, who are sovereigns upon a smaller scale, enjoying a great independence, warring perpetually against each other, pillaging and burning the houses of the common people. As these holy personages are essentially men of peace, they do not fight themselves. For this inferior work they have smaller lay rajahs, who command a sort of warlike caste of men, the Zinkabs, and to them these lesser employments are given. Below these governing and fighting classes, there is a hard-working and peaceful class of men, the labourers, who provide their masters with food. They are ever liable to be robbed, they are badly fed and badly clothed; they place all savings into the pious hands of the lamas, who naturally keep a portion of it for themselves.

In case of dispute it is the rule in the lama tribunals that the priestly party should never be condemned. Exactions, contributions, and hard labour are showered down upon the poor; the soldiers, the rajahs, and the lamas despoil them without mercy.

All this while caravans of pilgrims are ever bringing to the great dignitaries of lamaism, riches, offerings, bullion in silver and in gold, a sort of Peter's pence, in return for which are given to the donors scraps of paper on which pious phrases are printed, statuettes of terra cotta, rags of old clothes which had at one time been worn by some holy saint.

We can prove that the Mongolian race is inferior to the white race, for the Thibetans have never known a Rabelais.

The way in which the Mongolians and the Thibetans carry on justice among themselves shows a level of a low order. With them, as with most poorly-civilised people, in Sinai, in Burmah, and elsewhere, theft is considered a much graver crime than murder. At the time of Marco Polo's wanderings, petty theft and larceny were in Tartary punished by bastinado which often resulted in death. But if a man was condemned for stealing a horse he underwent capital punishment—his body was cut into two pieces. There are in the community no magistrates entrusted with the power of repressing this punishment. The individuals injured, or their friends, must bring the guilty man before the judge. But every sentiment of joint responsibility is not ignored, for the inhabitants of an inhospitable tent are punished if they have refused to give shelter for a night to an injured and forsaken traveller.

In Thibet there are lama tribunals, but their jurisprudence is still very coarse, and often iniquitous. Theft is always considered the great crime. After he has been in prison for six or for twelve months the thief is sold as a slave, his property is confiscated, and sometimes this punishment will extend even to his relations. In case of adultery the husband may kill the delinquents if he finds them *flagrante delicto*. But the rich man ransoms himself from murder by paying an indemnity to the rajah, to the high functionaries, and to the family of the deceased. In case of insolvability the body of the murderer is sometimes tied to the corpse of his victim, and both are thrown together into the water. Nor is the judgment of God unknown in the Thibetan Himalayas; a man there will prove his innocence by drawing a piece of money out of a pot of boiling oil, or in holding in his hand an iron made red hot in the fire. Sometimes each of the parties will poison a kid, and the kid who lives longest will show justice to have been with his master.

If the Mongolian race had not given rise to societies of a higher order than those seen in Thibet and in lama Tartary, it would hold only a very humble rank in the hierarchy of the human kind; but from what we know of China and of Japan, we are authorised to assign to it the second place in the order of human races.

II.

Social Organisation in Japan.

In speaking of Japan, we are enabled once more to make an encouraging remark for the future study of sociology. It may be taken for granted that there was no direct communication between our societies of the Middle Ages in Western Europe and that of Japan. In both countries, the human agglomerations have evolved separately, and yet the two organisations are in reality identical. The Japanese feudal system is now a faithful image of what ours once was, and like our own, it is probably the result of conquest.

In the topmost rank of society, the emperor, the mikado, sits in all his glory, surrounded with divine attributes. A superhuman mortal does not act like ordinary men; a most majestic etiquette governs each one of his daily actions. In the year 1788, an irreverent fire made him run away in all haste, and for two days he was obliged to eat rice, which had not been very scrupulously picked. This was an event in the annals of monarchy! The mikado, the successor and representative of the gods, is in principle the proprietor of the whole empire. The principals and nobles own their parcels of land only as feoffees, and they enjoy only the usufruct of the soil. One of these great personages, under the emperor, the siogoun, shogun, or taicoun, allows the mikado to rest torpid in his half-divine solemnity of existence; and then the high official is enlaced so tightly in a strait etiquette waistcoat, that he makes his annual visit to his sovereign only by proxy.

Below these great dignitaries, there is a whole hierarchy of nobles of various orders; they are landlords of the first degree (daimios), who have granted smaller feoffs to their vassals of a lesser rank; and they in their turn under-let to the peasants. And if we include other nobles who possess no other feoffs, but their two swords, the samourais, a sort of *condottieri*, men in the service of the princes, we shall then have mentioned all the orders in the governing classes in Japan.

The whole nation divides itself into eight classes: 1. The

princes and nobles of the first rank. 2. Nobles of the second rank, who owe military service to their suzerain. 3. The Sintoist and Buddhist priests. 4. The samourais, or the petty nobility, who have lost their rank. 5. A class of citizens, comprising the subaltern officers and physicians, having the seigneurial right to wear a sabre, but only one. 6. Wholesale merchants. 7. Retail merchants and artisans. 8. Peasants and day labourers, who, for the most part, are serfs of the nobles, and are crushed under the burden of heavy fines.

Then after these eight classes, or rather outside every class, come the category of pariahs, who are obliged to live in particular villages; men who may not enter inns or other public places, and who are not included in the census of the population.

These outcasts are tanners, curriers, and others; all those who live by the preparation and the manufacture of skins.

In Japan the feoffs are hereditary, but in case of disinheritance they revert to the sovereign.

This society, so accurately divided, in which every member of the governing classes pays homage to his lord, and receives homage also from his vassal, in which the feoffs are possessed merely as hereditary usufructs, is in fact the image of our European feudal system, but it differs from it in one particular feature.

Suspicion is, we know, a characteristic of all semi-barbarous societies. The superior is everywhere considered as a man of a different mould to his inferior; he pretends to govern him, and he usually distrusts him. But in Japan this want of confidence has come to be a principle running through every system of government. The families of the princes were retained as hostages at Yeddo, and the princes themselves were obliged to live in this town six months in every year, or one year out of every two. In Yeddo the smallest details in their daily lives were governed by severe etiquette. Princes who were the owners of two neighbouring feoffs were not allowed to live together upon their domains, except when they were notorious enemies, for these high personages were constantly at war one with the other.

When they were over rich the siogoun would ruin them by

inviting them to dinner, and this entailed the necessity of a most sumptuous banquet in return. There was a body of police who kept a very strict look-out upon these important personages, and the proudest of the nobility were always glad to be enrolled.

The system of spies was carried even into a lower scale. The houses were divided into groups of five, and the chiefs were held responsible for people in the other groups. Every head of a family was bound to watch over that portion of the street near his house; and if asked to do so, was obliged to draw up a report, under penalty of fine, of whipping, or of imprisonment.

No one was allowed to change his residence without first having obtained a certificate of good conduct from his neighbours.

A man was bound to remain always in the same class of life in which he had been born.

This policeman-like system had been pushed to excess with the peaceful islanders of Loo-Choo, for here a watchful and suspicious eye was ever prying into the smallest actions of private people. The peasants were fearfully oppressed by the nobles, and lived in a state of misery; and the sovereign mikado, or miniature taicoon, after death took his place among the gods or kamis of the country. The populace, whom he had caused to tremble while he was alive, still feared him after his death, and offered up sacrifices to his shade, not to obtain good things, but that it might not injure them from the bottom of the tomb.

Nevertheless in this society, where tyranny is so carefully organised, where the social edifice rests only upon force and upon cunning, the administration of justice is no longer carried on merely according to the good pleasure of the rich and of the strong. There are tribunals in which judgments are publicly and solemnly pronounced, and from whose decision there is no appeal. Crime cannot be ransomed by money; impunity to rich men cannot be guaranteed. The penalties are not numerous, but they are severe. They are: forfeiture of office, imprisonment, banishment, confiscation, and death, which is often accompanied by torture. Capital punishment always carries with it confiscation of property.

Penal joint responsibility was invoked in Japan as in ancient

Egypt. A man was adjudged guilty of crimes that he did not prevent, and was often punished for the faults of others.

It would be idle to deny the influence that the institutions and the government of a country have upon a people who live under them; but still every race, and every ethnical group of men, do not move in the same way. Despotism will for ever quell some people, others will resist it. It would seem that the Japanese belong to the latter category.

All travellers are agreed in allowing to the Japanese energy, pride, and independence of character; and the remarkable efforts that they are now making to introduce European civilisation into their country, show clearly that they at any rate have not been altogether enervated by feudal and inquisitorial despotism.

XII.

Society in China.

In our short sociological review we have already noticed many ethnical groups belonging to very different races. Some of these were savage, others barbarous or semi-barbarous, and some few more or less civilised. But in these societies, various as they are, we recognise one feature common to them all: the predominating quality is a shameless and undisguised egotism. In all these societies the social formula is nothing else than a more or less savage organisation, a more or less intelligent system of government by force.

We find in China, and for the first time, a society governed by higher and better motives. Not that here despotism is extinct; the people are still immersed in it. But the conscience of the governing classes has grown wider and has become more enlightened; the best of their members have felt keenly the desire for the general welfare, and to their lasting honour they have endeavoured to put the reins of government into the hands of the more intelligent men.

The ideal of the Chinese society is no doubt still far from just. The idea of government hitherto realised is no more than an enlargement of the system that prevails in every family. The mass of humanity is still regarded as a collection of children and minors, for whom guidance, protection, and chastisement are necessary. The general welfare is much considered, though the governing powers are careful not to allow to the individual any stretch of liberty of which they do not believe them to be worthy. They endeavour to put every man into his proper place in the social hierarchy. They wish that the nation may be governed by the most intelligent of its members, by mandarins, chosen after competition, no attention being paid either to caste or to birth, and to whom authority will be measured out in proportion to their merits.

As is customary, we find at the head of society a monarch hedged in with divine prestige, a son of heaven, raised immeasurably higher than other human mortals. He makes or unmakes the law; he chooses or dismisses the mandarins; he has the power over life and death. The forces and the revenue of the country are in his hands. Everything reverts to him; everything proceeds from him. Upon his accession to the throne the principal personages in the country bring their daughters before him, that he may choose from them five wives. One of these will be the principal wife, and her children, all other things being equal, will be preferred to the children of the other wives, when the son of heaven appoints his successor.

But the imperial power finds a barrier to his will in the vast hierarchy of scholars. In theory the emperor is the chief of an immense family, "the father and the mother of the empire." It is he who appoints all the administrative functionaries, but he has to conform to a law higher even than himself. In theory he is all powerful; but he cannot, in point of fact, choose his agents except from among men of letters, and according to the classifications established by graduated competition, divided into three series; all of which are open to every Chinaman, and which are the only means of opening to him an administrative career.

It is the idea of family that regulates all these social distinctions. Each functionary possesses a certain portion of paternal authority. The dissertations of moralists and philosophers, the allocutions of mandarins, the proclamations of the emperor, are ever exhorting, commenting upon, and exalting filial piety. Family sentiment is a fundamental virtue which is often pitched to the height of passion.

As a good father of a family, the emperor of China is strictly obliged to give assistance to his children. It is his duty to keep full the garners of plenty, where in case of famine the needy may find, at a small cost, barley, rice, and millet. An imperial edict of the year 1260 declares that aged men of letters, orphans, those who are abandoned and have no home, the sick and the infirm, are the population of heaven. But the theoretical father of 300,000,000 of Chinamen is himself watched over and lectured. During his lifetime "the son of heaven" is closely looked after by duly-appointed censors, who, under peril of their life, are bound to bring him back to his duty, if he departs from the right path. When the emperor Thsin Chi, 212 years before our era, in a fit of madness commanded that all books containing political or religious laws—in fact, all the historical traditions of China—should be burned, 460 scholars in Pekin alone preferred to be burnt alive than approve of such aberration on the part of their sovereign. The imperial censors have many a time honoured themselves by similar acts. When the emperor is dead, his life and his reign are judged, and a posthumous decision, laudatory or critical, is henceforth attached to his name. During his lifetime he is even sometimes held responsible for earthquakes; and on these occasions, as was the case in 1069, he is asked if there has not been anything reprehensible in his conduct, or any abuse in the government, that ought to be reformed. Even nowadays these censors will present to the emperor a critical examination upon this or that act in his public or private life, and these minutes are published in *The Pekin Gazette*. In a word, the emperor of China, who is also "the son of heaven," and before whose throne no one may approach without knocking his forehead nine times on the ground, cannot choose any petty government officer except from the list of candidates composed of the names of men of letters.

In China, the hereditary titles extend only to members of the imperial family, and to those who are the more or less probable descendants of Confucius; but these distinctions are purely honorary. In reality, in the heart of the empire, the governing classes are formed solely from the men of letters; and with the help of competition, these men are chosen from the whole population of the empire.

It may be worth our while to consider for a moment this curious organisation, of which the object is to place the reins of government into the hands of the most intelligent men.

The foundation of the class of scholars in China dates back as far as the eleventh century before our era, at a time when all Europe was still sunk in the most coarse barbarism. In principle the corporation was chosen by universal suffrage; the mayors only are now chosen in this way. Literary competition was substituted for general election in the eighth century.

There are minute regulations to prevent every kind of fraud in these examinations; for candidates crowd into them by thousands. Examination is the only means by which a man can obtain government employment; everywhere in the administration, from the top to the bottom, the offices are purely civil. The military chiefs have authority only over their subordinates; the troops are encamped a long way from the large towns, and do not come in except upon the invitation of the civil mandarins.

The idea of confiding power to the most intelligent men reflects honour upon the nation who first conceived the idea; but unfortunately, Chinese knowledge is very backward, and, as a rule, mnemonical exercises are the only tests required of the candidates. The greater number of quotations from ancient authors found in the examination paper of the candidate, the greater merit he is supposed to possess.

However this may be, the examination-rooms are crammed with candidates. At Ning-Po, Sinibaldo de Mas has seen three thousand aspirants try for thirty-seven places.

These trials are made gradual, and the ambitious scholar may have to wait in expectation for many years. There are four degrees that may be mentioned: bachelor, licentiate, doctor, and professor. These must be gained in order, and each successive step gives to the holder of the office a more important position. But for several years the capability of the individual is constantly tested by additional examinations. There is no competition for special functions, each mandarin may occupy any place corresponding with his rank.

In theory this organisation of a class of scholars is very remarkable; it is the greatest effort that has ever been made to realise the true social problem: to everyone according to his works. In actual practice, and in spite of rules, the institution has very much degenerated. Grades have been bought, judges have sold their opinions. In point of fact, like everything else, this wonderful organisation has become altered; but the largest ethnical agglomeration that has ever existed, after having founded, maintained, and governed three hundred millions of people for thousands of years, after having in this enormous collectivity so uprooted all idea of hereditary privilege—the Chinese have learnt, not without much astonishment, that in Europe there was an hereditary order of nobility. In China, the grades equivalent to our titles of duke, marquis, count, baron, and knight are given to those civil and military mandarins who have distinguished themselves in their administrative functions, but these titles are held only for life. They can never be transmitted to the descendants—rather to the progenitors—for it would be against all rule and principle to contradict in any way the fundamental maxim, that a son should be better qualified than his father.

Every mandarin of high rank convicted of negligence in the fulfilment of his duty is made to go down two degrees in the social hierarchy, and is for two years deprived of his salary. To obtain from a great personage by undue influence a flattering notice in a report made to the emperor is considered a crime both for the officer making the report and for him who is praised. If connivance is proved, the higher functionary is beaten and is exiled; his creature is punished by decapitation and by confiscation of his property. Instead of protecting its officers, the Chinese government is suspicious of them, and holds them aloof. If a popular sedition breaks out, the governor is invariably deprived of his situation. The people, it is thought, are harmless; there must therefore be either an abuse of power or great incapability.

It is also forbidden to functionaries holding jurisdiction to acquire land in their district. In principle the Chinese administration is conceived with much good sense and a great desire for justice; but nowadays, in actual practice the government departs very widely from the old ideal laws. In China, as in many other countries, the men are not so good as their institutions; but these institutions have so much prestige attaching to them that the Chinese army counts only 10,000 men, a very insignificant military force for such a vast empire.

China is certainly the only country in the world in which military glory is scoffed at, and in which the profession of arms is held in small honour. A known Chinese philosopher, and one quoted in the world of letters, has said: "Give to the conquerors only funeral honours; greet them with cries and wailings in memory of their homicides." In these latter years, after the injustice and insults that the Europeans have imposed upon China, these people have somewhat recovered from their disdain for war and everything connected with it. Europe may one day become more aware of the fact.

In the interior government of the country the authority of the mandarins is still mainly moral. In theory every government is as the father of a family; and in this way their administration is naturally over minute and harassing. Confucius has laid down that the conduct of children should be carefully watched: the officers "ought to look after the people as they would after a son." But this administrative tenderness cannot exist without jealous supervision. In the fourteenth century a functionary used to come every evening into each inn, and then bolt the door upon every traveller after first taking his name. The next morning he came back to make his call, and the travellers continued their journey under the care of a guardian, who at the next station entrusted them, giving at the same time a receipt into the hands of the next functionary. At the same epoch, a signboard fixed upon the door of each house indicated numerically all the inhabitants, and even the number and the kinds of animals that were kept.

But in China people are far removed from the intolerable tyranny practised in the countries where there are hereditary castes of men. Every Chinaman may follow the profession that he pleases without need of authorisation, and even (a most enviable thing!) without having to pay for a patent. It is lawful for anyone in China to print, to sell, to distribute books and pamphlets, to placard advertisements and loose sheets. The only associations forbidden are those which attempt to overthrow the dynasty. In this respect, as in many others, the Europeans have still a good deal to learn from the Chinese.

The imposts are not numerous in the Celestial Empire, and they are established upon a very simple basis. A tax upon salt, of which the Chinese are very fond, is the principal duty upon food. All the other taxes are unchangeable, they are laid upon landed property and affect the merchant and the artisan only by indirect means.

China has not, like some countries in Europe, pushed to excess the mania for centralisation; its 300,000,000 inhabitants scattered over a vast extent of country would be inconveniently lodged. Each province, under the authority of a governor, forms in itself a small state; it has its own customs, levies its own taxes, and makes its own contracts. In China, as elsewhere, the nation is but the result of the grouping together of small agglomerations of ethnical unities, each having dissimilar tastes, different needs, living under conditions essentially unlike, and all of which it would be absurd to congregate together under the rule of one uniform despotism. How many of our statesmen would do well to go to China to learn this very simple lesson!

In order to govern a country well it is first necessary to know it thoroughly and accurately. The reigning dynasty in China has ordered that a general report of the empire be drawn up, containing minute details as to the population, products, topography, the towns, the fortifications, the temples, the examination-rooms, etc.

This report fills from two to three hundred volumes. Monographs of each province, department, or district help to complete this huge collection.

Besides those rites and customs which age has consecrated in the country, there are written constitutions, modified from time to time, which determine the general basis of government.

The organisation of Chinese justice is also most interesting; like everything else, it rests upon the notion of a family upon an extended scale. The condemned man must thank the mandarin for the chastisement he has been kind enough to inflict upon him; for after all it is only a paternal correction, and the judge was very sorry to have been obliged to enforce it.

Attempts have been made to reduce the number of causes in the courts of law. Each group of a hundred families have a chief; he imposes the taxes and he is made responsible for a great many offences. Under penalty of bastinado he must look to the proper cultivation of the land. To give his people a dislike to lawsuits, one emperor went as far as enjoining the tribunals to harass the litigants, who would therefore be driven to have their differences and their disputes settled before the mayors in their own districts.

To prevent acts of personal violence, and abandonment of people in distress, the government has created responsibilities which often bear very unjustly. If for instance a man dies in a field, or upon the doorstep of a shop, the proprietor of the land or the merchant will undergo a penalty.

Every murder entails upon the murderer capital punishment and deprivation of funeral honours; for his relations it means ruin and dishonour. But as a man is held responsible for a suicide of which he has been the means or the cause, it is by a threat of suicide that the weaker will often punish the stronger adversary; it is by suicide that he takes his revenge. For a spoken word that has given pain, for an affront, etc., a man will hang himself, or throw himself into the bottom of a pit. Suicides are very common; they are thought to be honourable and glorious in public opinion.

The Chinese idea of justice is quite utilitarian; it is unrefined, and not at all metaphysical. Punishment is not measured according to the moral gravity of the offence, but according to the extent of injury caused. The penalty exacted for theft will be in proportion to the value of the object stolen.

The punishments inflicted are brutal; they remind one very strongly of primitive justice. They are: bastinado, which in China is very often used; slaps with thick broad pieces of leather; small iron cages, in which the prisoner is crouched up; imprisonment; banishment into the interior of the empire, or exile into Tartary; death by strangulation or decapitation. Formerly slavery was also a Chinese penalty.

A high judge forms the whole tribunal. He questions the prisoner, who remains on his knees before him; and there is always an executioner, who, upon the judge's order, beats the witnesses for the prosecution or for the defence, when their answers are displeasing to his lordship. There is no advocate; but the magistrate may, using the discretionary power given to him, allow the parents or the friends of the prisoner to plead his cause.

In other respects the criminal legislation of the Chinese is considerably advanced, at least in theory. It admits extenuating circumstances, a non-retroactive effect, the right of appeal, temporary liberty under the responsibility of the magistrates, merger of punishments, and the right of mercy, which is reserved to the sovereign.

On the other hand the duty of a judge in China is not always soft and pleasant. Measures are taken against him, and he is threatened with severe chastisement. If he marries or takes a mistress in his district he is liable to have eighty blows; and double the number if by chance the wife or the concubine is the daughter of a litigant over whom he has to pronounce a decision. For an erroneous decision given, even upon an appeal case, the judge is beaten.

Beatings here and beatings there, the stick plays a very important part in the administration of the Celestial Empire. It is however the old family maxim: "The wise father chasteneth his son."

But everything is not childish in this vast organisation. China shows the most ingenious and the greatest social effort yet made by the Mongolian race. It is the largest human agglomeration that has ever existed; it has lasted for thousands of years and has for a long time occupied the foremost rank in the human race in the perpetual march onwards towards a more perfect condition of life. It is the first large society which, to its everlasting honour, has for ever broken the mould of castes and abolished hereditary privileges. And it is still the only country in the world in which, in theory at least, a systematic attempt is everywhere made to give to individual merit its full reward. There is much grandeur in the idea, certainly much fairness; and it will assuredly form the basis of better societies which we may hope to see in future ages. China may have awkwardly realised her conception, but to that conception she has certainly owed her ancient prosperity. Modern Europe has slowly entered upon the same path, and for the most part clumsily enough. In Europe, as in China, candidates for examination are tested by the retentiveness of their memory more than by their powers of originality. And even if this method of trial were better understood, man's intelligence would count for much, but it cannot include everything. There is behind it that which supports it: character, generous sentiments, every moral force so absolutely necessary to one who has the care of souls. Now, all this side of the human creature, so difficult to appreciate accurately, has nothing to do with school or other examinations; and yet it is that which vivifies man's intelligence and his activity; it is that which we must first grasp if we wish really to test his value.

Another objection, quite as serious, must be brought against the Chinese civilisation; that is, its besotted infatuation for what is past and gone. After it had organised its class of men of letters, China found herself very superior to her neighbours; she sanctified herself in her own eyes; she denied the progress of which she was herself such a remarkable instance, and then decreed absolute immovability. Henceforward everything in China was congealed; over minute rites were made to govern the smallest details in social life. Before the war of 1840 a merchant in Canton bethought himself of putting a European-shaped rudder on to his boat, but before the skiff was launched a mandarin, a dragon in the performance of what he thought was his duty, ordered the boat to be destroyed, and imposed a fine upon the irreverent innovator. Facts like these betray the character of the people; it is the "laudator temporis acti," beyond which nothing shall go, held up as an eternal maxim. We may be quite sure that in pulling to pieces this haycock civilisation we shall find that the rough aggressions of the Europeans have rendered service to the Celestial Empire.

Nevertheless, the edifice of Chinese society has its good points, and if we make certain allowances, there is much truth in the eulogious judgment of one who is certainly well qualified to give an opinion on Chinese manners and customs. Milne says: "The political organisation of the Chinese empire, now more than four thousand years old, is the most philosophical, the most rational, and the most free from prejudices of all sorts that has ever existed in any age or in any country in the world. That is the reason why it has lasted so long."

However this may be, the Chinese form of government is the most advanced social system yet realised by any Mongolian race. It is despotism tempered by reason; and to find a political organisation of a higher kind, we must look among the most civilised branches of the white races.

XIII.

Societies among the White Races.

The different ethnical groups of the white races are now classed mainly upon a linguistic basis, in default of less uncertain knowledge, that history and anthropological anatomy cannot give to us. Ingenious as it is, the classification by means of modern languages is open to criticism, if we wish by it to determine to what ethnographical group or what race man actually belongs; nevertheless, all things considered, we may accept it. Moreover, it is convenient and it is generally allowed. We shall therefore follow it, dividing the white race of man into three great branches—the Semitic, the Iranian, and the Aryan; and we shall endeavour to describe what has been the social constitution in each of these races.

A. *The Semitic Societies.*—Like all the human races which have played a large part in this world's history, the Semitic race is divided into a certain number of ethnical groups, of which each has had a different fortune, and has become more or less civilised after its own manner.

But a notable portion of the Semites now extant have hardly, as yet, passed the pastoral and nomad phase; and if we may trust to induction and to tradition, we may reasonably suppose that such, in far-distant times, was the special characteristic belonging to the whole race.

At the present time the Semites are still grouped into small tribes, forming a sort of large nomad family under the orders of a chief. Some of their tribes have shown such a strong love for this wandering life, and so much aversion for every sort of progress, that in the tribe of the Nabatæi, according to Diodorus Siculus, it was forbidden, under penalty of death, to sow corn, to plant fruit-trees, or to build habitations.

But the Semitic and wandering tribe contains germs of the elements belonging to the ancient monarchies. Slavery is in full force; a special magistrate, a cadi, is entrusted with the administration of justice. This chief is a respected monarch, whom people do not approach without first kissing the ground, and whose power is transmitted hereditarily to his eldest son. This chief is especially a military chief; no razzia is ever undertaken without his authorisation. Such was also probably the primitive organisation of the Hebrews, who were divided into *tribes*, *families*, and *houses*.

Here and there the tribes would league together; here and there they were subjugated by a chief, who governed despotically over a certain number of clans. Then the people betook themselves to agriculture, they built towns, they founded monarchies like those of Assyria and Babylon. As far as we judge, from such confused and incomplete knowledge as has come down to us, these ancient monarchies were in nowise original, from a sociological point of view. The crown was then hereditary. The monarch enjoyed almost unlimited power. Society was, as it were, immovable; professions and trades were always handed down from father to son.

It is in the Bible and in the Koran that we must look for precise and detailed information as to the way in which the Semites, in gradually becoming civilised, planned their social organisation.

Among the Hebrews, when one chief of the bands, more fortunate or more artful than his rivals, found himself in the position as commander over a greater or less number of tribes, he was usually both priest and king; for the Beni-Israelites were super-eminently bigots. Melchizedek was the most high priest of God; Saul offered up a holocaust, etc. The government among the Hebrews was always essentially theocratical; the chiefs were the lieutenants of Jehovah.

In Judaea there was an hereditary aristocracy. The Pentateuch tells us of a council of elders, of judges, of scribes, who were the representatives of families and of houses. It was a council of this kind that went to Samuel to ask for a king; it was this council which, at the request of the father, caused the son to be stoned; it was this council which heard the plaint of the husband who pretended that he did not have the first offerings of his wife—a crime that also entailed lapidation.

The judges were the monarchs, and their power was not hereditary; but it became so after the days of Solomon. The despotism of the kings of the people of God was not less severely exercised than that of other Semite kings. We know how David acted towards Bathsheba; his son Solomon no sooner mounted on the throne than he committed fratricide and other murders. The fastuous splendour of this monarch, so celebrated it would appear by his wisdom, his dealings with foreign women, his crowded harem,—these are all facts which we may do well to recollect.

The theocracy which came from the great Arabian monarchs was much of the same kind. Before Mahomet, there was in Arabia only a few small confederations of tribes; but with the assistance of fanaticism, all these elements grouped themselves together around the authority of the prophet. Then servility, which, as we have seen, was customary in all tribes, soon developed in the most shameful manner. The subjects of Mahomet used piously to keep every drop of water which he had used, they used to collect the hairs which fell from his head; they used even to lick his saliva! But Mahomet was no more than a theocratical judge, as also was Moses. After him the first caliphs were still elective; but after Moawiah the caliphat became hereditary. Then a systematised policy of oppression bore upon the whole race, and still more heavily upon the conquered people, who paid to the master a heavy capitation tax in return for the great kindness with which he allowed them to cultivate their lands, which by right of conquest belonged to the conqueror.

The great majority of the Semitic race has never conceived, either in ancient or in modern times, a social ideal more elevated than that of a mute despotism, very similar to that which has been established all over the world in the larger primitive agglomerations of men. But perhaps we may make an exception in favour of Carthage, and of Tyre, the metropolis. This fact has its value, for it will suffice to establish the sociological perfectibility of the Semites.

If we may believe Aristotle, Polybius, Diodorus, and others, Carthage enjoyed a well-organised form of democracy. Notable families were established there—the Magos, the Hamilcars, the Hannos, etc.—but there was no hereditary aristocracy. A sort of senate was elected by the pentarchies or electoral colleges. With the exception of judicial authority, every other power was vested in this assembly. Between the sessions it was replaced by a commission chosen of its own members. This commission appointed a few delegates or ministers to exercise the executive power; their chief or suffetes was named president of the assembly. A military suffetes, or vice-president, had the command of the military forces. The two suffetes were chosen by the people out of a list drawn up by the assembly, and their command was only nominal, except in the case of prerogative. The military suffetes were also held in check by a committee named for this purpose by the general assembly. This committee supervised the operations of the suffetes, and decided upon his prorogation or upon his recall as they thought expedient.

The existence of a political constitution both so republican and so complex among a people of the Semitic race, very backwards in other respects, may naturally cause us some surprise; and we can hardly credit the testimony of ancient authors, inasmuch as this democracy rested, like every Semitic civilization, upon slavery, of which we must now say a few words.

The institution of slavery is inherent in every inferior civilization. Before men can understand that there should be in liberty, in freedom over his own actions, a right not resting merely upon the law of prescription, he must necessarily have first arrived at a higher degree of intellectual development; his heart must first warm his intelligence, and his intelligence must first warm his heart.

We find slavery at the bottom of every Semitic society, ancient or modern, nomad, monarchical or republican. The romance of Antar shows to us how in ante-Musulman Arabia the slave was absolutely submissive to the will of his master, who kept his life in his hands, who naturally had the right of violating every woman slave. But the male slave may be ennobled by the chief, and then enjoy the rights of a free man.

Nor do we find that the Koran thinks of protesting against slavery. The enemy conquered by the faithful are either put to death or else brought into servitude. This latter was usually the fate among women and children.

Slavery was common in the ancient Semitic kingdoms of Assyria and Babylon; it existed also in the republics of Tyr and Carthage. In the Homeric ages the Phœnicians used to make raids upon the slaves on the coast of Greece; they then took them into Egypt and sold them. Herodotus tells us how they ran away with Io, the daughter of Inachus, king of Argos. In Spain the Carthaginians used to make the Iberian slaves work in the mines; they overwhelmed them with blows, and they exhausted them with work. At Tyr there was a whole population of slaves, and we hear even of servile revolts.

The Jews, the people of God, had their slaves. When he was much in want of money the Hebrew father might sell his daughters as slaves. The wretched populace might even sell themselves

The thief, incapable of paying his fine, became the slave of him from whom he had stolen. Prisoners taken in war were slaves. But even the Jewish conscience had some scruples about slavery. We read in Exodus that the slave be set free at the end of six years, but the master has the right to keep his wife and children. There were at this early period certain laws protecting the slave; but the master might with impunity beat the slave to death, on the condition that the slave lived one or two days after his beating, for "the master had bought him with his money." We have now said enough to show that morality was at a very low ebb among the ancient Semitic races.

Justice, as administered by the sons of Shem, was also very coarse. Among the ante-Islamite Arabs the chief of the tribe was often invested with judicial power, and he decided the differences according to his own good pleasure. But still the poem of Antar speaks of cadis who served as arbitrators between the chief and his followers. Among the Hebrews, judicial matters were decided either by the elders or by the king. We find the talion law exacting "an eye for an eye, and a tooth for a tooth," to prevail everywhere; and during many ages it must have been the only code known among the nomad tribes. By degrees, as societies established themselves and became more civilized, legislation grew to be more complex. The Hebrew false witness was made to bear the same penalty that had been enacted against the innocent man whom he had accused. For every goat stolen a man had to restore four goats, for every ox five oxen. For voluntary homicide the talion law was in force; for involuntary homicide a pecuniary compensation was received. For blows and personal injuries the injured man might exact the same punishment upon the culprit. The greatest of all crimes among the Hebrews, the crime for which there was no remission, was idolatry. If an individual was convicted he was stoned; if a town was so convicted it was anathematised, all the inhabitants were put to the sword, and the town was burnt. Certain incests and sodomy were punished with death.

Capital punishment was ordinarily by lapidation, sometimes by fire or by decapitation. For smaller offences a bastinado or the rod was very common. That, however, was considered only as an admonitory punishment, and there was no dishonour attaching to it.

Like the Bible, the Koran allows the talion law, and also pecuniary compensation. For instance, twenty camels for an involuntary murder. But thieves may also be liable to having their hands cut. The penal code of Mahomet is not original, for its author declares that he drew it from the Bible. In addition to the penalties enacted by the Koran the modern Mussulmans have added punishment by impalement, often inflicted in Arabia on the Bedouins, who rob the pilgrims. But for the majority of crimes the social justice of the Arabs is slack. In Bagdad the punishment of a murderer is generally decided by the relations of the victim: it is a case of a private quarrel in which those not concerned are careful not to meddle.

On the whole, we see that the Semitic race, taken in a mass, have freed themselves from savagery, but not from barbarism. But on one point we must give them justice. Some of their ethnical groups—at least if we may believe the ancient writers—seem to have been the first in the world who have organised republican governments with deliberative assemblies, to have separated the legislative and the executive powers. That is a sociological progress of the first order, and one which the superior races only have realised.

II. *The Persians.*—In a treatise upon sociology, we may be brief in speaking of the Persians, both ancient and modern, for their institutions seem to have been very unoriginal.

In spite of a certain moral dignity, which we may discover here and there in the Zend-Avesta, amid a thousand inanities, the social ideas of the ancient Persians never went beyond the most absolute despotic monarchy, or the institution of castes.

Even under the legendary government of Djemschid, the population in Persia was divided into four castes, which we now find in India: the priests, the warriors, the labourers, and the artisans. We need hardly say that, as in every society of this kind, there was outside the legally-recognised classes a large number of slaves.

From what Greek writers, and Herodotus especially, tell us of the ancient Persian monarchy, we may easily imagine what it was at a still more remote date; for it is not in going back through the course of ages that we shall have most chance of finding, in the annals of humanity, any notions of justice and of liberty.

At the time of Cyrus, Xerxes, and others, Persia was under the most unalloyed form of despotic monarchy. The king of the Persians was, above every other, "the grand king;" his smallest caprice was law. When the Lydian Pythias allowed himself to ask of Xerxes, when he was preparing to invade Greece, the favour of keeping his eldest son by his side, offering at the same time as soldiers his four other sons, the king of kings caused the man for whom the intercession was made to be cut into two halves. During the march, as the Persian troops defiled before their master, the rod was constantly held over them.

In spite of centuries gone by, and in spite of the introduction of Mahometanism into Persia, the kings in this country still exercise a power quite as despotic as was that of Xerxes. They have the right of life and death over all their subjects, and over the members of their own family. Fraser saw a young Persian prince accustoming himself to go about blindfolded, for he said: "When our father dies we shall all be put to death, or else deprived of our sight, and I am trying to see what I can do when I am blind." The whole Persian code of laws is resumed in this short phrase: "It is the will of the Shah." Below this tyrant in chief, there is a whole hierarchy of sub-tyrants, men who are slaves on the one hand, and despots on the other. At the bottom of the social scale is the class of cultivators who have to bear a whole mountain of injustice. These men are ill-treated, robbed, ransomed at pleasure; they have to defend themselves as best they can against the tax-collectors, by every cunning device that they can invent.

In a country such as this, justice is an empty word. When there is a shadow of it, it is ever the same talion law that has existed for centuries. The thief may be pardoned by the man from whom he has stolen. The heir of a murdered man may commute with the assassin, or kill him, as he pleases. In Fraser's time, the murderer of a young man was given over to the mother of the man murdered; she drove a knife into him fifty times, and then cut his lips with the knife all stained with blood!

There are magistrates, cadis, or governors, who sit in judgment and inflict fines, bastinadoes, strangulation, extraction of the eyes, or decapitation. But as nearly all the Persian functionaries receive bribes, crimes are easily washed out for a few tomans.

This frightful state of things, which has probably lasted through all historic ages, from the earliest times, has produced a general debasement of character in the whole nation. The people are obsequious, servile, stealthy, cruel, and bigoted. For the institutions of a nation, when they have lasted through a long series of generations, at last mould the people according to their image.

C. *The Vedic and Hindoo Societies.*—If the Persian branch of the so-called Aryan race offers to us but little interest as regards the constitution of societies, it is very different when we turn to the Indo-European branch, properly so called. On the one hand, this fraction of the human species is the one of whom there is most record in past ages, and consequently we may retrace its evolution; and, on the other hand, it comprises the races of men who are the most richly endowed, the men with the largest brain, those who have struggled the hardest, who have been bolder than others, and who have had the gift of invention most strongly developed.

It may be doubtful to maintain the theory that the Aryan races have originally all sprung from one district, from one creative navel, situated in the Hindoo Koosh mountains, or in the plains of Pamir; but it is certain that in these regions man has soonest arrived at any large degree of moral, social, or intellectual development. This is what will now first claim our attention.

The Vedic hymns furnish us with much information upon the social condition of the Aryan tribes of Central Asia, before the time of any large emigration, either towards Europe, or along the valley of the Indus towards India.

There was then no caste known among the Vedic tribes. Certain social functions were marked out, but there was no impassible barrier line between them. The priest who was entrusted with grinding the sôma, or with lighting the divine fire, did not disdain to till the ground, to tend the cattle while they were grazing; he even went into the middle of the battle by the side of the Kshattriyas. There were the Brahmins, the Kshattriyas, the Vaicyas, but between these classes there was no hierarchy, no subordination was necessary. The servile class, that of the Çoudras, did not exist, for the Vedic Aryans had not yet become conquerors.

It is now somewhat a feudal organisation. The Aryans are divided into small groups, into tribes, who are obedient to a chief owning upon a hill a fortified habitation, from which he governs the Vaisyas, or Viças, agriculturists or shepherds, scattered over the plain. This chief is a wealthy man. Among the Vedahs richness is the condition of royalty. The master is surrounded with sumptuous splendour. He mounts an elephant, or a gilt chariot; his head is adorned with a tiara, or crest of diamonds; he is covered with precious stones; a troop of Kshattriyas compose his suite. In the Vedic society the spiritual power shows but poorly enough. The Brahmin seems generally to be reduced to performing the part of a chaplain, or the flatterer of the petty barbarous king. Nevertheless he clothes the king with religious investiture, he crowns him, he tries to extort from him presents by selling to him his prayers at exorbitant prices. He prays the aurora to "grant a wholesome plenty to the noble lords who have showered presents upon him." He extols in pious hymns the horses, the cows, the chariots, the jewels that the royal munificence has bestowed upon him, and he prays that Indra and Agni may reward the generous donor a hundredfold.

These sovereigns, gracious as they are towards the representatives of the gods, are invested hereditarily with the exercise of power, and sometimes they organise a complete feudal system, recognising a suzerain, the great king, the maharajah.

Among the Vedahs there is no question of slavery. It would seem that long before the conquest the people, the artisans, the labourers, and shepherds, grouped together in villages at the foot of the strong castles, were only vassals, and were treated mildly enough. There is nowadays a considerable degree of familiarity existing between the princes and the subjects in Nepaul, in Rajpootana, and in Peshawur. But when the Vedic Aryans became more established in India as conquerors, they altogether modified the constitution of their societies. The feudal kingling became a powerful and absolute monarch. Three Vedic classes were changed into hereditary and isolated castes. The conquered people formed a servile caste, that of the Çoudras, below which there were half-castes, or unclassed groups of men, who constituted a class apart in a still lower scale. Then the Brahmins became glorified creatures; they grew to be demigods.

The Manu Code describes minutely this theocratical society, in which the king is in principle the secular arm of the Brahmin, in which also the spiritual and temporal powers are so completely distinct as to give delight to any disciple of Comte.

In the first place the king receives from a Brahmin the divine sacrament of initiation. After the ceremony he is consecrated in every sense of the word. Anyone that shows ill-feeling towards him must perish. The principal duty of this sub-Brahmin monarch is to chastise, for the Brahmins have gone so far as to deify chastisement; it has become a genius created by God. From daybreak the king must present his homages to the Brahmins; he must especially give a great deal to the Brahmins. Every present made to a Brahmin who is a good theologian carries with it infinite merit; but to a man who is not a Brahmin it has only an ordinary merit. A king, even if he is dying for want, must never receive tribute from a Brahmin who is versed in Holy Writ. A king ought to be very careful not to kill a Brahmin, even though he be guilty of every possible crime, or to confiscate his property. There is no iniquity in the world greater than to kill a Brahmin; the king ought not to be able even to conceive the idea.

The pious monarch chooses seven or eight ministers, with whom he examines all affairs of importance; but when he has formed his decision, he ought always to obtain the sanction of a Brahmin.

The country is well organized, the communes are grouped according to a decimal system, by ten, twenty, a hundred, a thousand; and each group has a responsible chief dependent upon the chief of the superior group.

The king imposes a tax upon cattle, upon the annual savings, upon the produce of the soil. The Vaicyas pay annually a fine comparatively very small. The poor Çoudras, who have no property worth taking into consideration, pay their tribute by one day's labour in every year.

We can easily understand that the privileges and duties in every caste are very unequal. The three first castes only participate in the social advantages, but very variously. The Brahmin shines in his splendour at the head of all society; he is the chief and the proprietor of everything that exists. The Kchattriya must protect him, the Vaicya work for him, the Çoudra must serve him. The Brahmin may beget an adulterous child, if he undergoes a purification lasting for three days; but if his wife is unfaithful to him, the king must order her to be devoured by dogs in the public place, and her accomplice to be burnt upon a bed of red-hot iron.

The Brahmin has the right to compel the Çoudra to serve him; he may rob him with perfect safety of conscience. To collect riches, in any way, however honestly, is forbidden to the Çoudras.

In the Indian society the administration of justice is already organised; but in this, as everywhere else, the Brahmins have the upper hand. The king may deliver justice in person, but he must be assisted by some Brahmins. He may delegate his judicial authority, but only to the Brahmins. The cases are regularly conducted; witnesses are heard, after they have been harangued and after they have taken oaths. The offences are few in number. Theft and adultery, assimilated to theft in most primitive societies, are the offences which most concern the legislator. The fine to be paid for stealing will vary according to the caste of the thief; but here, strange to say, it increases in proportion to the dignity of the guilty person.

In spite of this peculiarity, which may cause us some surprise, the Indian society, as it is explained in the Manu Code, does not differ, fundamentally, from those of other larger primitive monarchies which we have already described. It is an organisation of the most complete despotism, for the benefit of a privileged class, in the hands of whom the king and the military caste are only the instruments. Because of the indispensable services which it fulfils, the working and commercial caste still enjoy certain rights; but the servile and the mixed castes are considered merely as domestic animals.

After they conquered India, the Vedic Aryans seem to have sociologically degenerated. They have, no doubt, founded vast empires; but they have exchanged their relative liberty, which was enjoyed by these primitive people, for monarchical servitude and the rigid immovability of castes. Nearly everywhere, as we shall see, the Aryan race has undergone a similar evolution. We may add that in its main features, this social organisation, inferior as it is, has in India withstood the weight of centuries. Even now, the system of castes and of despotic rajahs exists more or less over a great part of India. The English domination has tended to lessen it, though it has not abolished it. Even where the British functionaries and residents have taken their place beside the rajahs, they are powerless beside the system of castes. And quite lately M. Ujfalvy has found the same power in full force, even in Russian Asia, among the ethnical helots of the white race settled in Central Asia on the northern side of the Himalaya mountains.

D. *The Social Condition of the Afghans.*—If we start from the supposed country of the Vedic Aryans, and travel westward, and as we go glance over the societies founded in far-distant ages by the people belonging to the so-called Aryan race, we shall, without much difficulty, be able to disentangle from the ethnographical division of the different groups, and from out of the historical revolutions, a large and general evolution, similar to that which we have traced in the foregoing section. Everywhere as they have freed themselves from the savagery of prehistoric ages, men belonging to the Aryan races have grouped themselves together in semi-barbarous tribes, and then, whether they came together with the intention of forming feudal societies or aristocratic republics, making a servile class perform all the hard labour, they have always brought themselves into a state of absolute monarchy; and it is only in quite recent times that the monarchical power has become mitigated in some countries and broken in others.

The present race of Afghans, in those regions where neither the Musulman nor the Indian civilisations have taken root, have still preserved the organisation of warlike and semi-barbarous clans which has everywhere existed among the Afghan people.

Many of the Afghan tribes are only nominally governed by the sultan of Cabul, and certain of them choose their own chief from an aristocratic family. The power of these chiefs is very variable. Sometimes it extends to the right over life and death. These clan chieftains live as great feudal lords, they own fine castles, and they maintain in their houses a state of luxurious splendour.

The emir of Cabul, a suzerain more or less obeyed by these clan chieftains, coins his own money, declares war, levies taxes, mainly consisting of a tax upon landed property, custom duties, and fines. These taxes are collected by the chiefs of every village.

In a country so little centralised as is Afghanistan, the administration of justice is far from being uniform. In Cabul, and among tribes under the direct control of the emir, the Musulman customs have all the force of law. In law the judicial power has become a royal prerogative, and criminal matters are brought before the emir wherever he may have chosen his residence. Everywhere else he delegates his power to cadis, or to muzis, reserving to himself the right of appeal that may be brought from their decisions, though he never interferes himself spontaneously. Social justice is as yet unknown.

In the rural, far distant, and almost independent districts, the king's justice is almost nominal. When the people are under the care of the central government the village chiefs are held responsible. They are bound to deliver up the guilty persons, or else to pay their fines and indemnities, and to reimburse themselves by levying taxes. Among the most independent tribes the ancient Afghan customs are still preserved. The people have not yet come to that state of civilisation to be aware of any social interests. Public opinion holds it to be right and just for each individual to take the law into his own hands; the talion law is customary, and this is usually practised at once and without delay. Sometimes matters are brought before a council of old men, of mollahs, of notables, or before the chief of the tribe; but recourse to judges is always considered in public opinion as an act of weakness: the strong man ought to avenge his own cause.

We have already mentioned the curious Afghan custom which makes of the young girl a monetary unity. This custom also holds good in law. Among the western Afghans, as among the ancient German tribes, crimes and offences may be ransomed by giving over to the injured party, or to those in his place, a certain number of young girls, or their value in money. There is an established legal tariff: a murder costs twelve young girls; mutilation of the hand, of an ear, or of the nose, is bought by six girls.

Except for the matter of compensation, which sounds strange to our ears, we find here the old German Wergeld, of which we shall again have to speak. Many other sociological analogies also may be discovered between the present state of the Afghan tribes and that in the early days of Greece, of Rome, and of Germany.

E. *Ancient Greece.*—The mere mention of the name of Greece awakens in us feelings of respect and of gratitude; our recollections are at once surrounded with a halo of glory. We know, in point of fact, that in this privileged corner of the earth the human understanding did for the first time command itself, that there was here an intellectual harvest unique in its kind. The Hellenic genius, the most equally balanced that has yet existed, has enlightened everything, it has attempted everything, and shown everything. It personifies truly the virile age of the human kind, for after having become extinguished, enervated by Asia, enslaved by Rome, it has left behind it such a ray of light, that merely by glancing at it all feudal Europe, benighted by a thousand years of mental and political slavery, awoke again, dazzled at the new spring, and began afresh to enrich the human mind.

But, as we have more than once had occasion to remark in the course of this book, the Hellenic civilisation was not born spontaneously. Like every other human effort, its beginnings were very humble. Between the social condition of the primitive tribes in Greece and that of the age of Pericles, there is all the difference which separates the Venus of the Capitol from the rude forms in terra cotta found in the tombs of Mycenæ.

The early Greeks were divided into small independent tribes, who obeyed their military chiefs, a sort of cacique, often at the mercy of caprice and the violence of his subjects. It was only under fear of common danger that the people came round their guardian shepherd and were obedient to his orders. Important affairs were often decided by a majority of votes. The name of king is much too pompous for the leader of one of these clans, whose precarious power was exercised only over a handful of inhabitants. Cecrops, wishing to reckon the number of his subjects, ordered them, tradition tells us, each to place a stone in a certain spot. The stones were afterwards counted; there were found to be 20,000. But Cecrops was a great king. The small kinglings did not dare to decide upon anything without first consulting the chiefs of the tribe, or perhaps all the members.

Ordinarily they were rich enough; they possessed flocks of cattle, and a special domain cultivated by the slaves. The members of the tribes, and especially of the conquered tribes, paid to them subsidies in kind. Their office, which was ordinarily transmitted to their eldest son, was not merely civil; they exercised also at the same time a sort of priestly duty, they presided over the sacrifices, etc. The religious duties of the people were heaviest upon those who had to perform the office of chief in each family.

When the Hellenic clans all came together with the object of discussing some common project, such as the siege of Troy, for instance, the power of the chief of the confederation was always considerably enshackled. We all know how with what liberty of speech the king of men is apostrophised in the councils of war related in the Iliad. At a much later date, in the established monarchies, as at Sparta, for instance, five ephors, named annually by the people, were chosen to watch over the kings, the queens, and to guard the public treasury.

Tradition relates that twelve Hellenic towns united to found Athens. These twelve cantons for a long time preserved their autonomy. Thucydides gives to Theseus the honour of bringing them together in one body. But the Grecian mind was essentially prone to particulate, and in spite of the amphictyons, Greece never succeeded in founding a real federation.

In their intimate organisation, the smaller states of Greece, whether they were republican, or more or less monarchical, resembled in many ways the primitive societies in every country. In all of them slavery was common, and many did not shrink from the institution of castes.

In almost legendary times, the people of Athens were divided into four classes: labourers, artisans, priests, and warriors. Later, the sacerdotal class disappeared, or rather it became merged into the aristocratic class, into the class of the well-born, the eupatrides.

Then, by degrees, the social evolution ever going on, the archonte, who was chosen for life, replaced the king. Then the archonte lost all political power; the church separated itself from the state. When the archonte was deprived of his temporal power, he was chosen by lot. To the gods was left the care of choosing their terrestrial vicar; but for all other functions, notably for determining the strategic places of the executive power, universal suffrage became the law of the state.

These were the golden days in Greece; but even in these glorious republics, which have left their mark upon all the ulterior development of civilised humanity, the social edifice was based upon slavery. Aristotle laid down the principle that, in a well-governed state, the citizens ought not to be obliged to trouble themselves with the first primary wants. Serfs, which in Sparta were called helotes, in Thessaly, penestes, in Crete, perioeki, performed all the hard social labour; and all this servile mass of people were considered beneath the law, because the slave could be driven out as game. It was the materialist Epicurus who seems first to have discovered in Greece that the slave was a human being. "The slave is," he said, "a friend in a lower condition;" and he recommended owners not to beat their slaves.

Among the Greeks, as everywhere else, slavery had its origin in war and in conquest. The Thessalian penestæ were no other than the ancient inhabitants reduced to the state of domestic animals; the helotes of Sparta had the same origin; for the right of people, all over the world, began by placing the conquered man at the absolute mercy of his conqueror. There were, however, other motives for slavery. Until the laws of Solon, a man might in Athens borrow by hypothecating not only his own liberty, but also that of his wife and children, which it was always lawful for the father to sell.

From the time of Solon the castes of primitive Athens became classes founded more or less upon the gifts of fortune. From the first of these classes, that of the proprietors, having an annual revenue of 500 medimni, the people chose the higher city functionaries; the number eligible was restricted, and, after all, the reform introduced by Solon did not change much in the social hierarchy.

The partitioning of imposts was more equitable than that of the political rights, and the privileges were very costly. To be entitled to be ranked in the first class a man paid a tenth part of his revenue, in the second a twentieth, and a sixtieth only in the third or last class. This was the proportional system of taxation that our modern states have not yet had the courage to introduce.

Much money was brought into the public treasury from the succession duties upon immovable property, and upon many indirect forms of taxation which affected mainly the commercial transactions. The scale of taxation in Athens was never definitely established, not so, at least, during the time of the full bloom of the republic. For laws, after having been for some time immutable, and considered as of divine origin, became simply measures of public utility, changeable according as necessity required.

There were also many heavy charges upon the public treasury. In Athens, each man was not, as in our modern societies, left to himself, surrounded by every chance in the struggle for existence. The state considered itself bound to give portions to poor girls, to distribute at very low prices, or perhaps quite gratuitously, corn in times of need, and to amuse the people by theatrical representations. The modern egoism, which has for its maxim "Every man for himself," is very far removed from this organised form of joint responsibility. In sociology, as in many other things, Athens has shown to humanity more than one good example.

In the same way, from barbarism to civilisation, the conception of justice evolved itself among the Greeks. The talion law sufficed to the early Hellenes as to so many other races. Aristotle and Diodorus Siculus furnish us with proof of the fact. The talion was seen in the laws of Solon, and hence sprang a whole theory of jurisprudence. Men perceived that the law "an eye for an eye" was insufficient when the aggressor had taken the second eye of one who before had only one; and that in this, if equity was to be observed, instead of "an eye for an eye," the rule should be "blindness for blindness." This barbarous form of justice was a long way from that of the Areopagites.

But when a penal code, public tribunal, and rules of procedure had been established in Greece, there was no magistrate entrusted with the office of prosecuting the murderers. The right of initiative lay with the relations of the deceased. The accused was not even made to undergo preventive imprisonment. He might take flight and so avoid prosecution, but then his property was confiscated and sold by auction. The legislator had not yet arrived at a clear idea of social justice. In his eyes murder was a wrongful act of a peculiar kind; and for a long while the homicidal man might assure himself from punishment by disarming the vengeance of the relations of the deceased by a pecuniary compensation. For a long time also the accuser had the right of being present while the guilty man was being punished, who, at his request, was condemned to capital punishment. In a word, the arm of justice was only a substitute for the vengeance of the aggrieved person.

More elevated notions of general justice came to be formulated in the minds of a few philosophers. In practice the Grecian legislation confined itself to applying the talion law; but Aristotle wrote: "Justice is perfect virtue, not only in itself but in connection with others; that in justice every virtue is contained." The same philosopher also says: "The city rests upon love even more than upon justice, and that supreme justice is love."

What would have befallen the Greek societies if nothing had come to trouble the course of their evolution? We are authorised to suppose that this spontaneous development of Greece, already so glorious, would have been but the prelude to a more full, a more intelligent, and a more perfect blossoming. The Macedonian brutality put an end to the most interesting sociological experience that has been attempted in humanity, to the regular progress of the most intelligent of the human nations. Philip and his glorious son cost dear to the Greeks. After them a deluge of Asiatic despotism invaded the whole country, and an abject Demetrius, a Nero upon a small scale, was at last deified in the city of Pericles.

V. *The Roman Society.*—As our object is not to write an anecdotical history, but to note the main features in the social development of the human races, we need not dwell at length upon the sociological evolution in Rome in so far as it was analogous with that of Greece.

In Latium, as in Attica, there were at first very large families despotically governed by a chief—*pater familias*; then these families grouped themselves into curiæ, analogous to the Grecian phratries. Later, these curiæ, each having their own worship, their own tribunal, their own government, associated together to form themselves into tribes. By degrees the confederation of the tribes grew into the city.

Hence resulted a great number of small federations, often at war one with the other, under the command of a military chief, whom they called their king, though his power was, like that of the chief of the Hellenic tribes, very much confined.

As everything in these small Roman societies was hereditary, so was also the royal dignity often hereditary, though not always, as was the patriarcat, and as was slavery. For in Italy, as in Greece, the system of castes was at the basis of the social edifice. It is not our object to rewrite Roman history; we all know how the Roman monarchy became a republican aristocracy; how this republic obtained the mastery over the greater portion of the Aryan people; how, enervated by its successes, overweighted by other less-developed nations coming into and forming part of its own vast territory, it dwindled away and became lost in the slavery of the Lower Empire. Such is the habitual result of conquest—the conqueror and the conquered are both usually impoverished.

From our present point of view the judicial organisation in Rome concerns us more fully than its historical facts. In Rome, as in Greece, the talion law was the primordial custom. The earliest written laws—the Laws of the Twelve Tables—still allowed it unless in case of agreement with the injured party: "Si membrum rupit, ni cum eo pacit, talio esto." The Theodosian Code mentions the talion law.

For a long time there were in Rome neither penalties nor well-defined forms of procedure. The magistrate having the *imperium* exercised it at his discretion. The permanent commissions were instituted for the recognition of certain crimes.

Caius Gracchus formed one of the commissions to take cognizance of cases of murder and of poisoning.

Sylla took similar measures against arson and spurious witnesses in testamentary dispositions.

The Cornelian laws established, without the slightest care of justice, penalties graduating in an inverse ratio to the dignity of the social classes. The judges were nearly always chosen from the governing classes; the list was published every year.

There was no civil or judicial hierarchy; appeal was unknown; but the judicial functions are otherwise specialised. There was a judge of law, the prætor, and a judge of fact chosen by the parties from an annual list drawn up by the prætor. The judge of fact was held responsible for any bad decision.

The Roman mind passes for having been essentially judicial; but it was rather cavilling, litigious, very jealous of its judicial formalities and careless enough about real justice. We know how in modern Europe, we who have inherited from Rome much of her legal procedure, that the most solemn and most minute judicial forms have often sanctioned monstrous iniquities. And even nowadays, though the Anglo-Saxon jury in nearly every country in Europe holds a check upon the tormenting jurisprudence of the legists, the slightest matter is still crushed under the weight of useless formalities, and only too often the judges, crammed to overflowing with their knowledge of the law, astonish us with their partiality and the iniquity of their judgments.

G. *The Celtic and German Societies.*—The Celtic, German, and other European tribes, brought by Rome under her subjection, began, socially speaking, much in the same way as the primitive classes began in Greece and Rome.

The Germans had their slaves, their nobles, their priests, to whom was extended the right of inflicting punishment, of beating, and of imprisonment. In Germany the families used also to form kinds of curia, and they remained united together even upon the battle-field.

These tribes had their chiefs, generally hereditary; often they were generals chosen merely on account of their valour.

The German slaves were a sort of colonists, having their habitations and paying to the master a fine in kind; but the master had the right to beat them or to kill them with impunity. The free man might become a slave by playing his liberty at a game of hazard.

The chiefs discussed together the important matters, and decided the smaller ones *proprio motu*; but every decision bearing upon the general interests of the community could be decided only in a general assembly, in an open field, where everyone was under arms. This is the direct form of government as we see it now practised in some of the Swiss countries, in Uri, Schweitz, Glarus, Appenzell, and Unterwalden.

Crimes entailing capital punishment were also debated in public assemblies; offences of a lesser importance were adjudged by the chiefs of the village, who were chosen by the assembly.

Every family was held to join in the friendships or the enmities of their chief. The vendetta was a duty; but commutations were allowed; everything might be ransomed, even homicide. This is the Wergeld, that we find again better organised in the fifth century of our era, among the Franks on the borders and the Burgundians established in Gaul. In the border Wergeld everything was fixed according to the tariff, each portion of the body had its price, which varied also according to the social category to which the injured man belonged.

We see here the same social iniquity, the same shamefaced candour, which we have also pointed out in Rome, in India, and elsewhere.

Like the conscience of every race of people, the conscience of the Indo-Europeans was at first far from delicate. It is only quite recently that a desire for equality, for true justice, has shown itself strongly in the human mind, and there is still ample room for very much wider development.

CHAPTER VII.

THE POLITICAL AND SOCIAL EVOLUTION OF THE ARYAN TRIBES.

The long analytical exposition continued in the last chapter may enable us to determine by comparison the social value of the Aryan race. It is beyond all doubt that this race now holds the foremost place in the competition for progress among the various human types. But such has not always been the case. We can hardly do more than make inductions as to the primitive origins of the Aryan societies, because, as far as history, legends, and even language will take us back into past times, we find the Aryans already emancipated from the very early state of savagery. Not that the Indo-European, considered individually, has always been superior to primitive bestiality. We have seen Europeans adopt most completely the sort of life of the Tupinambas of South America, or of the Melanesians of Fiji, and go back even as far as anthropophagy. Cook has met in certain small Malay states in Batavia the descendants of the Portuguese, reduced to the condition of the servile classes; and if we threw the sounding-line down into the depth of squalor and misery which the civilised varnish of our own contemporary societies have but badly covered, we should find people no further advanced than the Red Skins, or perhaps than the Fuegians. However, taken as a mass, the least civilised groups of the Indo-European branch—for instance, the Kafirs of Afghanistan—have already acquired a certain degree of moral and intellectual development. They may be still barbarians, but they are no longer savages. There can be no doubt that the antique progenitors of the race were originally savages of the coarsest type. On this point prehistoric archæology will amply instruct us.

Following the early stages of civilisation, the Aryans evolved, as did their brethren, the Semites, even also as the best-endowed branches of the great Mongolian race. They passed from the horde to the tribe, from the tribe to the town, from the town to the great oligarchical or monarchical states. But at a very early date some of their groups distinguished themselves by a certain moral strength, a spirit of initiative strongly marked. Among some of the tribes in ancient Germany also might be seen an individual independence, a spirit of equality. Even the loiterers in the Aryan race, the Slavs, if we may believe Procope, began by forming small democratical tribes, and in the Russian mir we see the remains of this primitive organisation. But the oligarchical or monarchical servitude is already in germ in the mir, for the individual is thus deprived of his liberty. Also, after the passage from the nomad to the agricultural life, when the social functions became determined, slaves of the labouring and fighting classes were formed in the small communities; the former were soon brought into subjection by the latter, and reduced to the condition of the Egyptian fellahs. Another cause ought to be taken into account to explain the transformation which the other branches of the Indo-European race have not felt, not at least in the same degree; that is, the influence of the Tartar blood, so largely mixed with the Slav blood all over eastern Europe.

As a matter of course, in the Indo-European race, as in every other, equality, when it was thought of, was at first conceived after a very narrow fashion. Inequality, sometimes, of the most unjust kind was common between castes or classes subordinate one to the other. Equality was never supposed to exist but among men in the same social category. But below the privileged classes there was an enslaved multitude of people for whom there was no legal right; and these men were made to sweat, to suffer, and to die, in order that their masters might live and enjoy themselves.

We can understand that a social organisation such as this should end in a monarchical despotism; and in fact such was the social destiny of every primitive state founded by the Indo-Europeans, even the republics of Greece and Rome. The impulsion once given made itself felt a long time after the fall of the Roman empire; the European feudal system was nothing more than the subdivision of the unitary despotism in imperial Rome. The fragments of the great Latin monarchy became smaller monarchies, which received the titles of kingdoms, principalities, duchies, counties, etc. Bonds of feudal suzerainty tied together more or less closely all the remains of the great crumbled edifice; but the institutions of privileged classes were everywhere carefully preserved; the noble profession of arms was everywhere glorified; the right of force was everywhere held to be the first law; the degradation and the slavery of the labourer and of the villain were everywhere at the basis of the social organisation.

If the Indo-European races had stopped there, there would be no need to make further crowns of laurels. The political constitution of feudal Europe is certainly far inferior to that of China. But during the last few centuries of European history a very great social transformation has taken place. In England a parliament has put a check upon monarchical power; in France the sub-despots have in their turn been conquered by the most powerful members of their own body, and the multitude, who have now but one master, may think of setting themselves free.

At the same time, the general intelligence has developed, reason has made itself felt, the scientific mind has become enlarged. In the seventeenth century religious persecution drove to the other side of the Atlantic fanatical men whose character was sombre and severe, and these exiles have shown to the world how easy it is to live without aristocracy and without a king. And, finally, the great upheaving of society in modern times, the French Revolution, has maimed for ever the feudal system. From that time a new era in humanity began, an era of redemption and of emancipation, and though we are still only in its early days, it is possible to prejudge roughly the general course. But before attempting to do so it may not be useless to resume very briefly all the series of social changes through which humanity has been made to pass.

CHAPTER VIII.

THE POLITICAL AND SOCIAL EVOLUTION IN HUMANITY.

Is it possible for us to formulate sociological laws? The universe is assuredly not governed at haphazard. Chance is but a word, and everywhere, not only in the small human world, but even in the cosmical community, effects are very closely linked to causes. But to ascertain the laws of all this concatenation is a most difficult task. At what expense of labour have we, even up to this time, been able to trace out in the inorganic sciences a few general fundamentary notions worthy of the somewhat pompous name of laws? Is there, even in physics, even in chemistry, one single true law, "a law of brass," which will admit of no exception? It is very doubtful. Even in biology, where the intricateness of the phenomena becomes very embarrassing, the so-called laws are quite free from any rigorous constraint, for at every instant we find whimsical exceptions frisking about and jumping merrily over the legal ditch. But in sociology the apparent confusion is far greater, for there the skein of causes and effects is so entangled, they result from so many other antecedent or concomitant phenomena, that there would be some boldness on our part to talk of laws. We maintain, however, that there are sociological laws, and that it is our duty to discover them. We shall use our best endeavour, enumerating, examining, and classing together the facts, but we must leave to our grand-nephews the honour of crowning the edifice. For our own individual part we are obliged to confine ourselves to pointing out a few main features in the general outline, which is itself not always very distinguishable. We will now try to do so, resuming, upon the basis of the facts already given, the progressive social development in humanity.

Quite at the commencement, the human mammalia, weaker and less well armed than many of their competitors in the animal kingdom, instinctively joined themselves into small groups, endeavouring so to gain a little collective force by putting many weak bodies together into one bundle.

At this time the ideal of human existence was far from being high. To eat, and not to be eaten, to satisfy their amorous passions like beasts in the thickets, as do now the Papuans, the New Caledonians, and the Andamans: such were in this primitive stage of the social development the only objects in human life. During this early and very inferior phase, man wandered about through the forests, naked, almost unarmed, devouring everything that was at all eatable; and in case of great need he would eat his wives and his children. These were small hordes of people totally ignorant of what was meant by family, by morality, or by laws; they were almost equally ignorant of industry. Such were the ethnical unities at that time. Each little group lived promiscuously, huddled up together, under the command of the strongest male, after the manner of the chimpanzees. The word *society* is too eulogious a term to apply to these small agglomerations; we must turn to the zoological vocabulary to find a fitting qualification. Man then lived in a flock, in a *gregarious state*.

Every human race began from the gregarious stage; some groups of men have not yet passed it.

It was, there can be no doubt, the vital, inveterate, and never-causing competition among the human hordes which pushed them more or less quickly into the way of progress. The more fully the members of any group of men assisted each other in times of danger, the better chance had that group of endurance, of outlasting its less prudent rivals. Therefore, by force of things, the association ameliorated itself by degrees; the family sprang up; industry made progress; customs which experience had shown to be useful became instinctive and obligatory. When the social instinct became developed, even to a small extent, the ethnical unity then increased; several hordes came together; the phase of men living in tribes then began.

But the more numerous the tribe was, the more need was there for a complex form of organisation; a certain division of labour, a commencement of hierarchy soon established itself. In the plundering or warlike expeditions men found it to be advantageous to allow themselves to be guided by the boldest, by the most experienced; but as the daily life among these tribes was but a long succession of warfare, it soon became customary to obey a chief who was chosen temporarily. When men thus lived together in hordes, a sort of submission was naturally forced upon them; for the strongest men enforced his will or his caprice upon his companions. Therefore, the tendency to bend before a master had become very early implanted in the human brain. By degrees a sort of hierarchy became established in the tribe; the best warriors or the most skilful huntsmen formed a rudimentary aristocracy, obeying often the advice of the chief, or sometimes governing the community with no chief at their head.

At the same time these chosen people of the tribes had some thoughts as to their genealogy; as the family became constituted the descendants of the chiefs and the notables had more chance of inheriting either directly or collectively certain social advantages. This was especially the case among the more advanced tribes, among those whose means of subsistence were not entirely dependent on the more or less fortunate results of a hunting expedition among the tribes whose habits were pastoral or agricultural. As soon as men came to own flocks of cattle and fields whereon to feed them, the social condition became less unstable. These were their properties, or, in other words, their collective strength: but these properties, even when the ownership lasted only for life, were very soon divided into most unequal shares. It was not likely that the chief, the foremost warrior, after a successful attack, should not take for himself the largest share in the plunder. Why should not he claim the lion's share in the flocks and fields of the conquered party? But the conquered people themselves became also a property, generally individual, sometimes collective. In this way slavery was established.

The institution of slavery was a fact of the highest importance; it has left its mark upon every ulterior development of society. It was at first a progress; it has ever since been a stumbling-block. In principle, when the small human groups had neither fixed dwelling-place nor sure means of food, the prisoner taken in battle was ordinarily killed and often eaten upon the battle-field. As we have seen, such was the custom with the New Zealanders. But when hunting and warfare were no longer the only important occupations, when it became necessary for men to work—either tending the flocks or digging the soil—the lives of the conquered prisoners were often spared on condition that they performed the hard labour. They were turned into domestic animals, over whom their conqueror had every right, even that of life and death.

Henceforward the social organism became much more complex. There were chiefs, there was an aristocracy, there were slaves. It became necessary therefore to draw up a code indicating the rights and duties of everyone, a code which should pass traditionally from one generation to another. At the same time the same reasons of social utility which had more or less civilised the hordes, and afterwards grouped them into tribes, brought about aggregations of a still more complex kind. Spontaneously or not, the tribes came closer together, they formed themselves into larger associations; villages grew into cities; the tribes as they joined themselves together constituted monarchical or oligarchical states, always resting upon the basis of servitude. From this point in the social evolution a new class or caste was formed. The intellectual torpor of the first ages was dispelled, but ignorance being still very great, metaphysical or religious speculations played an important part in the mental condition of every people. The humble sorcerer in the primitive tribe acquires a sacred character; he is in time raised to the dignity of priest. Often also the chiefs take upon themselves both the political and the sacerdotal power; they command men, and they hold intercourse with the gods. Sooner or later the functions become distinguished; the priest has his own separate office. But even then, the chiefs and the priests find it worth their while to understand each other; and the government, when not purely theocratic, is at least impregnated with theocracy.

In societies so constituted, whether the government be oligarchical or monarchical, the classes are clearly defined, each having unequal rights and duties. There are the aristocrats, the priests, the labourers, and the slaves.

This system of castes has ever, either owing to a natural evolution or to violence, terminated in absolute monarchy, in a centralisation of despotism—a most natural result, for slavery is the essence of every society established upon the basis of castes. The privileged classes at last felt more or less the yoke which they thought lawful to impose upon those below them, and against which the conscience even of the slave did not protest. As a matter of fact, nowadays, in Africa, the negro slave thinks that his condition is quite a natural one; he would impose it upon others were he able to do so. The idea of there being injustice in his lot in life never occurs to him. In all the great races, the vast agglomerations of men were first constituted under this system of castes and of despotic monarchy. Ancient Egypt, Manu India, the Semitic kingdoms, even the Roman empire, and, nearer to us, Peru and Mexico before the Spanish conquest, show us this social phase in all its splendour. And in our own times, in Africa, there are examples, perhaps not so wonderful, but quite as characteristic.

Whole races of people—the African negroes, the Indians in Central America, the Semites—taken in the main, have never succeeded in reaching a higher social ideal; and all the efforts of the most advanced Mongolian groups—the Chinese, for instance—have had no other result than substituting for the castes a large corporation of men of letters, restricting to some extent the caprices of a monarch, who is at the same time worshipped as though he were a god.

The feudal system, which has existed in a very crude form in Polynesia, which came to its full bloom in the Europe of the Middle Ages, and which we still find in Japan, is only a particular form of the system of castes. In feudal countries the ethnical group always comprises slaves and nobles by the right of birth; it is nothing more than a hierarchy of petty despots, one controlling the other, and all governed by him who is the most powerful.

Under such a system the mass of the people are made submissive to a perpetual guardianship. The governing classes have decreed everything, have ruled everything beforehand. Everyone must remain during the whole course of his life exactly in the same position in which his birth had placed him; as far as possible he must carry on his father's trade. He is told what he must do, what he must say, what he must believe. As long as the human mind is minor, all these trammels are borne docilely enough. But among the better-endowed races, reason tries to free itself. Man inquires if all this oppression is lawful; if the more privileged classes, if those who are better born are essentially superior. Science, and philosophy, its offspring, plant revolutionary ideas in the best-endowed brains; religious myths, which had so long bolstered up political abuses, are shaken at their foundation; those who are oppressed dare to think of freedom; the oppressors themselves begin to doubt the legality of their rights. A social transformation then becomes necessary, and in one way or the other it is effected. The Renaissance, the Reformation, the foundation of the United States of America, and last, the great thunderbolt of the French Revolution, have been, in the so-called Indo-European race, the principal stages in the metamorphosis, which is still far from being completed.

We find also, corresponding to this evolution, profound modifications in the organism of societies. At first the enslaved masses, by whose labour everyone else lived and was happy, possessed nothing which they could call their own. The soil was the particular property of the privileged classes. In Sparta the helotes laboured for the free men; in India the Çoudras sowed and reaped the corn which the glorious Brahmins condescended afterwards to eat; in Rome the slaves and the colonists cultivated the *latifundia* of the patricians. We have seen that in Polynesia the chiefs had the right to appropriate to themselves anything that they wished to possess. This was at first the custom all over the world. Then the good pleasure of the strong man had its limits; fixed fines and taxes were substituted in the place of universal grasping. We need not say that these taxes were levied very capriciously and most unjustly. Even when there was a monarch, the powers and the saints paid to him very little, and the burden fell altogether upon the shoulders which had from long ages past been accustomed to support it. The desires of the governors and the patience of the governed being both unbounded, the paying classes paid for everything. They paid to work, to be allowed to carry on their business, to travel, to salt their food, and even to breathe fresh air. We do not now speak of taxes capriciously laid, of fines ten times commuted and ten times re-established, simply by the right of force. It would be useless to insist upon the fact that past ages have bequeathed to us many vexatious imposts unjustly established; but the great social evolution that has taken place during the last few centuries has introduced a certain degree of justice. Whether we have succeeded well or ill in placing the burden upon the proper shoulders, the strong man must now feel that his hands are not quite unfettered.

In the same way, too, law and justice have evolved. The first laws were no other than customs traditionally preserved. How did these customs arise? There is no trace of them kept in the memory of the people; and nearly everywhere laws were supposed to be clothed with divine attributes. They were the orders dictated by the gods, or sent by the gods. To change them in any way would have been sacrilege. But as it developed, the human intellect grew so audacious as to contest even these divine prescriptions. Laws were despoiled of their religious prestige: people saw in them nothing more than the commands of the most powerful man, established more often upon force than upon right. At last, laws came to be merely measures of social utility, debatable and changeable according to our progressive wants.

It is especially in the evolution of the penal laws, properly so called, that one sees how the care for the general welfare has gradually established itself. At the commencement of every society, man is just like his brothers in the animal kingdom, quite ignorant of every idea of justice; the right of a stronger force is his only law. But by the very fact that his brutality gives rise to conflict, a vague instinct of justice is at last awakened in the human brain. He thought it just to repress violence by violence; then the idea came to him of establishing the balance between wrongs and revenge. Hence grew the talion law: "An eye for an eye, a tooth for a tooth." The care of applying this primitive law was at first left to the parties interested, because for a very long time the governing classes had other matters to think of than the administration of justice. When they did interfere, it was only with the object of putting the offender into the hands of the injured party, who then carried out this talion law as well as he was able. By degrees manners grew softer; the spirit of foresight and of calculation increasing, the injured party renounced his right of vengeance, accepting in exchange compensation in kind or in money. At the same time, the judicial functions became in some measure determined; codes were drawn up, at first very simple and very cruel. Repression began by bearing only upon a very small number of actions, often remissible enough, according to our modern ideas. In the opinion of the Hebrews, the greatest, the most inexpiable of crimes was idolatry. The want of respect to masters was everywhere one of the main offences. As a general rule, the early penal laws punished theft more severely than murder. For in primitive societies life was held cheaply enough, and also the right of vengeance was left to the friends of the deceased. Not until a comparatively late date did the community take upon itself the right to punish murder, to institute for this purpose tribunals, a form of procedure, a penalty soundly and equitably established. Do we not find, even in Athens, that the tribunal, in case of murder, confined itself to appointing "an avenger of blood!" In the spirit of our modern codes, the spirit of vengeance is completely taken away from the individual, but only to be handed over to society. Our penal laws aim principally at punishing the guilty person, at making him suffer. In the future, justice will think only of putting the criminal beyond the possibility of doing harm, to correct him of his faults, and to make of him, if possible, a useful citizen. In the future, justice will guard herself from showing anger, she will break her sword; to establish her balance weights and measures she will be guided only by social utility; she will become thoughtful, and will lean upon the observation and the experience of mankind.

If, after the manner of the god of the metaphysicians, for whom every bygone age is but as a moment, we were to endeavour to envelop in one short formula the slow progress already accomplished by poor humanity on its long journey in search for improvement, we might say that all social evolution is but a gradual emancipation of the individual, both in his mind and in his body. But we should have to close our eyes very determinedly, if we wish to make ourselves think that this social renovation is completed.

It is difficult to predict the future, except for those who may have the gift of divination. But as we have now followed the evolution of societies from their cradle down to these our present days, we may perhaps, without over rashness, if we confine ourselves to generalities, hazard a few conjectures as to the future destinies of humanity. No doubt, in the Indo-European societies, the weaker creatures have been very greatly relieved, but they do not yet walk upright. The mountain of oppression which has weighed upon the shoulders of the humble has been much lessened, but it will one day be altogether removed. Many privileges have been wiped out, but there are still others that must be abolished. Liberty has already enlarged the brain of him who was once a slave, and instruction must now come to give it greater power. Depths of suffering, of misery, of vice still remain, and these must one day disappear. In a word, we must succeed in equalising, as far as it may be possible, the chances of the combatants who enter into the arena of life.

If, as in a fairy tale, a magician could bring before us the picture of a future epoch, perhaps not too far distant, we should see the superior human races constituted into republican federations, and their social organisation radically modified. The confederated ethnical unities would then be small groups of men, governing themselves in everything that did not manifestly touch the general interests of the republic. In each of these groups, the social activity would be altogether spent in useful occupations. The physical, moral, and intellectual education of the young generations would be watched over with greater care. Men would endeavour to lessen gradually at fitting opportunities the organic inequalities, the only ones which still exist in our happy times.

"To everyone according to his works" has become the great social law; the inequality of conditions rests merely upon the differences of individual worth, and upon services rendered. All the useless trammels have been broken; nothing is forbidden but that which is manifestly injurious to the social body. Kings, priests, and standing armies have all disappeared into the dismal limbo of the past. All artificial inequality has vanished. The fairy of inheritance no longer lays riches into any cradle, and society now holds out her helping hand to give succour to the weak. The individual is aided as much as possible, and is governed as little as possible. The Brahmins had made of chastisement a divinity. The Europeans in the future will punish little; they will anticipate and reform a great deal. Without crushing anyone, they will, by intelligent selection, ever ameliorate the poor human kind, confiding the social government to the wisest. Their device will be: knowledge, justice, joint responsibility.

BOOK V.

INTELLECTUAL LIFE.

CHAPTER I.

THE DIFFERENT DEGREES OF PSYCHICAL LIFE

I.

Psychical Evolution.

In spite of many reveries a hundred times imagined and printed on the subject of the souls of plants, and even of minerals, we are obliged to recognise, if we wish to keep clear of positive biology, that except in the nervous cells—in certain of the nervous cells—a conscient life does not exist. But this conscient life is a privilege inherent to the aristocratical castes of cells. In many of the radiates or inferior molluscs the nervous ganglions are no other than centres totally unconscious of reflex action. A little higher in the zoological scale we find a more complex nervous system already provided with conscience, but this conscient life is still very rudimentary. The reflex actions no longer silently pass over the nervous cells, but they provoke nothing more than impressions of sorrow or of pleasure. Then in proportion as, in the superior kinds, the organic commencements distinguish themselves and become complicated, in proportion also, the nervous centres develop—special sensibility and intelligence (which is dependent upon it) join themselves to movent power and impressionable feelings. The conscient life is then provided with all its principal means; it has only to ripen into perfection.

From the fœtal period every human being goes through the whole psychical series, and shows to us again upon a small scale his ancestral evolution, of which he is the final term. During the last months of pregnancy the human fœtus is, like many invertebrate, susceptible only of movent power; it acts and reacts under the shock, but unconsciously. At the time of birth the new-born child is susceptible of painful impressions, but his special sensibility hardly exists; not until afterwards, and then slowly, is intelligence born in the child and begins to develop itself.

In a word, the embryology and the taxinomy of the animal kinds, enlightened by the doctrine of transformism, show to us the slow acquisition of conscient life in the animal kingdom. A whole organic history is unfolded out before us. Once formed, the nervous cell, which is primarily a registering measure, has stored away impressions gradually becoming more delicate and more complex. In principle this cell has been no other than a centre of movable reactions, of reflex inconscient actions. Then the transmission has ceased to be silent; it has been accompanied with impressions of grief or of pleasure; by a new progress a peculiar sensibility is born, and various sensations have been tested and definitely registered. And from the conflict of all these impressions, more or less revivifying, the intelligence is born. Henceforward there is a conscient me, which, very awkwardly at first, more artfully later on, is able to compare and classify the recorded ideas, to foresee future events more or less clearly, to imagine, to invent, to deduct, to induct, and at last to reason.

From the fœtal life to the adult age every human being goes through this series of evolution, becoming at every stage a simple machine showing a reflex inconscient action, then afterwards a sensible register, and finally a creature more or less intelligent and reasonable. He is all the more reasonable in proportion as his impressions and sensations allow themselves to be governed, and as they lose their primitive excess of intensity and of natural colouring.

We see here a whole graduated scale, which all individuals and all races of men succeed in climbing more or less well. Intellectual progress will ever show itself in man by a greater subordination to the automatic life, by his having a greater command over himself. But in this respect men and nations are very different. In the inferior degrees of intellectual development the human being is but the plaything of outside circumstances. Impressions and desires exercise such a violent hold upon such a man that they rarely allow him to be master of his own actions. His nervous centres are as yet but very poor registering-machines; his impressions are quickly wiped out; his memory is short; the present, always very vividly before him, soon eclipses a future hardly thought of; he has no sort of idea of foresight. Neither can he fix his attention. The present is to him a tyrant which the hazy gloom of the future cannot dispel. Also the nervous conscient cells, the psychical cells, are in a state of perpetual instability; the equilibrium of their molecules is constantly troubled. The conscient me cannot therefore become fixed. Every effort of attention tries him; he is essentially versatile. In the opinion of all travellers, fickleness, want of foresight, incapability of fixing the attention, are psychical characteristics inherent in every inferior race of men.

II.

Comparative Psychology of the Human Races.

To explain with any sort of fulness the short review of comparative psychology, on which we must now say a few words, we should require a volume, and we have only a few pages at our disposal; we must therefore confine ourselves to a very short enumeration of the most characteristic details.

In this, as in most respects, the Melanesians, and especially those of Tasmania and Australia, are at the bottom of the social ladder. Cook was struck with their apathy, with their want of curiosity. They barely paid any attention to the English or to their ships, to all the novelties which they saw before them. D'Entrecasteaux bears the same testimony. "The Tasmanians," he says, "express a wish for every kind of trifle, but drop them immediately afterwards; everything seems to distract them, nothing can occupy their minds." At a later date, in the schools in the English colonies, the Tasmanian children gave proof of strong memory as to persons, places, and objects; but it was with difficulty they could be made to understand grammatical constructions, and they were especially extremely rebellious to arithmetic.

The "soul" of the Australians was very like that of their Tasmanian brethren. Some of them did not even condescend to look at Cook's vessel; others left in a corner the stuffs that had been given to them as presents. They seemed to be more interested with a few turtles that they saw upon the deck of the vessel than with anything else. Leichard says that, for details, for particular objects, they have a tenacious and a photographical memory; but that they cannot understand even the simplest drawing. They mistook a portrait of one of themselves for a ship, or for a kangaroo. That which struck them most in a book was that it could be opened and shut like a mussel; they called a book by that name. It has been attempted to teach the Australian children to read and write; but civilisation has modified their natures only on the surface. Their early years of manhood are often marked by an explosion of savage-like instincts, which irresistibly drive them back into the kind of life from which they had sprung.

The New Caledonians, more civilised in appearance, are extremely changeable; their intelligence is at the lowest possible ebb. They know of no name designating their own island.

The Bushmen are not more developed than the Australians. They have no proper names; they despise an arrow which has failed in striking its object; they think that of two chariots the smaller is the child of the larger; they do not trouble themselves about any future meal, but when they are finishing the one which they are actually eating.

The Hottentots, barely more intelligent, but already a pastoral people, have a strong, exact, and tenacious memory for everything that concerns their cattle. Like the Australians, they are susceptible of receiving a certain dose of European education; but also like them, they return immediately afterwards to their savage mode of life. Like the Australians also, they are quite ignorant of common prudence; they will without hesitation consume all their provisions in one day, turning themselves, like Falstaff, into a "bolting-hutch of beastliness," without any thought of the morrow.

The very inferior savage, like our own infant children, does not know what to-morrow means.

The Kafirs, the neighbours and the hereditary enemies of the Hottentots, barbarians rather than savages, and belonging to the superior black races, have learnt agriculture; and they are prudent enough to store away the produce of their harvests. But still they do not kill an elephant without making excuses "to the great chief elephant, whose trunk is his hand," and without declaring to him that his death was merely an accident. In the same way the negroes of near the Gaboon river, when they have killed a leopard, compliment him upon his beauty. But these same negroes, who commercially are very crafty, never think of provisioning themselves with food before they are in actual want of it. In the centre of Africa, a Tibboo chief could not understand a landscape drawing, any more than one of our own children three or four years of age.

All travellers are agreed in saying that the majority of the black races in Africa may be compared to our young European children. They have all the light-headedness, the capriciousness, the want of prudence, the volubility, and the same quick and confined intelligence, as a child. The negro child is precocious; sometimes he will surpass the young white child of the same age. But his progress very soon stops short, the forced fruit does not ripen. In the same way, the small Andaman Negritos learn their letters very quickly, they repeat them like parrots; but they have great difficulty in joining ideas to the words they have learned.

Not counting the Esquimaux, who seem to have come from Mongolian Asia, there is everywhere in the American man, from the north to the south, a certain unity. It is the same human type which has progressed, more or less, according to the regions in which he has lived, and which we may study in the different phases of his development from extreme savagery to semi-civilisation.

The Fuegians, a people without any industry, without any prudence, not having got beyond the Age of Cut Stone, show to us the primitive American man. Their saddened mind is not capable of astonishment or of curiosity. On board Cook's vessel they saw many objects, but without looking at them; and these objects they surely could have never seen before. They are still in a state of the coarsest animism. A missionary, who complained of the heat, was answered by a young Fuegian that he was wrong to reproach the sun, for if the star hid himself they would soon have an icy cold south wind.

From the time of the Spanish conquest the history of the Indians of South America is most instructive. It shows to us how very slowly has the intelligence of man developed in the different races. Except certain tribes of the Grand Chaco and on the Chilian plateaux, who have become more or less pastoral, the aborigines in South America have mentally undergone very slight modifications. All the tribes which have lived exclusively upon hunting and fishing are still in the lowest degree of savagery. Such are the Atipones, the Botocudos, many tribes in Columbia and others. In the heart of a country of exuberant fertility the majority of the native Brazilians are still grovelling in a state below that even of the Esquimaux. Though they have become pastoral, and live upon their flocks and their horses, the Araucanians of Chili, the Puelches, the Patagonians of the pampas are still savages, untamable, and are more wandering now than ever. On the other hand, nearly all the people who have become agricultural and pastoral, or who have at least joined agriculture to their desire for the chase, have submitted themselves to the Spaniards, have become Christianised, and have accommodated themselves to Christian habits. All the Chiquitos have been reduced by the missionaries; the Peruvian branch, a long while ago curbed by civilisation of the Incas, is now quite submissive. The true savages have resisted or have died out. The ancestral instinct is so strong in them that a Botocudo man who had become a doctor, and had received a diploma from the University of Bahia, has been known to throw off the clothing and the life of civilised men in order to go back and wander about naked in his native forests.

Facts of the same kind have also been observed in North America. In spite of the Jesuitical missions, the Californians still live principally upon acorns and upon what they can kill either hunting or fishing. The Red Skins prefer a fish-hook which has taken a large fish to a handful of hooks quite new. Their imprudence is very great. Like the Caribs, who will sell their hammocks in the morning for less money than they could have done the evening previous, the Red Skins will destroy a whole herd of bisons and take only their tongues, without thinking that in two days' time they may again be hungry. The intelligence of the Ahts is so deadened, that in order to fix their attention one must repeat to them several times the same question. As a rule the Red Skin is like a child in his thoughtlessness, and like a decrepit old woman in his obstinacy and in his want of understanding. In North as well as in South America, the only tribes which have more or less conformed to European civilisation are those who had of themselves arrived at a certain degree of intellectual development, the Choctahs, the Cherokees, and others.

Among the Esquimaux intelligence seems to be more generally awakened than among their enemies the Red Skins. They could understand the maps that Ross showed them, and they even drew maps upon the sand, marking the principal objects, the hills and mountains, by a stone or by little mounds of sand. But the Esquimaux are not real Americans.

But of all savage races none are more childish than the Polynesians. Their thoughtlessness and their light-headedness are extraordinary. It is impossible to fix their attention upon anything for two minutes. The most civilised, the Tahitians, had no idea as to their age; to recall the date of an event that had happened two or three years previously was altogether beyond their power. But some of them could speak from memory of the old traditions in their race; that was a practice which in certain families the children had been taught to observe. But they held generally that memory and knowledge were gifts of the gods, quite spontaneous; and at the death of one of their habitual story-tellers they always placed the mouth of a child over the mouth of the dead man, so that he might so catch the spirit of the deceased in its flight into the next world.

Like the children of all inferior or backward races, the Polynesian child is precocious, for in inferior civilisations a rapid development is necessary; man has not the time to remain long a child, but his intelligence is as narrow as it is quick in coming to perfection. According to Lieutenant Walpole, the little Sandwich islanders brought up in the English schools showed at first an excellent memory, but they were incapable of receiving any kind of superior instruction. In the same way the New Zealand children are at first more intelligent than the English children, but they are very rarely capable of any sort of high education. Sometimes, also, among the Polynesians who have been brought up in European civilised manners, the tenacious ancestral influence gains the upper hand; and when the neophyte has come to an adult age he feels the constraint of civilisation to be so irksome to him, that in spite of himself he casts aside the yoke and returns wild and savage into the woods. Mr. Marsden has noticed a fact of this kind in a young Tahitian, brought up in the school at Port Jackson, where he had been taken when he was eleven years old.

So far we have been concerned only with the inferior races; but the early stages of the other races have not been more brilliant, as indeed is proved to us by the want of development in the most humble of their species, and even in certain of their ethnical groups. We need not despise the large civilisation of Japan or of China, but the nomad Mongolians are as yet very backward. Some of them, the Ostiaks for instance, never kill a bear without making their excuses to him afterwards. The Mongoloids of Kamtschatka are still in the age of the coarsest cut stone; and certain Malays who live from day to day only upon the produce of their fishery, and take no heed as to their next meal before they have digested the last, are still in a state of extreme savagery. In Europe, at the time of Tacitus, the Finns had not yet learned the art of agriculture.

Taken as a whole, it would seem certain that the white race are now well clear of savagery. That is the general result; but in the heart of societies apparently the most civilised, how many creatures do we not find as little intelligent as the lowest savage! We may also notice that our civilisation, still imperfect, is of comparatively recent date. Before the days of history cycles of savagery slowly unfolded themselves, as is now shown to us by prehistoric archæology; and even the Latin historians may well have seen some of these European ethnical groups. If we may believe them, the Breton Celts, before the Roman conquest, were not more civilised than the present Polynesians.

III.

It is a commonplace observation that the moral and intellectual development do not always go together hand-in-hand; the fact constantly strikes us in our so-called civilised societies, and we remark also the same thing in comparing races one with the other. Wallace tells us that the Dyaks are more sincere, more frank, more honest than the Malays and the Chinese. But nevertheless, as a very general rule, intellectual activity in any society is the prime mover in all great progress, industrial, moral, or social. On this point there can be no possible doubt as regards industry. In the moral and social development the relation of ideas is less evident. But in a word, morality will depend strictly upon the kind of education imparted to a series of generations, and the value of this education is closely dependent upon the intelligence of the ethnical group. In the same way the social condition is higher and more consonant with justice in proportion as the governing classes are more enlightened as to their true interests, and especially as they can foresee more clearly into the future.

Now, foresight is especially the result of intellectual development. To have this gift man must be endowed with keen observation; he must be capable of concentrating his attention to enable him to group together and compare facts, to deduce future events from present and past history. But the observation of the inferior man is very restricted; he is concerned only with that which relates to his most urgent needs; his memory is short, events gone by do not dwell in his mind: no savage race has a history. It is almost impossible for the inferior man to relate an event exactly without changing anything; even the Hindoos do so with the greatest difficulty, consequently in their rich literature the historic element is wanting. The power of concentration of thought, especially upon any intellectual object, is weaker even than his memory. The Abis of South America can answer accurately only a very few questions. Near to the lake Tanganyika, in Africa south of the equator, Burton, endeavouring to notice in each tribe the names used for counting from one to ten, succeeded only after an infinite deal of trouble. After a few minutes, the negro, when questioned, became vague and stupefied; his answers were incoherent; he was obliged to sleep in order to refreshen himself. Burchell, too, relates the same thing of his teacher of languages in Kaffraria. As regards any complex reasoning, combining together any number of observations and ideas, the dull intelligence of the primitive man can never grasp it. This would be as completely impossible for him as strong force of will, capable of overcoming and controlling his natural desires.

To confirm these general ideas we must now consider the principal manifestations in the human intelligence, starting from the most elementary, from those seen most constantly in daily life, to the most abstract, to the industry of languages, and to science.

CHAPTER II.

INDUSTRY.

We should fill a thick volume were we to describe all the industrial inventions of the human species. We must, therefore, rest content with a sort of enumeration, and even then confine ourselves to certain primordial industries: arms, the invention of fire, pottery, metallurgy, and agriculture.

We see here the principal manifestations of the inventive genius of humanity on the side of industry; and it is owing to them that man has been able to conform to nature's laws, and to increase in number and in strength.

Behind every conscient activity in man there is a secret agent, which marks it with its stamp; that agent is the intelligence, very unequal, both in races and in individuals. From what has been said in the last chapter, and from the long exposition of facts given throughout this volume, we are now enabled to classify the human and the sub-human races in the order of psychical development. Between the lowest and the most glorious types of man there is an abyss; it is not unfathomable, though it must be crossed very gradually and very slowly. Everything in social life works together, one thing into the other; the races, or the people, which from a nutritive, sensitive, affective, or politic point of view occupy the inferior degrees in the human hierarchy, are also backward in industrial progress.

A. As regards the manufacture of arms, the Tasmanians, when they were extant, were of all the Melanesians the most inferior. They were content with their club, their bone spear, and their wooden javelin, the point of which they hardened in the fire; but the cleverest of them made their weapons more deadly by fastening on to them, with the *xanthorhœa* gum, coarsely-cut stones. In addition to these primitive instruments, the Australians had their *boomerang*, a weapon more curious perhaps than dangerous, in the form of a bent stick, and fashioned in such a way that it flew through the air and whirled round in an opposite direction to that in which it was thrown. These boomerangs remind us of the magic arrows mentioned in some Sanskrit poems, which came back of their own accord into the quiver of the warrior. The bow, the most universal of all propelling arms, was known to the majority of the Papuans, and even to the northern Australians; but it is certainly a Malay or a Polynesian importation, for the New Caledonians had no bows, and we may take this as a significant fact. The manners and industry would surely seem to indicate that all the branches of the Melanesian race have originally come from one and the same settlement.

But as regards arms, there is little variety throughout all the human kind. The club, the javelin, and the lance are in fact the foundation and the complement of the primitive arsenal; and they are little more than a perfected form of the stick or broken branches of trees, the ordinary weapons of large monkeys. To these we must add, excepting in a portion of Melanesia, the bow, the most serviceable propelling instrument known to savage humanity, and one of the great inventions of the old races. Another missile also, not a very learned one, was the sling. This too was common, for it has been seen in New Caledonia and in Polynesia, where they preferred it to the bow, then also in use. The sling was used, too, in South America, in the Pelew islands, in the Marianne islands, and among many Asiatic and European people.

These offensive weapons were no doubt different in each race, in each group, according to the greater or less degree of skill exercised in making them, and in the nature of the materials employed. We may say as much of the defensive armour. The Australians warded off the blows with a long and narrow shield made of bark. The majority of the Africans, and of the Red Skins, and of other races, used shields made of leather. The coats-of-arms of wadded cotton worn by the Mexicans and by some of the half-civilized people on the shores of the lake Tchad, and the cuirasses of skin worn by many hordes of men, differed only in substance from the coats of mail, cuirasses of steel, so lately worn in Europe, and which we still find in Central Asia. We see therefore that until the invention of gunpowder, the art of killing, which has been, and is still amongst the majority of men, the most necessary of all arts, rested only upon a few simple and uniform ideas. Firearms are no more than one of the many applications of metallurgy and of empirical chemistry; and their manufacture, even including all the science of modern projectiles, has only put into operation a number of industrial and scientific methods, forming part of one very elementary idea. The Indian of Guiana had need of as much inventive genius to imagine his nicely-contrived air-cane (sarbacane), into which he puts a light arrow with a few rolls of cotton round the top end, moulding itself on to the sides of the cylinder, as the sheeting of lead of the modern shell on to the rifling of the cannons. The inventor of the sarbacana also must have remarked the poisoning properties of certain substances, he must have learned that which we ignore, how to mix them so as to kill, to calculate the effects of this subtle poison. In all that a great effort of intelligence and imagination was first necessary.

No doubt the inventors of the bomb and the explosive shell have been encouraged in their industry by the greatness of the results which they were endeavouring to effect. They have anticipated that their future projectile would burst in the roof of a house, make holes in walls, cause a general explosion, tear human creatures into rags; and to bring about so much destruction they have not been niggardly in using their labour. In the same way the Indian or the American, constructing his venomous projectile, has, in his imagination, seen his prey, animal or human, overcome by the poison, fall lifeless into his hands, in spite of his own weapons of defence and in spite of all his strength. The invention of the sarbacana, which we also find between the Amazon and the Orinoco rivers and in the Malay archipelago, is equal then, in a psychical point of view, to the nicely-perfected firearms. The Indians know full well what they are about when they say of their poison: "It is our form of gunpowder!" But if we consider that the envenomed arrow is used mainly for the chase, we shall then have a right to consider that the invention of the savage is quite as intelligent as that of the civilised man, and that it is also more moral.

B. Of all primitive inventions the greatest and the most fruitful has surely been that of fire.

As the animal genesis of the human kind is now beyond all doubt, we may therefore conclude that man has not always known fire. And this induction is also confirmed by the traditions of the Egyptians, the Phœnicians, the Persians, the Greeks, the Chinese, and others. According to Pigafetta, who wrote the history of Magellan's voyage, the Mariana islanders ignored the use of fire in the year 1521; at first they imagined it to be some all-devouring animal. Even nowadays the Australians are not very skilful in the art of lighting a fire. Some of these tribes do not know how to keep it burning, and they will make long journeys to procure fire, when by accident the women have let it go out. All over Australia, one of the main duties of the women is to keep small sticks of *banksia grandis* constantly burning. These switches are like a wick, they have the property of burning slowly.

The numerous pyrolatric religions, the legend of Prometheus, etc., prove abundantly that in many primitive civilisations man had a full knowledge of the importance of his conquest when he first discovered fire.

The early pyrogenic methods are not numerous; they are often similar among very various races. Nearly all of them may be reduced to one of the three following methods: gyration, friction, or percussion. Although primitive men little suspected the modern principle of the unity of physical forces, they always endeavoured in their pyrogeny to transform movement into heat.

The most widely spread pyrogenic method was that of gyration. The operation was making the point of a very dry stick turn round with sufficient rapidity in a small hole bored into another stick. That was the means employed for producing fire by the Australians, the Bushmen, the Nubians of Sennaar, the Fuegians, many of the tribes of the Red Skins, the inhabitants of the Caroline islands, the Kamtschadales, and others. The Arâni of the Vedic Aryans was nothing else; but it had given rise to a whole mythological conception, for the race was intelligent. Two pyrogenic sticks had endowed two different races. The turning stick, Pramanta, was the father of the god of fire; the immovable stick was the mother of the adorable and luminous Agni.

The mode by friction was merely moving rapidly backwards and forwards the point of the male branch in the furrow of the female branch. That was the means used by the Polynesians, and also by some of the Malay hordes, notably by those of Atelima.

The method by percussion was by striking sparks from the sudden contact either of two stones, or of two pieces of mineral, or of a silex and a piece of metal. The Algonquins used two stones. The Esquimaux preferred two pieces of ferruginous pyrites. The Semites used flint and steel; which were still used in Europe not so many years ago, for we are all still impregnated with the relics of past times.

The use of fire when it had been once discovered gave rise to many industrial inventions, specially to pottery and to metallurgy, which have both played very important parts in primitive civilisations.

C. Most of the groups in every human race have been potters. We must, however, except in past times the majority of our European ancestors of the Age of Stone, and in modern times, the Australians, the Tasmanians, and what is much more singular, the Polynesians, already free from primitive savagery, and arrived at the Age of Polished Stone.

The history of pottery alone would prove that in proportion to the greatness of progress, the onward march of the tortoise has been astonishing. It is but slowly, painfully, and ever protesting himself that man changes his industry and his usual mode of life. In the far-distant prehistoric ages, all over the world, and also now in all primitive pottery that has been burnt with fire, we do not find hollow or scooped-out vessels. This invention is comparatively recent; it is nearly contemporary with the invention of polished, ornamental, and artistic pottery. The turning of the pottery was also a rare art and is almost modern; simple as it may be, it was unknown to all prehistoric humanity, and to all savage people in contemporary humanity. Only the most civilised nations of the old continent have adopted it, or have invented it.

On different spots of the globe, even in Europe, old practices, anterior even to the invention of pottery, have subsisted down to modern times. In the sixteenth century the islanders of the Hebrides used skins for boiling; in these skins they heated water in the same way as the Shoshonis of North America do now in their wicker-work saucepans.

In terminating this short sketch on ceramic art, we may remark that the potter's skill has always been disdainfully left to the women. It was doubtless because this industry, essentially primitive, was invented during a social phase in which hunting and warfare were the only manly occupations, and at a time also when the care of the kitchen was left entirely in the hands of the weaker sex. And even nowadays the fabrication of pottery among the Papuans, the Niam-Niams, the Guaranis, and others, is a labour exclusively feminine, and in many fragments of prehistoric pottery we see the impressions of small fingers which in all probability were feminine.

D. No doubt the potter's art has played a very important part in the development of civilisation, especially as without it the culinary art would hardly exist; but primitive metallurgy was an invention still more fruitful. In his perpetual fight against nature man never had really a fair chance until he had learnt how to employ mineral substances; how to extract the hard and useful metals, how to cast aside the alloy, to melt or forge the metal into arms, into utensils, and into instruments.

With arms man could, without being at too great a disadvantage, conquer his terrible adversaries in the animal kingdom. Also, the use of metals multiplied in many ways all kinds of primitive industry. With his stone hatchet a Carib took a month to cut down a tree, or a year to scoop out a canoe. Everything was in the same proportion, and the smallest industrial result was not obtained until after very long and laborious effort. The use of metals increased the capabilities of man by tenfold. But every people had not their Vulcan. In the far-distant times the Ages of Cut and of Polished Stone have been prolonged over geological periods; and even now there are whole races of man who ignore or who have ignored the use of metals. Some people have not got beyond the use of copper or of bronze.

Except certain Papuans of the eastern part of New Guinea, who have been somewhat civilised by the Malays, and who forge their iron in the Malay fashion, by means of a double wind-bag, metals were ignored all over Melanesia and so also in Polynesia.

In America the majority of the savage tribes were or are still in the Stone Age, but among the Americans on the north-western shore, more or less in communication with northern Asia, copper and sometimes iron were not absolutely unknown, but they were procured only by way of exchange. The half-civilised nations of Central America, before the time of the Spanish conquest, knew how to melt and extract gold, silver, lead, and brass; they had copper mines, which they worked after a clumsy fashion by digging horizontal galleries into the side of the mountains. They knew how, by mixing brass and copper, to fabricate bronze; but they did not know what to do with iron, which was a common mineral in the country. Slabs of pyrites of iron, cut and polished, served the Peruvians as looking-glasses; they had not yet thought of making any other use of their ferruginous metals.

In some of the antique mounds in North America bracelets of copper have been found, and near the lake of Erie mines of copper were worked in prehistoric ages.

Not going outside European archæology, a so-called necessary law of succession has been established between the Ages of Stone, of Bronze, and of Iron; but we cannot say that this law has everywhere and always been perfect. No doubt, we may find it in Egypt, where stone instruments were used at a comparatively modern date in certain religious ceremonies, and where, during many centuries, bronze was the metal most frequently employed. But all over negro Africa, iron has been known from time immemorial; the people know nearly everywhere how to extract it from other mineral substances, how to forge it, and how to mould it. Instruments of iron, probably belonging to the Berbers, have been found under the dolmens in northern Africa; and all over the rest of the continent, very rich in ferruginous substances, there is no trace of an age of bronze, nor of copper.

Iron is worked everywhere in Africa from the country of the Kafirs, as far as Senegambia and the valley of the Nile. Nearly everywhere, also, the means employed are a double wind-bag; this is used also in New Guinea, in Malay, and in Madagascar. The accounts of modern travellers in Africa have made us acquainted with this ingenious contrivance, consisting in principle of two bags, from which by alternate pressure the air is driven into two tubes; these two tubes unite into one, and the air is so carried down into the middle of the pit. This system is too complicated to suppose that its use can have been spontaneous nearly everywhere. It is probably a Malay invention, and has been afterwards borrowed from them. There has certainly never been any regular communication between Malay and Madagascar, but many authentic examples will prove that contrivances, far inferior to those of the Malays, have been transplanted to enormous distances. Coincidences such as these do not of course constitute ethnical migrations. They are insignificant occurrences which do not modify any race of men, though they may conduce to diffuse very widely industrial inventions.

All the Africans know how to work iron, but they do not all know how to extract it from other mineral substances; such was at least the case with the Kafirs, according to Levaillant. It is in the middle portion of Africa that we find the great metallurgic zone. In Senegambia, in the valley of the Niger, in the basin of the Upper Nile, the large furnaces, seven feet and a half high, are constructed in clay, and at a very trifling cost. The mineral substance is placed in the upper portion of this small contrivance; the coal burnt is in the lower part; the draught is caused by the air from the double wind-bags. The construction of these primitive high furnaces is better adapted as one gets nearer to the basin of the Upper Nile. The Bongos, for instance, a poorly-civilised people, know how to build furnaces with three compartments, and these are often permanent constructions. The relative geographical closeness to Upper Ethiopia and to Egypt may perhaps partly help to explain these facts.

But a general diffusion of the use of iron, both in ancient and contemporary Africa, is one of the most notable facts in the history of the human kind. In negro Africa we can, in fact, follow step by step the evolution of the metallurgy of iron, from the simple hole dug in the clay, and round about which burning coals and mineral substances are piled up, as is the practice at Mandara, until we come to the almost scientific furnace as we see it in use among the Bongos. All this certainly authorises us to conclude that Africa is one of the great homes for the primitive metallurgy of iron.

If we except the Esquimaux tribes, who are still in the Age of Stone, all the Asiatic people know the use of hard metals. But some of the Mongolian tribes in eastern Asia are more familiar with copper than with iron, though iron has been known for a long time past by all the civilised Mongolian nations, in Japan, in China, in Thibet, and elsewhere. The Aryan nations of Asia also discovered iron at a very early date, because we find mention of it in the Vedic hymns. The Greeks knew it later, for all the Iliad is an epopee of the Age of Bronze. Iron is mentioned in the book of Genesis; but the Bible tells us that it was with fetters of brass that the Philistines bound Samson. Gaul before the time of the Roman conquest was nearly everywhere in the Age of Bronze.

We may say, therefore, that there has been no civilisation at all advanced which has not more or less known the use of metals; and the most intelligent, the great Aryan, Mongolian, and Semitic civilisations, and even that of ancient Egypt, though at a comparatively late date, knew how to extract and to forge iron.

But we should fall into error if we thought that the use of iron was everywhere and always the chief sign of a high civilisation. The metallurgy of iron is more or less common all over negro Africa; and yet the Kafirs, the Bongos, and others are very much less civilised than were the ancient Peruvians and Mexicans, the Greeks of Homer, the Hebrews in the book of Genesis, and other people. The intellectual development of a nation cannot be tested by one single characteristic. Any partial discovery may no doubt give a great impulse to progress; but before the progress can come man must first have been in a condition to profit by it. The most important sign is cerebral progress, the development of the brain and of the intelligence. The worth of the tool will depend upon the use that the workman can make of it.

E. The remark at the end of the last paragraph will apply to agriculture as well as to metallurgy. In the progressive movement of humanity, no industry has played such an important part as agriculture. According to Sully's famous expression, agriculture is the "breasts" of the people. Every great civilisation, all those which have grouped themselves together and produced large agglomerations of men, all those which have become real homes and hearths, giving to humanity warmth and enlightenment—they have all been based upon agriculture. But there are, and there have been, agricultural attempts amongst the wildest savages.

The inferior Melanesians, the Tasmanians and the Australians, in all respects the lowest of men, have doubtless never thought of agriculture. They might collect well enough certain fruits, certain vegetable substances, but the idea of sowing had never taken root in their bestial brain. But their cousins, the Papuans, rather more intelligent, and less isolated, who have received some notions of it from different quarters, are all more or less agricultural. Even the New Caledonians, savages as they are, have known how to root up the ground with a hatchet and with fire; they have learnt how to cultivate the taro (*arum esculentum*), the igname, the sugar-cane, and to water their plantations.

In Africa, if we except the pastoral Hottentots, the nomad Arabs, and the Tuaricks of Sahara, who despise the labourer and the townsfolk, all the races are agricultural. The rudimentary civilisation of the Kaffirs rests principally upon agriculture; whereas the savage tribes of the Gaboon, less skilful in cultivating the soil, look to the chase as a means of supplying them with a greater proportion of their food. In all the middle zone of the African continent agriculture is very highly esteemed; the people know how to cultivate sorgho, rice, cassava, bananas, etc. We may make one observation upon African agriculture: nowhere are domestic animals employed; all the labour of digging the ground, sowing, etc. is done by women and by slaves.

With the exception of the Fuegians, the majority of the Pampas races (the Patagonians, the Puelches, the Charruas, the hordes on the Grand Chaco, and others) to whom we may add the

Esquimaux of North America, all the native tribes of America are more or less employed in agriculture. The Indians of the pueblos naturally understand it very well; the majority of the Red Skins have an agricultural season lasting for a few months. In speaking of food we have said what were the plants cultivated by the American aborigines. We have seen, also, that among the ancient nations of Central America agriculture was already fairly advanced. Besides maize, the Mexicans used to cultivate cocoa, the agave, tobacco, etc.; they were familiar with the science of the distribution of crops, and of irrigation. They had upon their lakes, like the Chinese, established floating gardens. The Quichuas of Peru were still more clever. With them agriculture was the grand affair of the community. They cultivated the quinoa (*chenopodium*), the potato, maize, the *oxalis*, the *oca*; they knew the fertilising properties of the guano, the consumption of which was regulated by law; they knew how to dig watercourses down to their lakes, to execute great works of irrigation determining the quantity of water to which everyone had a right. They made the sides of their mountains valuable by cutting gradually-inclining paths; they knew how to cultivate the different plants, putting each one at the most suitable altitude.

Excepting the New Zealanders, all the Polynesians were more or less agricultural; but the most clever were the Sandwich islanders. They knew how to build aqueducts, to construct terraces upon the mountain slopes, like the Peruvians, and they carried their terraces up as far as the snow line.

Various as are the populations in the Malay peninsula, they are nearly all agriculturists; and we have seen that in many districts the culture of rice, which requires the assistance of a great many hands, had given rise to the system of holding land in community.

On the vast *Asiatic* continent, if we except the Esquimaux, there is no ethnical group at all important who are quite ignorant of agriculture. But La Pérouse says that the Mongolians of Saghalien, and also on the western side, were not yet agricultural, and confined themselves to collecting eatable bulbs from a sort of lily. In western Mongolia the Tartars have become quite agricultural from the contact with the Chinese, and the nomad people even sow small fields of millet.

We know well enough to what degree of perfection the Chinese have carried the art of agriculture.

The Aryan and the Semite branches of the white race have always been more or less agricultural from time immemorial. The Vedic Aryans were agricultural, and to find a purely pastoral state, even among the Arabs, we should have to go back to the pre-Islamic ages sung by Antar.

In short, the greater part of the human race is agricultural; but the different human groups can give to it only as much intelligence as they possess. In the early stages the distribution of crops and dressing are unknown. After each harvest the bit of land cleared is forsaken. Tools are also very rough and rudimentary. Man first digs the soil with a pointed stick; he then generally makes a few holes, into which he puts the grain. The wooden pickaxe is the only agricultural instrument known to the New Caledonians, to the Caribs, to the Nubians of Darfur, and to many others. The ancient Peruvians made use of a stake, through which they drove a piece of wood horizontally, and upon this they placed their foot. The Kafirs and the Bambarrans have now a sort of rudimentary spade. The plough is unknown throughout all negro Africa; but it was known in ancient Egypt, where cows were put into the yoke. This was a most important innovation. But it would appear that the plough is an Asiatic invention. Its primitive model is still found in Celebes island. It is a machine with only one arm, with a wooden socket made of the palm-tree; it is drawn by buffaloes. No doubt slaves, and even women, were formerly harnessed to these light ploughs; and this custom still prevails in China.

If we may believe Hesiod, the first Greek ploughs were no more than a long wooden hook; the curved end was driven into the ground. The Hebrews apparently knew of the plough, for in Deuteronomy we find that it is forbidden to yoke together an ass and an ox. In Job, too, we find the harrow mentioned, an instrument long unknown to the Greeks.

This savage form of agriculture is no doubt very far indeed behind the agricultural industry of modern ages, and even of the old civilisations in Asia and in Egypt; but putting all technical considerations aside, that which we may call the psychology of agriculture we find everywhere to have been the same. We have seen that the quite inferior races are incapable of any power of foresight, and agriculture demands necessarily an insight into the future. Every form of agricultural labour demands thought as to what the morrow may bring forth. The Australian, whose only delight in life is to stuff himself with the flesh of whales, and the Red Skin, who thinks of little but destroying game merely *for* the pleasure of killing, are utterly incapable of considering events four-and-twenty hours in advance.

And also the history of agriculture will show to us the perfectibility of men, even of those of the lowest kind. The Australian cannot conceive that to-morrow he may be hungry; but the New Caledonian has this care upon his mind. The Red Skin is essentially a huntsman and is most improvident; but certain tribes in New Mexico have learned the use of the harrow and of the wooden plough. In 1836 the Cherokees became an agricultural people, at the expense, it is true, of their slaves and of their wives. The Peruvians knew how to establish a sound and far-seeing organisation, based mainly upon agriculture, and in which the only thing wanting to them was liberty. Progress, therefore, is not a dream, as we have more than once been asked to believe.

V. But though it has been both the cause and effect of the development of civilisation, industrial labour has had upon human societies many unhappy results: among others we may mention the system of castes and of the enslaved classes. Industry must certainly be ranked among the principal reasons for slavery, which indeed, in quite primitive societies is a progress, for it replaces slaughter. Manual labour, however little sustained, supposes a stationary life and a persistent effort of will; and this is alsolutely opposed to the savage man, or even to him who is barbarous. Except the fabrication of arms, all industrial labour was at first imposed upon the woman and upon the slave. Man kept for himself what he thought was the more noble task: that of killing animals and his human competitors. His work in life was hunting and warfare.

We know well enough that this disdain for manual labour has lasted through all the historic phases in humanity, down to our own times. But even after the abolition of slavery and of bondage, the labourers were often divided into castes and into corporations of a more or less degraded kind. We must, however, make an exception in favour of certain people in Africa, among whom the blacksmith is held in very high honour. The Kafirs call him "the iron doctor," and among the Bambarras the blacksmiths form an aristocratic class. We find nothing of this sort in modern Europe, where the governing classes hold in poor estimation the artisan and the peasant, though they do not always themselves reach to that degree of intellectual development which, even though it existed, would hardly excuse their disdain.

In primitive humanity industrial labour weighed very heavily upon the weak and the forsaken, and in our so-called civilised societies this same state of things has not disappeared so fully as many of us imagine. It is true that we have no longer slaves or serfs among us, but we have men who work for wages, who are constrained to perpetual toil, often excessive and dangerous, and whose only advantage over the ancient slave is that they can change their master.

From the perfectioning and development of large industries in our modern societies there has also resulted for the working classes a baneful consequence. The industrial consumption has grown to be enormous, and to suffice for its own wants it has been obliged to inaugurate upon a vast scale labour in common and the most minute divisions of this labour. In this respect, as in many others, the old Chinese civilisation was in advance of us. Before it is completed a Chinese flute passes, as does a pin with us, through a very large number of hands; the labour, therefore, of each workman becomes purely mechanical and stupefying in its simplicity.

Even in savage or in barbarous societies man is in some way an artist. He manufactures completely the objects of his own industry, he can perfect them according to his own taste, he can interest himself in his work, and he exercises his own intelligence. There is nothing of this sort in our large manufactories, where the human creature is reduced to a mere automaton, executes every day of his life a small number of movements which are ever and always the same. Hence are born our large class of pariahs, who emaciate themselves with enervating labour. We might also mention homicidal industries, decimating the hard-working victims in our large societies, to their infinite detriment.

We hold it to be all important to remedy those flaws in our civilisation. We ought to transform the wages system into a free association, and above all things lessen the hours of daily work. Certain industries, also, which are necessary but unwholesome, ought to be considered as a social drudgery, existing because of their public utility, and therefore divided out among those interested. In this case, and in this case only, the extreme division of labour would be salutary. Fourier's industrial armies would gradually replace the armies of men now turned into soldiers.

The future will place civilised societies in an inexorable dilemma: justice or death.

CHAPTER III.

PURE INTELLIGENCE.

THE effects of intellectual power, properly so called, are so widely spread, that we cannot now even attempt to give a sketch of their extent. We will confine ourselves to saying a few words upon the most primordial acts of the human intelligence, mentioning them only as specimens, as standard measures, whereby we may characterise the different races, and assign to each a place in the hierarchy of the human kind.

I.

Languages.

We no longer lose our time in trying to argue that language is a thing of human origin; there are only a few belated minds who endeavour to contest the fact, by using arguments belonging to another age. No doubt, among the superior races, the luxuriant complexity of flexible languages may dazzle us, but a comparison of the idioms spoken by the whole medley of the human kind, their hierarchical classification, a study of their rise and of their evolution, will inevitably lead us to connect the articulate language with the animal cry, in which also there is no divine element. As a matter of fact, every cerebral impression that is at all exciting may reflect itself upon these or those movent nerves, and the cry is but a reflex action of the mind. It is an automatic sign of the vocal organs, especially of the larynx. In man, and in many animals, certain sentiments will provoke cries, inflections, modulations of the voice, as expressive as they are spontaneous; it is a mechanism showing its elasticity.

To this primitive foundation of language are immediately added the imitative onomatopœia. With more or less conscient feeling, the primitive man, and the child, who resembles him so strongly, endeavour to reproduce the noises which most frequently strike their ears. But they will copy differently, according to their race; for imitation is of necessity imperfect, and each human type has a mode of auditive impressionability peculiar to itself.

In order that a real language may free itself from this very rudimentary verbal utterance, a social life of some sort is necessary, and a social life with all the attendant incidents, the conflicts, and the fruits of liberty. We know the story of the two children brought up by Psammouthis in silence, and away from the rest of the world: the conclusion is false, for experience has shown, beyond any doubt, that a child brought up in silence cannot learn to speak. Father J. Xavier, nephew of François Xavier, when he was a missionary in India in 1594, heard from the lips of the Emperor Akbar a curious history similar to that of Psammonthis. The all-powerful monarch had an idea to make an experiment as to the origin of language. He therefore caused thirty children to be brought up together, in a confined place, under the care of nurses and guardians, who were enjoined to keep absolute silence, under penalty of death. The children imprisoned in this way grew up, and became, as was natural, dumb, and absolutely stupid, having no other language at their command than a few gestures, to express their bodily wants.

Even when they live freely in the middle of our societies, children do not speak unless they hear, and this is irrefutably proved by the speechlessness of children struck with congenite deafness at a time when they receive no special education. In order that articulate sounds may fix themselves in their memory with any determined meaning, it is absolutely necessary that they should hear the same sounds repeated several times. But childhood is of all others the age most suitable for learning to talk. During the first years of life, vocal imitation is always the easiest; as Itard has remarked, it is very often automatic, unconscient. Now, this mental condition of the child probably shows to us, with more or less accuracy, that of our ancient ancestors, hardly yet human beings, who created articulate form of speech. But this early language was very poor, as may be seen nowadays from that of the inferior races.

To whatever linguistic family the idioms of the poorly-developed human types may be connected, there are two characteristics common to them all: the extreme indigence of their vocabulary, and the want of abstract and general terms. A few facts chosen from many others will show the truth of this statement.

The Weddahs have none but the most usual words at their command, necessary to express the simplest acts of their daily life and the things that they see around them.

In the Tasmanian language there were no adjectives; the

people could qualify a thing only by comparison. The Tasmanian vocabulary had words to designate this or that kind of tree, but the general word "tree" was wanting. We need not, therefore, be astonished to find that the Australians, so similar to the Tasmanians, have no expressions to say "justice, crime, fault," etc.

The language of the Bushmen is so poor that they have constantly to make gestures one to the other; they cannot therefore talk in the dark. The Bechuana dialect has no word corresponding to "conscience, spirit." The Fulah has neither masculine nor feminine; it classes beings into two categories: those which belong to humanity, and those which belong to animality.

The American idioms, numerous as they are, are not more philosophical. According to Spix and Martius, the aborigines of Brazil have no words to say "colour, kind, sex, mind." The expressions "time, space, and substance" are wanting in most of the American dialects. The Choctahs have words to designate the black oak, the white oak, but not to say "oak tree;" and the Californians have only one expression to say "toad" and "frog;" their moral qualifications are all taken from their sense of taste. The same word among them will designate a good man and a savoury aliment. We may remark by the way that a similar confusion exists in our European languages wherever there was originally a mental condition of a very coarse kind.

The Malays cannot say "red, blue, grey, or white;" they have no word signifying colour. The vocabulary of the people on the banks of the Drave is wanting in expressions to convey ideas of God, of the soul; the word "will" is also wanting. There is no Basque word that has the large sense of our denominations "tree, animal," etc.

But without going into savage countries, we may observe similar facts even among the races in Europe, who have a rich vocabulary at their command. Whatever be the total opulence in any language, it becomes poor when it is handled by an unintelligent or poorly-educated man. Great writers have at their command a thousand expressions, corresponding to each delicate shade and turn of their thoughts; but for the peasant or the unlearned man a very modest supply, containing perhaps a few hundred words, will be amply sufficient for all his wants.

If the richness or poverty of a vocabulary will convey to us a fair idea of the degree of intellectual development in any race, the quality of the articulated sounds will also instruct us as to the general character of a people. The Papuans of the New Hebrides, a most ferocious nation, spoke a language all bristling with consonants and with harsh articulation. On the other hand, every sharp hissing sound was banished from the Tahitian dialect; there were very few consonants, a constant repetition of syllables, and the language had generally a very infantine character. The Polynesian idioms, very different in their vocabulary, became more phonetic in their sound as one approaches nearer to Melanesia. The language of the warlike New Zealanders was guttural; the k, the v, the hard consonants were predominant. This was also the case among the anthropophagous islanders of Pomotou.

Another feature, common to the languages of the inferior races, is the extreme variety of their dialects, however close may be their linguistic relationship, however identical their general construction. In Australia, between neighbouring districts, the vocabulary will change, and the natives are often obliged to speak English if they wish to understand each other. In the same way, in New Caledonia, the people belonging to tribes at all distant do not understand each other. The Red Skins, in the Rocky mountains, who have the same tokens but who belong to different tribes, are obliged sometimes to converse by signs; and in the exogamic tribes the men and the women will often speak different languages. Something of the same kind still exists in Europe in the valleys of the Caucasus, a region which the Persians call "the mountain of languages."

This diversity of dialects, in languages having no written literature, owes its origin mainly to the isolation of the tribes into small groups, which are still barbarous. Each one of these small ethnical unities lives separately, troubling itself little about its neighbours except to fight them, and its idiom, not established by any law, soon acquires its own special form.

Thanks to modern linguistic science, which has succeeded in

unravelling from the infinite diversity of verbal forms general characteristics enabling us to classify languages into large families, it has become easy to trace the evolution of the most complex idioms. Every language first began by monosyllabic utterance, and many have not yet succeeded in getting beyond the first stage. The language of primitive Egypt was monosyllabic, as are also the majority of the Mongolian languages: the Chinese, the Thibetan, the Burmese, and the Indo-Chinese languages. The genealogy of Sanskrit will lead on to an ancestral foundation of monosyllabic roots, upon which the whole growth of the language entirely rests.

Linguistic progress first began by the juxtaposition of monosyllables, coupling into one element, which kept its primitive value, other elements playing the part of suffixes or of prefixes, and determining the moods of the invariable element. In this way monosyllabic words grew into agglutinations of words. The agglutinative languages are still very numerous. We may mention among others, the Japanese, the Corean, the Malayo-Polynesian languages, the American and the African languages, the Drave and the Basque languages.

The third and most ingenious form of articulate language is the flexible form. In flexible languages man has not been content to agglutinate together the roots; he has modified these roots, and, therefore, we have now dialects that may be bent and twisted as we please, capable of rendering every possible shade of thought. The class of flexible languages will comprise all the Semitic and Indo-European dialects, but those only. The flexible languages, therefore, are the most dignified, belonging to the superior races.

Nothing is more interesting than this evolution of articulate language; but we can by no means say that the degree of development in any race or people will depend accurately upon the hierarchical order of their language. As regards humanity, as we now see it, that which shows in any language the measure of its mental energy is much less the place it occupies in the general classification than the richness or the poorness of its vocabulary. There are certain monosyllabic languages which very early became permanently fixed, as regards their construction, though they were ever evolving new forms of expression. For instance, the majority of the people of the Mongolian race have not gone beyond the monosyllabic phase of language; but that has not prevented the Chinese from creating a great and wonderful civilisation. On the other hand, the Australians, the native Americans, the Africans, from the Hottentots as far north as the Fulahs, speak agglutinative languages, though all these people have never yet emancipated themselves from savagery or from barbarism. We may perhaps infer that these races have not created their own languages, and that during the very long prehistoric period foreign initiators brought to them idioms which had taken root and grown elsewhere.

May we be allowed, after having spoken of the forms of language now past and gone, to make a few conjectures as to their future; at least of the Aryan languages, of which we know the past evolution, and which we see still constantly changing? If, in spite of their common origin, these languages have so distinguished themselves one from the other, that has certainly resulted from the dispersion of the nations, from the separation of the people who formerly spoke them. But European civilisation is evidently now evolving towards fusion; interests so long hostile are now becoming joined together. Manners and laws are becoming uniform; the work of advancing towards a common end is gradually progressing. If these wide transformations, of which we now see the commencements, go on without hindrance for a sufficient number of centuries, the Indo-European languages must necessarily form themselves into one synthetic language, into one future Aryan tongue, as complex and as rich as the primitive Aryan language was once coarse and indigent.

II.

Mathematical Aptitudes.

The idea of numbers, as understood by the educated man, is essentially an abstract idea, but of the most rational and logical kind; for totally separate from concrete objects, it rests altogether upon objective reality. Mathematics partly appertain to metaphysics, for they seem to soar above the ideas which had prompted them, but they are in fact a scientific form of metaphysics. We may soon become convinced of this in studying their early stages.

When man tries to picture to himself a certain quantity of similar objects, all at the same time, and to keep in his mind a recollection of them, he very soon becomes confused, and his confusion will come all the quicker in proportion as he is less intellectually developed. To enable him to retain his ideas he has recourse to mnemonical signs. When man gave a particular, an ordinal name to a few of these signs, when he succeeded in abstracting the denomination of the material object from that to which it was fastened, numbers were then established.

Mnemonical objects used at first to give support to the number are ordinarily of the most simple kind: they are small pieces of wood or of stone. Men reckoned by calces or stones (hence our word "calculate"), or more often, and nearly all over the world, first by the fingers and afterwards by the toes.

The Weddahs of Ceylon, who seem to be the least intelligent of men, have still no mathematical faculty whatever; they have no name for any number.

The Tasmanians, a little more advanced, used to be able to say "one" and "two;" for a higher number they had to say "many." Sometimes they could succeed in saying "two and one," or even "two and two." In order to say "five" they lifted their hand as high as a man's head. They had, therefore, the idea of the number five, but the expression was wanting to them.

The Australians have only two numerical expressions, but in putting them together they can count as far as ten. The most intelligent of them when they want to express the number five say "hand," and for the number ten they say "two hands."

The majority of the tribes in New Caledonia have only four nouns of number. For five they say "a hand"; "two hands" will mean ten. If they wish to go beyond ten they begin to count again as far as five, and after that they put forward a foot, or five toes. When they have got as far as twenty they say "a man." That means all a man's fingers and all a man's toes. Some few clever calculators can continue in this way; but the most skilful mathematicians in all New Caledonia cannot get beyond two or three hundred. Beyond this colossal extent of numeration people make use of the expressive saying, "the grains of sand could not count it."

The digital form of numeration is, as we shall see, very common among primitive races, and it certainly must have been the basis of the decimal system.

Like the Tasmanians, the Bushmen have only two nouns of number, which they continue more or less by juxtaposition: 2 + 1; 2 + 2; 2 + 2 + 1, etc.

The mathematical faculties of the Kafirs are hardly more advanced. Three nouns of number are sufficient for the Dammaras; beyond that they make use of their fingers. They are obliged to sell their sheep one by one, as they are incapable of counting them or of reckoning the price. The numerical expression among the Zulus for saying six is, "take the thumb of the other hand."

In the middle part of Africa, where people are more civilised and especially more commercial, the mathematical aptitudes are greater. The Yarriba children amuse themselves by counting with cauris, shells which are used as small money coins all over Africa. If one man wants to reproach another for his ignorance he says to him: "You can't add nine and nine together." But the Arab influence has now become widely spread over these regions.

The numeration of the Indian Americans resembles very much that of the Melanesians and the Africans. Among many of the native tribes in South America, in the subarctic regions, the numeration does not go beyond the first numbers, afterwards it becomes digital.

To express a number higher than four the Guaranis say "innumerable." Many tribes, whom D'Orbigny has called Moxeans, the Itonamas, the Canichanas, the Movimas, are very deficient indeed in the power of reckoning; they cannot go beyond two or four.

The Abipones have really only three nouns of number. To signify four they will say "the fingers of an emu"; five, "the fingers on a hand"; twenty, "the fingers on the hands and on the feet." We find the same plan of enumeration among the Guiana Indians, among the Caribs, among the Tamanacs of Orinoco.

The Esquimaux and the Ahts of North America can count only a very few numbers upon their fingers. Ross has seen Esquimaux who could not get beyond the number ten.

Many Indians in counting are obliged to raise their fingers and then to put forward their feet. Very often, when an Indian wishes to signify twenty he will say "an Indian," or "a man."

It is evident that the decimal system of counting, which we see used now among many civilised people, and which was used also by the Mexicans and the Peruvians, sprang from this digital numeration common among the primitive tribes. But the Mexicans had a very extended numeration. The numeration used by the Quichuas of Peru was employed also by the Araucanians, the Puelches, and the Patagonians, who for the higher numbers still make use of the same denominations as the Quichuas. We know that these latter people used to count their numbers by the knots in their quipos, which to them were real string registers; whereas the Americans, to unite their numbers, had invented a whole system of points.

Something similar to the Peruvian quipos was also customary in the Sandwich islands; here the messengers used to carry with them thin pieces of cord on which they made their knots, and these knots were their numerical signs. In most of the islands the people used to count with stones or with small pieces of wood, each of which was meant to signify a dozen. This same primitive method was used almost everywhere. The Polynesian numeration is decimal, and in principle it would seem to have been borrowed from the Malays; the Malay word *rima* or *lima*, which is meant to signify both a hand and five, is used as a name of number all through the Malay territory, in Madagascar, and in Polynesia as far as Easter island. But in the Pomotou archipelago we do not find this word *rima*, which is common for some other names of number, probably of Malay origin; and this is a fact not without importance when we come to determine the origin of the Polynesian islanders.

The decimal numeration of the Polynesians would, one would think, have allowed them to count as far as the higher numbers; but in practice they could scarcely get as far as two thousand, which indeed for savages is going a very long way.

The decimal system is common also in Mongolian Asia, or in those portions of Mongoloid Asia which are more or less civilised. But it would seem that mathematical aptitudes have as yet progressed very little way with the Siamese, for the tribunals will not accept the deposition of any witness who cannot count as far as ten. It is very different with the Chinese, for they have written books upon mathematics, and they have recently adopted the geometry and the logarithms of Euclid. In mathematics, as in everything else, the great Mongolian race holds an honourable position in the general competition in the human hierarchy.

But it is in India more especially that mathematics have early acquired a scientific development. In the fifth or sixth century of our era the Hindoos had invented a system of trigonometry. In the fourth century, according to Colebrooke, the algebra scholar Arya-Bhatta used to resolve equations from several unknown quantities. This precocious development in mathematical science, and especially in the very abstract branch of algebra, in a race more prone than another to metaphysical ramblings, would seem to denote a more or less narrow connection, a psychical relation between these two methods, the one rational and the other irrational.

Hindustanee is also very rich in numerical expressions. There is a word, "lak," to say a hundred thousand, and another word, "krar," to express ten millions; whereas our European languages have not, as had the Greek, a special expression to say ten thousand.

It is almost needless to say that the European nations did not all at once arrive at the higher mathematics; the barbarous nations in ancient Europe were assuredly not more advanced in mathematics than the Polynesians. Strabo tells us that the people in Albania did not know how to count beyond a hundred. In the Basque language there is no original word to express a thousand. The ancestors of the Indo-European races must certainly have begun to count upon their fingers, as is still the practice with many savages. This was also the mode in which the Abbé Sicard, at the close of the last and the beginning of the present century, used to teach his deaf and dumb pupils before giving to them special instruction. They were made to count upon their fingers as far as ten; beyond that they cut notches upon a piece of wood.

For aptitude in mathematics, as for everything, the human brain has evolved very slowly, and it also started from a very low ebb. Quite at first, man, incapable of the slightest abstraction, knew of no numerical term; like certain animals, in seeing objects before him he may have had a vague idea of number. By degrees he invented for himself a system of enumeration, at first very rudimentary, using his fingers as mnemonical pins. Then, freeing himself altogether from the objective world, he succeeded in stringing together an infinite number of purely abstract qualities, as he found it no longer necessary to strengthen his memory by graphic signs, figures, letters, or lines, which indeed are nothing else than a rather more ingenious way of counting on one's fingers, by stones, or by pieces of wood. This was the only practice known to the early reckoners, who followed their rude mnemonical system.

III.

Computation of Time.

If the idea of number is abstract, that of time is perhaps still more so. The commencement of it may be simple enough. In proportion as the phenomena of the exterior world unfold themselves, they strike the man who witnesses them, and they engrave themselves in his brain in the form of sensations and impressions. But these mental impressions are successive, as are the facts which have provoked them; and they ever tend to efface each other gradually. They have therefore a very different degree of colouring, according to their age and their intensity; the conscient me can therefore compare them, and assign to each a relative date. That is the fundamental basis upon which, by slow elaboration, the human mind has succeeded in abstracting an idea of time; a purely subjective idea, though it was glorified and placed by the Greeks on the top of Olympus.

But this idea of time is conceived with less fulness in proportion as a race is less intelligent. Among the most primitive races, a man is incapable of saying how old he is, and even of giving approximately the date of any event more than a few months old; the chronological operations come very slowly to perfection.

In very early times the chronological periods observed are very short. Day and night are the only unities of time. Preferring the subjective to the objective, some of the Esquimaux whom Parry saw used to count by so many sleeps.

For distant dates the people would reckon from a remarkable event: a storm, an epidemic, an emigration, or the capture of an elephant, as is the custom among the Hottentots. This, we see, is but an accidental kind of chronology, having no sort of regularity. Many negroes, especially the Mandingos, calculate more or less the recurrence of years by that of the wet seasons. It is probable that, by observing the periodical return of seasons, man was enabled to arrive at some degree of precision in the computation of time.

The course of the stars, when man was able to notice their regularity, established landing-points that were still far more exact. The first astronomical phenomena observed were everywhere the apparent movements of the constellations, and especially the changes of the moon.

But this observation had at first a special character. We have already seen how great is the animism of the savage, who throws his own life on to the outer world. In the eyes of the primitive man, the sky is not dull and dead, as it is to the mind of the educated European. The Patagonians regard the Southern Cross as an ostrich (*nandou*), and the surrounding stars are dogs following it. To them the moon is a man, the sun is a woman etc. The Abipones think that the Pleiades come down upon earth, and when this constellation becomes invisible they say that their grandfather is ill. During this epoch in his evolution man interests himself in the stars, as in living creatures, and each phase in their revolutions are to him as so many dramatic incidents. This same train of thought outlives the period of mythological corruption, even in societies that are actually becoming civilised. The Bambarrans, for instance, still count their different seasons by the periodical return of certain constellations seen beyond the horizon.

But it was mainly by the revolutions of the moon, nearly everywhere imagined to be a living creature, that gave to man the first idea of time in any way definite. The lunar month seems to many people to have fulfilled the same position that the solar year now holds in our scientific chronologies; then the periodical observations of the risings and settings of the moon at different points of the horizon suggested the idea of a longer period in the lunar year, adopted among all half-civilised people. This lunar year is still adopted by the Bambarrans in the middle of Africa, by the Polynesians, by the New Caledonians. It happened often enough to run twelve months when they were able to reckon accurately the time of the conjunction. The four phases of the moon, the quarters, gave the idea of the week, which has been adopted from time immemorial by nearly all the people of the white race, Semitic as well as Indo-European, and which has also been adopted in New Caledonia, and by the Bambarrans in Africa.

But as the difference between the duration of the lunar year and the solar year is considerable, it naturally followed that with this system the order of the seasons was soon inverted. People introduced what remedies came most natural to their minds; they made alterations, they suppressed so many days; and so they continued until they discovered the solar year more or less exactly, unknown to nearly every primitive race. The Polynesians, however, had some rude idea of the solar year, but the discovery was not amply and effectually made until the days of a more advanced civilisation.

In ancient Peru, where the system of reckoning by the lunar year was still current, the people rectified it by solar observations, made by means of rudimentary gnomons, by columns, the length of the shadows enabling them to determine the epochs of the solstices and of the equinoxes. The Mexicans, more skilful astronomers, had altogether adopted the solar year, each year being divided into eighteen months of twenty days, and to complete the year of 365 days they added in five supplementary days. And every fifty-two years they interpolated twelve days and a half, so making up for the loss of six hours not comprised in the 365 prescribed number of days in the year. The Mexicans had also improved upon the rude gnomons of the Peruvians. From out of one of the Mexican plains has been dug a colossal solar dial-plate, upon which was engraved a calendar, indicating the hours of the day, the epochs of the solstices and of the equinoxes, and that of the passage of the sun, according to the zenith in Mexico.

Egypt, India, and China also discovered, each in their own way, the solar year, and with more or less precision. The Chinese almanack, adopted by the Mongolians, the Thibetans, the Indo-Chinese, divides the year into twenty-four divisions, marking the passage of the sun in the twelve signs of the zodiac; and, according to Bentley, the Indians had, as early as the year 1442 of our era, known how to divide the ecliptic into twenty-seven lunar stations.

The ancient Egyptians were already far advanced in their chronological reckonings. Their year was divided into 365 days, or twelve months, with thirty days in each, and five supplementary days. But as the solar year does in fact exceed this by about a quarter of a day, the Egyptians invented a long chronological period, a sothic or cynic period of 1460 years.

Many other nations have also grouped their years into long periods. The Thibetans had their cycles of 250 years. The Chinese counted by cycles of sixty years, and they began to do so three centuries before the commencement of the Christian era. But the Mexicans and the ancient Egyptians seem to have been the only people who thought of making these cycles agree with the irregularities of the civil year. The Mexicans had imagined a cycle of fifty-two years, which they called "sheaves," or "bundles," and in their hieroglyphs they used to represent them by a bundle of roses. Because of its very long duration, the Egyptian civilisation, which in all probability the people thought was destined to endure to the end of time, had adopted in its sothic period a cycle enormously long in comparison with the ordinary life, not only of individuals, but with that of nations. No other people have as yet shown so much confidence as to future ages.

The study of chronometrical reckonings, invented by the different races, will give us a very good indication as to their natural intelligence, the strength of their memory, and also of their powers of observation. Quite at the bottom of the social scale we see man living a hand-to-mouth existence, like the animals, from whom he is not far removed. Then he endeavours to recollect the notable events; he observes the regularity of the seasons, the most striking astronomical phenomena. He succeeds in imagining first a lunar year, afterwards a solar year, which by degrees he conceives with more perfect accuracy, every now and then correcting his errors. And the old civilisations which saw a long succession of years that had evolved behind them in bygone times, and which hoped for a still longer chain of years in the future, the short solar year was not sufficient. They framed their annals into long cycles. Of all these ambitious periods the sothic period is in our present opinion the most inordinate. But we may nevertheless believe that future humanity will imagine a term of years still more imposing, established by long astronomical revolutions, the mutation of the terrestrial axis, or the precession of the equinoxes; for there is reason to hope that the human kind is but at the commencement of its painful journey along the course of ages.

CONCLUSION.

THE AGES OF HUMANITY.

I.

To anyone who has taken the trouble to read this book, it will be superfluous to refute the pessimist doctrines which have recently been current among the *blasé* men and women of our own time. Some writers, whom we should imagine to be sound, plump men, ruddy in health, well provided with this world's goods, have endeavoured to make us believe that to live is the worst of all evils, and that henceforward every human effort ought to tend towards suicide. Job also sang this anthem very many years ago, but he might have pleaded extenuating circumstances. His bed was a dung-heap, he was eaten up with leprosy, and he was worn out by commonplace morality showered down upon him by his well-to-do friends. Owing to his unfortunate position, his ideas of the development of humanity were necessarily incomplete. But now, in our time, when the history of the human kind is known to us—at least in its broad outlines—now, when we know the evolution through which man has passed, we must either be blind or be wedded to some chimerical system of our own, if we dare to deny the law of progress.

Man, no doubt, is very weak; he is still a long way from being perfect. No doubt the coarse instincts of the beast are still alive in him, for he has freed himself from brute-like existence only by long and constant efforts, and animality has by no means lost its hold. But by a long course of steady progress, ever more and more conscient, he has improved his condition, and in future ages he will still do so to a much greater extent.

The nature of man, like that of all the superior animals, is complex; and in the preceding chapters we have endeavoured to note the main features, separating them one from the other.

In the mental life we have seen that the nutritive appetites were predominant; they roar so loudly that their voice drowns all others. In every race, primitive man is a sort of wild beast; his chief thought is to satisfy his hunger, to capture and devour his prey, which is often human flesh. But even as regards the nutritive appetites man makes progress. He learns how to vary his food, to modify it by means of cookery; and in this way he augments the niggardly fare with which nature had first provided him. As he learns to rear domestic animals, and more especially by his acquired knowledge of agriculture, he replenishes his larder with greater facility; his culinary skill becomes more perfect, and enables him to taste delicacies, of which, crude and inferior as they may seem now to us, he had at first no idea. Still moving onward, his invention always at work, he discovers that which we have wrongly called the *nervous aliments*: substances which act directly on the cerebral life, either to excite it, to disorder it, or to deaden it.

Man's artistic notions develop also at the same time. He no longer uses his sense of sight and of smell merely as a means of perceiving the phenomena of the outside world; he endeavours to picture to himself, to realise the objects which his senses present to his imagination. From this moment he can, at will, with ever-increasing skill fix and give birth to a number of impressions, sentiments, and even ideas. He becomes musician, painter, and sculptor.

And parallel with the sensitive side of human nature the affective aptitudes have also developed. At first man confined himself to the satisfying of his genesic desires; but in proportion as his sensibility has grown finer and his hunger has slackened, his power of loving has increased. The manifestation of affective sentiments were at first short and rare; his love for his wife or his little ones was only temporary, like that of animals for their females or for their young. And sometimes, when hard pressed by fierce and pitiless hunger, man has eaten his own children without much hesitation or twinge of conscience. It was not until after many ages that this early savagery gave him disgust, or provoked in his mind any feeling of horror. It was by very slow degrees that man's heart became so enlarged as to embrace his cares for his wife and children, his neighbours, his friends, his fellow-citizens, and at last humanity at large.

An ascending gradation of the same kind may be seen also in our social institutions, from the animal horde of men, in which the strongest reigned as brutal and absolute master, down to the clan, to the tribe, to the city, to the nation. The ethnical group has ever grown larger; social ties have multiplied themselves and become more complex; man's interests have become more dependent one upon the other; we have passed from anarchy to a rigid despotism, and at last have made for ourselves an individual independence, always increasing, and limited only by the real interest of the community. The government of human societies is now a science, with particular forms of procedure, all carefully studied; and the object is progressive amelioration of our kind from a threefold point of view: physical, moral, and intellectual development. Mythology is also unfolding itself, it is growing pale and gradually dying out. Man no longer ignores that the terrestrial world is the only one to which he can pretend to belong; his constant endeavour is to make his sojourn here more and more tolerable.

Each of these branches of progress supplies another more important than all the rest—the progress of intelligence. Primitive man could only live in the present, so weak was his memory, his imagination, his power of combining ideas. But the mental life gradually expands itself. The impressions marked in the nervous centres by the incidents and accidents of daily life become deeper and more tenacious. Many associations, change of sentiments, of ideas fix themselves in the nervous centres, and constitute a large capital of accumulated experience. Upon this solid basis man's intelligence spreads itself out and gradually widens; he learns to observe accurately, and to assimilate his observations, to draw deductions and inductions. Science is then born, and by degrees it finds its way into our everyday life.

In truth, all this very complex progress is above all things intellectual. The taste for refinement in our sensitive pleasures, in art, in humanising feelings, in the inauguration of justice and liberty, in our social relations, is all expressed in and proceeds from the gradual growth of intelligence.

II.

We have been obliged, for the purposes of our analytical study, to isolate one from the other the different modes of human activity; but in reality everything hangs together, is fastened, and works harmoniously. In the brain of the dullest savage there is some intelligence; and there are nutritive appetites in the man who, morally and intellectually, is the most highly developed. But these energies, though they are simultaneous, are so dissimilar, that in the normal course of its evolution the human kind passes through a succession of phases which we may rightly call *the ages of humanity*. Each one of these phases is characterised by a number of wants which govern and lead individuals and the ethnical groups, and these wants are less elevated in proportion as civilisation is less advanced. No doubt all these wants, all these principal aptitudes, exist in all races of men, but in very different proportions. In this respect the different human races may class themselves together according to periods of time, and according to the countries in which they live. They are more especially nutritive, sensitive, affective, or intellectual. We may say at once that a very small minority of the human population deserve to be classed in this last-named division.

Such is the general law, but we should strangely abuse ourselves if we expected to find it hold good always and everywhere; that every individual man, and all the human groups of men, necessarily evolve in a progressive scale merely by raising themselves more or less high. Retrogression is quite possible. We have given abundant evidence to show that the human races are very unequal, that they may be set in order in a long series; but it may also very well happen that individuals, or even small groups, belonging to a superior race should undergo a degeneration which will degrade them to the level of the most inferior races. In Mexico, in South America, in the Fiji islands, Europeans have returned to savagery and even to cannibalism. Cook has seen, in the Malay peninsula, Portuguese reduced by the Malays themselves to the condition of a servile caste of people.

To find other examples we need only look underneath the deceitful varnish of our civilised societies. No doubt that on its good sides our civilisation is prodigiously superior to that of primitive societies; but the greater part of those signs of progress which blind our eyes is the work of exceptional individuals,—men who have been innovators, and generally to their own prejudice. The innovator's calling is often a dangerous one. It is not the less true that among the dregs of our modern societies there are thousands and thousands of persons who in moral elevation and intelligence are hardly superior to the New Caledonians.

It is mainly to our industry that we owe our greatness. Now, our industrial productions, so exuberant and so complex, result principally from our ingenious implements and from our extreme division of labour. But this crumbling of mechanical labour has a most disastrous effect upon the general development of the intelligence. It has come from the formation of an ever-increasing class of modern workmen who have no time to think or to instruct themselves. Owing to this state of things we see crying inequalities in the various conditions of our social welfare and of our knowledge. These are fearful plagues in our civilisation; they are blots which all free and intelligent societies of men ought to endeavour to remedy.

Many vast reforms have yet to be accomplished, and all those who have the care of souls must work, each according to his strength, towards the fulfilment of the great task. As men neglect to do so they will assuredly be paving the way for future social convulsions which will endanger even civilisation itself. We have already felt the first attacks.

We must endeavour that justice, enlightened by superior intelligence, shall take the helm and guide us in our social dealings.

INDEX.

N

Q

R

CHARLES DICKENS AND EVANS, CRYSTAL PALACE PRESS.

www.ingramcontent.com/pod-product-compliance
Lightning Source LLC
Chambersburg PA
CBHW020852210326
41598CB00018B/1643
* 9 7 8 3 7 4 2 8 2 0 1 3 6 *